Collins

Combined Science

Higher

AQA GCSE

Revision Guide

AQA GCSE Revision

Combined Science

Trilogy

Ian Honeysett, Emma Poole
and Nathan Goodman

Contents

	Revise	Practise	Review

GCSE Combined Science Revision Guide   HT **Higher Tier Content**

Contents

HT Higher Tier Content

Contents

HT Higher Tier Content

Contents

HT **Higher Tier Content**

Contents

Contents

HT Higher Tier Content

Contents

HT Higher Tier Content

Contents

HT Higher Tier Content

Review Questions

Recap of KS3 Biology Key Concepts

1 What is the basic unit that all living things are made of? *cells / atoms / particles* **[1]**

2 The diagrams show an animal cell and a plant cell.

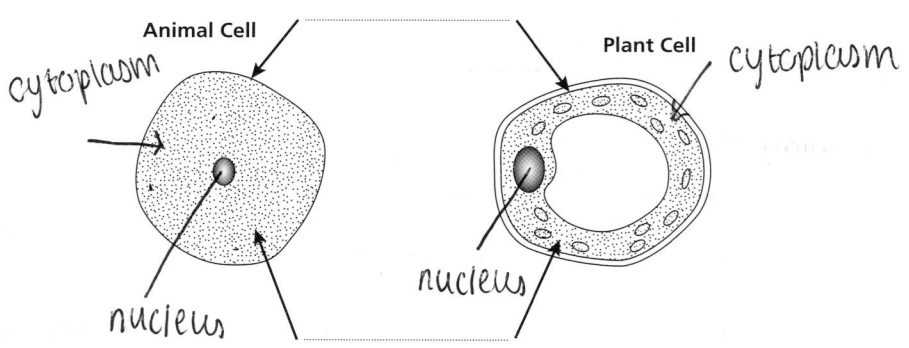

a) Some structures are found in both types of cell.

   Label **two** of these structures, shown by the arrows, on the diagrams. **[2]**

b) Name **two** structures that are present in plant cells but **not** in animal cells. **[2]**
   chlorophyll, vacuole

3 Human ova (eggs) and human sperm have important roles in reproduction.

a) What are eggs and sperm?
   Tick **one** box.

   tissues ☐   cells ☑   organs ☐ **[1]**

b) What does a sperm use to swim towards an egg? *the tail / flagellum* **[1]**

c) Name the male reproductive organ where sperm are made. **[1]**

d) In reproduction, a sperm fuses with an egg.

   What is this process called? *Fertilisation* **[1]**

4 Carbon monoxide, nicotine and tar all get into the lungs when a person smokes.
   turns lungs black   *breathing problems*
   Write down **one** harmful effect on the body of **each** of these substances. **[3]**

5 What is the process called by which plants produce glucose and oxygen? **[1]**
   photosynthesis

6 What is the process called by which the human body releases energy from glucose and oxygen? **[1]**
   anaerobic respiration

7 Give **two** functions of the human skeleton. **[2]**
   hold the shape, store fat and nutrients.

8 The chart shows a way to group living organisms.

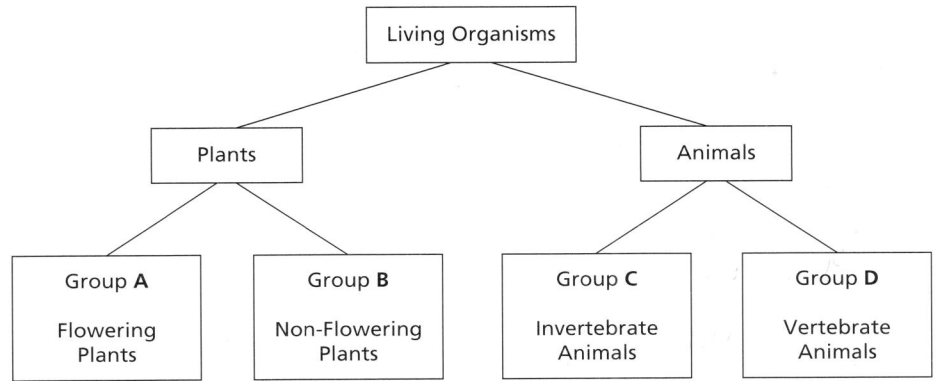

a) To which group, **A**, **B**, **C** or **D**, do the following living organisms belong?

i) Flatworm   ii) Mouse   iii) Daffodil   iv) Human   [4]

(C)   (D)   (A)   (B)

b) Some flatworms are parasites.

What is a parasite? [1]

an organism that feeds on ~~other~~ another living organism.

9 Rachel cut her knee playing netball. A medic put a plaster over the cut.

a) A plaster helps to stop a cut getting infected.

Give the name of **one** type of microorganism that can infect a cut. [1]

bacteria

b) While she was cleaning Rachel's knee, the medic wore rubber gloves.

Why is wearing rubber gloves important for the medic's health? [1]

So that harmful organisms stay away from the medic.

10 The human lungs contain millions of alveoli.

a) i) Which gas passes into the blood from the air in the lungs? [1]

~~more~~ oxygen

ii) Which gas passes out of the blood into the air in the lungs? [1]

CO_2

b) The walls of the capillaries and the alveoli are very thin.

Why do they need to be thin? [1]

to allow gasses to pass through.

c) There are millions of alveoli in the lungs. They provide a very large surface area.

Why is a large surface area important? [1]

So that ~~more~~ the rate at which the ~~pass~~ gasses pass is quicker.

Total Marks .... / 27

Review Questions

Recap of KS3 Chemistry Key Concepts

1 A match is used to light a candle.
The wax melts and then moves up the candle wick where it burns.

  a) What is the change of state when the wax melts?
Name the state **i) before** the change and **ii) after** the change. **[2]**

    Solid   melting   liquid

  b) The match contains carbon.

Write a word equation to sum up the reaction that takes place when carbon burns completely in oxygen.

carbon + oxygen   →   carbon - dioxide **[2]**

  c) Candle wax, wood and coal are all types of fuel.

Which of these statements is **true** for all fuels?
Tick **one** box.

Fuels are substances that can be burned to release energy. ✓

Fuels are always solid at room temperature.

Fuels are always black.

When fuels are burned, they react with carbon dioxide in the air. **[1]**

  d) Name the poisonous gas that can be produced when fuels that contain carbon are burned in a limited supply of oxygen. carbon ~~monoxide~~ monoxide **[1]**

  e) Coal contains a small amount of the element sulfur.

What is the chemical symbol for sulfur? S ✓ **[1]**

  f) Name the gas produced when sulfur burns. **[1]**

    sulfur dioxide

2 Potassium nitrate has the formula KNO_3.

  a) How many different elements are shown in this formula? **[1]**

    ✓ 3

  b) Which element is represented by the symbol K? Nitrogen **[1]**

  c) What is the total number of atoms shown in the formula for potassium nitrate? **[1]**

    5

3 A student uses a pH meter to measure the pH of four solutions.

  a) Tick the correct box to show whether each of the solutions tested is **acidic, neutral** or **alkaline**.

Sample	pH	Acidic	Neutral	Alkaline
A	12			✓
B	10			✓
C	7		✓	
D	2	✓		

[4]

b) The student placed the pH meter in a beaker of distilled water after testing each of the solutions.

Why did they do this? *To check if the water is neutral.* [1]

c) The student tested a sample of soil from their garden. The soil was slightly acidic.

What could the student put onto the soil to neutralise it? *an alkalinal solution or fertilizer* [1]

4 Edward and his family visit the seaside.
Edward has a bucket that contains a mixture of sand and sea water.

a) How could Edward get a sample of salt from the sea water? *distillation* [1]

b) How could Edward get a sample of sand from the mixture? *filtration* [1]

c) Mixtures, compounds and elements are different types of substance.
Sea water is a mixture, sand is a compound and oxygen is an element.

Draw **one** line from each type of substance to its definition.

Type of Substance   **Definition**

Mixture		Contains atoms of two or more elements, which are chemically joined.
Compound		Contains two or more elements or compounds, which are not chemically joined.
Element		Contains only one type of atom.

[2]

5 Rose placed some blue copper sulfate solution into a test tube.
She added a spatula of silver-coloured iron filings and stirred the mixture.
After 10 minutes the solution was green and the solid was orange-brown.

a) Why did Rose stir the mixture? *To mix the atoms together* [1]

b) Suggest a safety precaution that Rose should take during this experiment. [1]
wear gloves.

c) How did Rose know that a chemical reaction had taken place? [1]
the colour changed

d) Name the element produced during this reaction. [1]

e) Write a word equation to sum up this reaction. [2]

iron sulfate

Total Marks .... / 27

Review Questions

Recap of KS3 Physics Key Concepts

1 The wooden truck in the diagram is held so it does **not** move.

Magnet fixed to truck

Magnet fixed to wall

Wall

Wooden truck   S N   N S

Table

a) Describe the motion of the truck when it is released. *left* [1]

b) What effect will friction have on the motion of the truck? [1]

it will slow it down

2 During the summer, people use small hand-held fans to keep themselves cool.

a) Draw a series circuit containing a battery, switch and motor to show how one of these fans could be wired. [3]

b) Which component provides energy for the circuit? [1]

c) A bulb is added in series to the circuit.

How will this affect the motor? [1]

3 A series circuit contains three bulbs.

What will happen to the other bulbs if the bulb in the middle breaks?
Tick **one** box.

Both bulbs will go out. ☐

Both bulbs will get brighter. ☐

The one after the broken bulb will go out. ☐

Both bulbs will stay the same. ☐ [1]

4 A parallel circuit contains three bulbs.

What will happen to the other bulbs if the bulb in middle breaks?
Tick **one** box.

Both bulbs will go out. ☐

Both bulbs will get brighter. ☐

The one after the broken bulb will go out. ☐

Both bulbs will stay the same. ☐ [1]

5 Oil is described as a non-renewable energy resource.

   a) Explain what is meant by 'non-renewable'. **[2]**

   b) Give **two** other non-renewable energy resources. **[2]**

6 Choose words from the list to complete the paragraph.

chemical   **electrical**   **gravitational**   **kinetic**   **light**   **sound**   **thermal**

A bicycle light uses a generator powered by the turning of the wheels. As a cyclist pedals,

.... energy in her muscles is changed to kinetic energy.

When the generator turns, kinetic energy is changed to useful .... energy in the wires.

This energy in the wires is changed to useful .... energy by the bulb.

When the light is on, some of the energy in the bulb is wasted as .... energy. **[4]**

7 When a bird flies at a constant height, there is a downward force of 30N on the bird.

   How large is the upwards force on the bird? **[1]**

8 The diagram shows four forces acting on an aeroplane in flight.

Direction of flight

   a) Which arrow represents air resistance? **[1]**

   b) When the aeroplane is descending, what can be said about forces **A** and **C**? **[1]**

   c) When the plane is flying at a constant speed in the direction shown, which **two** forces must be balanced? **[1]**

   d) Just before takeoff, the aeroplane is speeding up along the ground.

      What must this mean about the size of forces **B** and **D**? **[1]**

9 Two cyclists are riding along a dark road at night.
One is wearing black clothes and the other is wearing light-coloured clothes.
A car approaches the cyclists with its headlights on.

   a) What happens to the light when it reaches the light-coloured clothes? **[1]**

   b) Explain why it is more difficult for the driver to see the cyclist in the black clothes. **[3]**

Total Marks .... / 26

Cell Structure

You must be able to:

- Describe the structure of a typical animal cell
- Describe how a plant cell differs from an animal cell
- Recall the main differences between prokaryotic and eukaryotic cells
- Describe the structure of a typical bacterial cell.

A Typical Animal Cell

- All cells have structures inside them – these are called sub-cellular structures.
- In an animal cell, the sub-cellular structures include:
  - a nucleus, which controls the activities of the cell and contains the genetic material
  - cytoplasm, in which most of the chemical reactions take place
  - a cell membrane, which controls the passage of substances into and out of the cell
  - mitochondria, where aerobic respiration takes place
  - ribosomes, where proteins are synthesised (made).

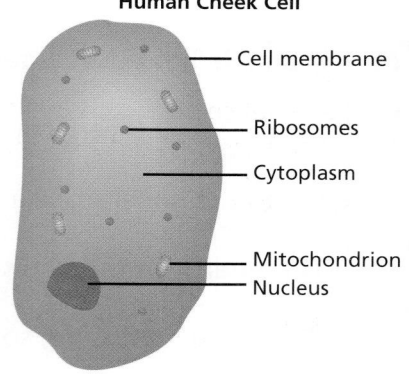

Human Cheek Cell

Cell membrane
Ribosomes
Cytoplasm
Mitochondrion
Nucleus

Plant Cells

- Plant cells and algal cells contain all the sub-cellular structures found in animal cells.
- They also have:
  - a cell wall made of cellulose, which strengthens the cell
  - a permanent vacuole filled with cell sap, which supports the plant.
- Plants need to make their own food, so some of their cells contain chloroplasts.
- Chloroplasts absorb light to make food (glucose) by photosynthesis (see pages 46–47).

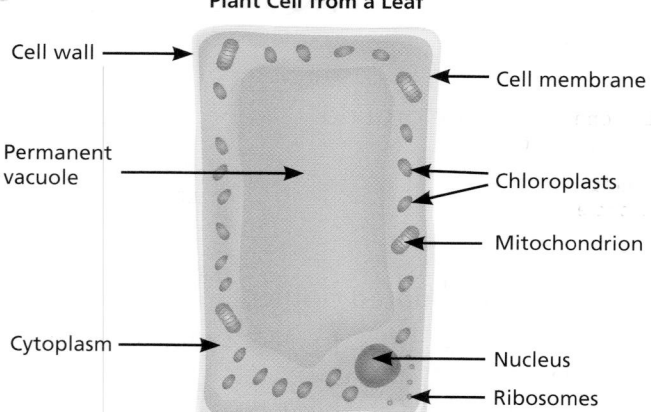

Plant Cell from a Leaf

Cell wall
Permanent vacuole
Cytoplasm
Cell membrane
Chloroplasts
Mitochondrion
Nucleus
Ribosomes

> ### Key Point
>
> Not all plant cells have chloroplasts. For example, they are not present in root cells because those cells do not receive any light.

Prokaryotic and Eukaryotic Cells

- There are two main types of cell:
  - prokaryotic
  - eukaryotic.

  Pro = NO!
- Plant, animal and fungal cells are all eukaryotic.
- Bacterial cells are prokaryotic.
- There are a number of differences between the two types of cell.
- Prokaryotic cells are much smaller in size and:
  - the genetic material is not enclosed in a nucleus
  - the genetic material is a single DNA loop and there may be one or more small rings of DNA, called **plasmids**
  - they do not contain mitochondria or chloroplasts.

A Typical Bacterial Cell

- Bacterial cells have many different shapes – some are round, some are rod-shaped and some are spiral – but they are all prokaryotic cells.

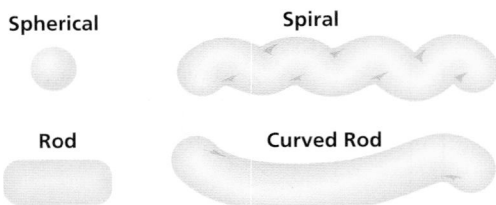

Spherical   **Spiral**

Rod   **Curved Rod**

- In bacterial cells, the roles of mitochondria and chloroplasts are taken over by the cytoplasm.
- There may be one or more flagella, which are tail-like structures that move the bacterium.
- Plasmids are present, which are loops of DNA that can be transferred from one cell to another.
- Plasmids allow bacterial cells to move genes from one cell to another.

A Typical Bacterial Cell

Plasmid DNA: a small, commonly circular, section of DNA that can replicate independently of chromosomal DNA

Chromosomal DNA: the DNA of bacteria is not found within a nucleus and is usually found as one long, looped chromosome

Cytoplasm

Cell wall: provides structural support to the bacteria (is not made of cellulose)

Flagella: tail-like structures that rotate to help some bacteria move

Key Point

Prokaryotic cells are much simpler in structure than eukaryotic cells. That is why scientists think that they developed before eukaryotic cells.

Key Point

Plasmids have become very useful to scientists. They allow genes to be inserted into bacteria in genetic engineering (see pages 70–71).

Key Words

sub-cellular structures
nucleus
cytoplasm
cell membrane
mitochondria
ribosome
cell wall
cellulose
vacuole
chloroplast
prokaryotic
eukaryotic
plasmid
flagella

Quick Test

1. Which sub-cellular structure controls the activities inside the cell?
2. Where are proteins made in a cell?
3. Write down **three** structures that are found in plant cells but **not** in animal cells.
4. What is the function of cell sap?
5. Where is DNA found in a bacterium?

Investigating Cells

You must be able to:

- Use different units of measurement to describe the size of cells
- Describe the advantages of using an electron microscope to view cells
- Analyse images of cells and perform calculations involving magnification, actual size and image size.

The Size of Cells

- A typical plant cell may be about 0.1mm in diameter and an animal cell 0.02mm in diameter.
- Prokaryotic cells are smaller – often about 0.002mm long.
- To describe the size of cells and sub-cellular structures, scientists use units that have different prefixes.

Unit	Number of Units in One Metre (1m)	
centimetre (cm)	100	1×10^2
millimetre (mm)	1 000	1×10^3
micrometre (µm)	1 000 000	1×10^6
nanometre (nm)	1 000 000 000	1×10^9

Using Microscopes to Look at Cells

- It is not possible to see cells as separate objects using the naked eye.
- The ability to see two or more objects as separate objects is called resolution.
- The light microscope was developed in the late 16th century and gave a greater resolution than the human eye.
- It allowed scientists to see plant, animal and bacterial cells.
- Some sub-cellular structures are even smaller than the resolution achieved by a light microscope and cannot be seen using this method.
- In 1933, scientists first used an electron microscope.
- An electron microscope passes electrons, rather than light, through the specimen and can give much better resolution.
- Cells can be seen in much finer detail, e.g.
  - the structures inside mitochondria and chloroplasts can be studied and this has helped scientists to find out how they work
  - ribosomes can be seen and their role in making proteins can be studied.

A Light Microscope

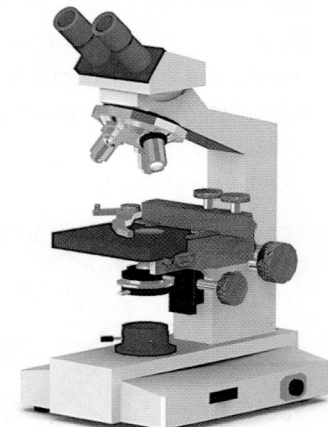

An Electron Microscope

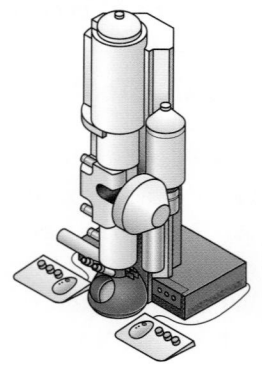

REQUIRED PRACTICAL	
Use a light microscope to observe, draw and label a selection of plant and animal cells.	
Sample Method	**Considerations, Mistakes and Errors**
1. Place a tissue sample on a microscope slide. 2. Add a few drops of a suitable stain. 3. Lower a coverslip onto the tissue. 4. Place the slide on the microscope stage and focus on the cells using low power. 5. Change to high power and refocus. 6. Draw any types of cells that can be seen. 7. Add a scale line to the diagram.	• The scale line can be added by focusing on the millimetre divisions of a ruler.

An Electron Microscope Image of a Bark Beetle

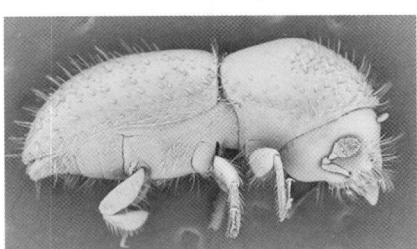

Calculating Magnification

- When a microscope is used to look at cells, scientists will often take photographs or produce drawings.
- These images are many times larger than the real cell or structure.
- The **magnification** is how many times larger the image is than the real object.

$$\text{magnification} = \frac{\text{size of image}}{\text{size of real object}}$$

A student looks at a plant cell under a microscope using a magnification of ×120.
The image of the cell is 1.5cm in height.

Calculate the height of the actual cell.

$$1.5 \times 10 = 15\text{mm}$$
$$120 = \frac{15}{\text{actual height}}$$
$$\text{actual height} = \frac{15}{120}$$
$$= 0.125\text{mm}$$

Convert the measurement from cm to mm.

Substitute your values into the equation.

Rearrange the equation to find the actual height.

Quick Test

1. Arrange these structures in order of size with the largest first:
   bacterium   liver cell   nucleus   ribosome
2. A nucleus is measured as 0.005mm in diameter. How many micrometres is this?
3. A student draws a cheek cell. The cell in their drawing is 50mm wide. In real life the cell is 0.025mm. What is the magnification of their drawing?

Key Words

resolution
electron microscope
magnification

Cell Division

You must be able to:

- Describe the arrangement of chromosomes in a cell
- Explain the importance of mitosis in the cell cycle
- Describe the properties and functions of stem cells
- Explain the use of stem cells.

Chromosomes

- The nucleus of a cell contains **chromosomes** made of **DNA**.
- Each chromosome carries hundreds to thousands of **genes**.
- Different genes contain the code to make different proteins and so control the development of different characteristics.
- In body cells, the chromosomes are found in pairs, with one chromosome coming from each parent.
- Different species have different numbers of pairs of chromosomes, e.g. humans have 23 pairs and dogs have 39 pairs.

A Section of One Chromosome

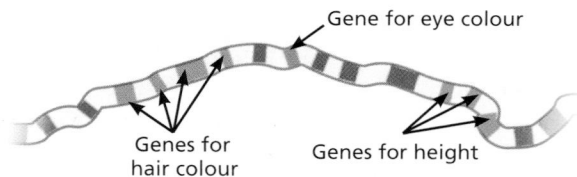

Gene for eye colour

Genes for hair colour

Genes for height

Mitosis and the Cell Cycle

- Cells go through a series of changes involving growth and division, called the **cell cycle**.
- One of the stages is **mitosis**, when the cell divides into two identical cells.
- Before a cell can divide, it needs to grow and increase the number of sub-cellular structures, such as ribosomes and mitochondria.
- The DNA then replicates to form two copies of each chromosome.
- During mitosis:
  1. One set of chromosomes is pulled to each end of the cell.
  2. The nucleus divides.
  3. The cytoplasm and cell membranes divide to form two identical cells.

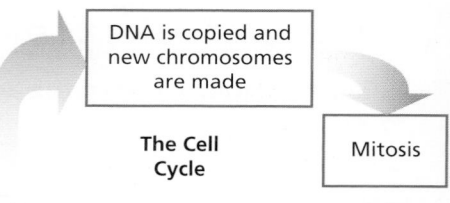

DNA is copied and new chromosomes are made

The Cell Cycle

Mitosis

Each cell grows and makes new sub-cellular structures

Mitosis

  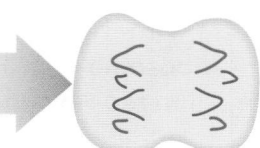 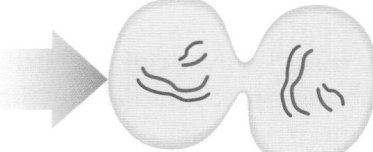

| Parent cell with two pairs of chromosomes. | Each chromosome replicates (copies) itself. | Chromosomes line up along the centre of the cell, divide and the copies move to opposite poles. | Each 'daughter' cell has the same number of chromosomes, and contains the same genes, as the parent cell. |

- Cell division by mitosis is important because it makes new cells for:
  - growth and development of multicellular organisms
  - repairing damaged tissues
  - asexual reproduction.

Stem Cells

- Some cells are **undifferentiated** – they have not yet become specialised.
- This means that they can divide to make different types of cells. They are called **stem cells**.
- Stem cells are found in human embryos, in the umbilical cord of a new born baby, and in some organs and tissues.
- Stem cells from human embryos are called **embryonic stem cells** and can make all types of cells.
- **Adult stem cells** are found in some organs and tissues, e.g. bone marrow. They can only make certain types of cells and their capacity to divide is limited.

Uses of Stem Cells

- Stem cells may be very useful in treating conditions where cells are damaged or not working properly, such as in diabetes and paralysis.
- They could be used to replace the damaged cells.
- A cloned embryo of the patient may be made and used as a source of stem cells. This is called **therapeutic cloning**.
- Stem cells from the cloned embryo will not be rejected by the patient's body, so they could be very useful in treating the patient.
- Some people are concerned about using stem cells from cloned embryos:
  - there may be risks, such as the transfer of viral infection
  - they may have ethical or religious objections.
- In plants, stem cells are found in special areas called **meristems**.
- These meristems allow plants to make new cells for growth.
- The stem cells can be used to produce clones of plants quickly.
- This could be useful for a number of reasons:
  - rare species can be cloned to protect them from extinction
  - large numbers of identical crop plants with special features, such as disease resistance, can be made.

Key Point

Although there are adult stem cells all over the body, they are very difficult to find and isolate.

Potential Uses for Stem Cells

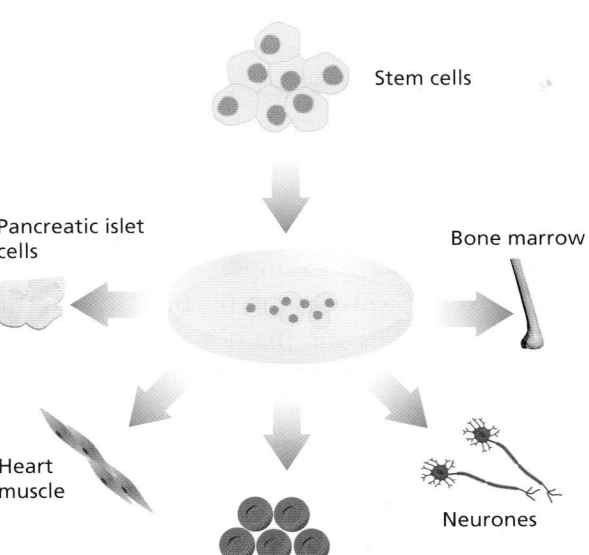

Stem cells

Pancreatic islet cells

Bone marrow

Heart muscle

Neurones

Blood

Key Point

Stem cell use (and the different views for and against it) is a good example of an ethical argument concerning a new technology.

Key Words

chromosome
DNA
gene
cell cycle
mitosis
undifferentiated
stem cell
embryonic stem cell
adult stem cell
therapeutic cloning
meristems

Quick Test

1. How many chromosomes are there in each human body cell?
2. Why is it important that the chromosomes are copied before mitosis occurs?
3. What is a stem cell?
4. Why does a plant have a meristem at the tip of the shoot and the tip of each root?

Transport In and Out of Cells

You must be able to:

- Describe the process of diffusion
- Explain the factors involved in moving molecules in and out of cells
- Describe how water can move by osmosis
- Explain why some substances are moved by active transport
- Compare diffusion, osmosis and active transport.

Diffusion

- Many substances move into and out of cells, across the cell membranes, by **diffusion**.
- Diffusion is the net movement of particles from an area of higher concentration to an area of lower concentration until they are evenly spread out.
- This happens because the particles move randomly and spread out.
- There are many examples of diffusion in living organisms:
  - oxygen and carbon dioxide diffuse during gas exchange in lungs, gills and plant leaves
  - urea diffuses from cells into the blood plasma for excretion by the kidney
  - digested food molecules from the small intestine diffuse into the blood.

Factors Affecting Diffusion

- The factors that affect the rate of diffusion are:
  - the difference in concentrations, known as the **concentration gradient**
  - the temperature
  - the surface area of the membrane.
- A single-celled organism has a large **surface area to volume ratio**.
- This allows enough molecules to diffuse into and out of the cell to meet the needs of the organism.
- In multicellular organisms, there is a smaller surface area to volume ratio. However, surfaces and organ systems are specialised for exchanging materials.
- The small intestine and lungs in mammals, gills in fish, and the roots and leaves in plants, are all adapted for exchanging materials:
  - they have a large surface area
  - the surface is thin so that molecules only have to diffuse a short distance
  - surfaces are usually kept moist so that substances can dissolve and diffuse across cell membranes faster
  - in animals, a rich blood supply maintains the concentration gradient
  - in animals, ventilation occurs to speed up gaseous exchange.

Osmosis

- Water may move across cell membranes by **osmosis**.
- Osmosis is the diffusion of water from a dilute solution to a concentrated solution through a partially permeable membrane.

Diffusion

Some particles

High concentration — Many particles → Low concentration

Net movement of particles

The Effect of the Size of an Organism on its Surface Area to Volume Ratio

Organism A

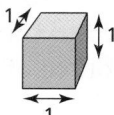

surface area = 6
volume = 1
surface area to volume ratio = 6

Organism B

surface area = 24
volume = 8
surface area to volume ratio = 3

Key Point

A dilute solution contains lots of water. A concentrated solution contains less water.

Osmosis

| Dilute solution (high concentration of water) | Concentrated solution (low concentration of water) |

Partially permeable membrane

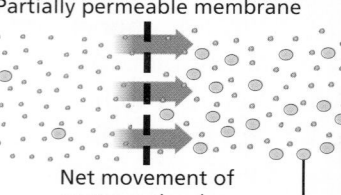

Net movement of water molecules

These molecules are too large to pass through the membrane

REQUIRED PRACTICAL	
Investigate the effect of a range of concentrations of salt or sugar solutions on the mass of plant tissue.	
Sample Method Potatoes can be used to measure the effect of sugar solutions on plant tissue: 1. Cut some cylinders of potato tissue and measure their mass. 2. Place the cylinders in different concentrations of sugar solution. 3. After about 30 minutes remove the cylinders and measure their mass again.	**Considerations, Mistakes and Errors** • The cylinders need to be left in the solution long enough for a significant change in mass to occur. • Before the mass of the cylinders is measured again, they should be rolled on tissue paper to remove any excess solution. • If the cylinders change in mass, they have gained or lost water by osmosis.
Variables • The independent variable is the one deliberately changed – in this case, the concentration of sugar solution. • The dependent variable is the one that is measured – in this case, the change in mass of the potato. • The control variables are kept the same – in this case, the temperature, the length of time the cylinders were left in the solution and the volume of the solution.	**Hazards and Risks** • Care must be taken when cutting the cylinders of potato.

Active Transport

- **Active transport** moves substances against a concentration gradient, from an area of low concentration to high concentration.
- This requires energy from respiration.
- Active transport allows mineral ions to be absorbed into plant root hairs from very dilute solutions in the soil.
- Active transport also allows sugar molecules to be absorbed from lower concentrations in the gut into the blood, which has a higher concentration.

Comparing Processes

	Diffusion	Osmosis	Active Transport
Allows molecules to move	✓	✓	✓
Movement is down a concentration gradient	✓	✓	✗
Always involves the movement of water	✗	✓	✗
Needs energy from respiration	✗	✗	✓

Quick Test

1. A person opens a bottle of perfume. Why do people in the room smell it faster on a warm day?
2. What is required for substances to be absorbed against a concentration gradient?
3. In addition to a large surface area, name **one** other feature that makes an exchange surface more efficient.

Key Point

If the potato cylinders do not lose or gain water, then the sugar solution must be the same concentration as the potato tissue.

A Cell Absorbing Ions by Active Transport

Root hair cell with high concentration of nitrate ions

Soil with low concentration of nitrate ions

Cell uses energy to 'pull' ions against the concentration gradient

---

Key Point

Anything that stops respiration occurring, such as lack of oxygen or metabolic poisons, will stop active transport.

Key Words

diffusion
concentration gradient
surface area to volume ratio
osmosis
active transport

Levels of Organisation

You must be able to:

- Explain how cells become specialised for particular roles
- Describe examples of specialised cells
- Explain how cells can form tissues, organs and systems.

Specialised Cells

- Cells are the basic building blocks of all living organisms.
- As an organism develops, cells differentiate to form different types of cells. They become **specialised**.
- Most types of animal cell differentiate at an early stage, but many types of plant cell can differentiate throughout their life.
- As a cell differentiates:
  - it may change shape
  - different sub-cellular structures develop to let it to carry out a specific function.
- Specialised animal cells include sperm, nerve and muscle cells.

> ### Key Point
>
> If cells are specialised, they become more efficient at their job but may lose the ability to do other jobs.

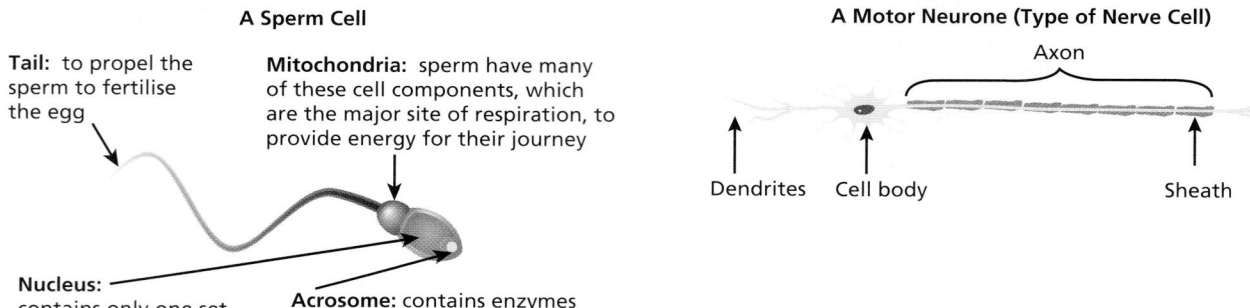

A Sperm Cell

Tail: to propel the sperm to fertilise the egg

Mitochondria: sperm have many of these cell components, which are the major site of respiration, to provide energy for their journey

Nucleus: contains only one set of the genetic material

Acrosome: contains enzymes to allow the sperm to penetrate the outer layer of the egg

A Motor Neurone (Type of Nerve Cell)

Axon

Dendrites   Cell body   Sheath

A Muscle Cell

Nucleus

Mitochondria

Protein fibres that can contract

Many mitochondria for energy

- In plants, root hair, xylem and phloem cells are all specialised cells.

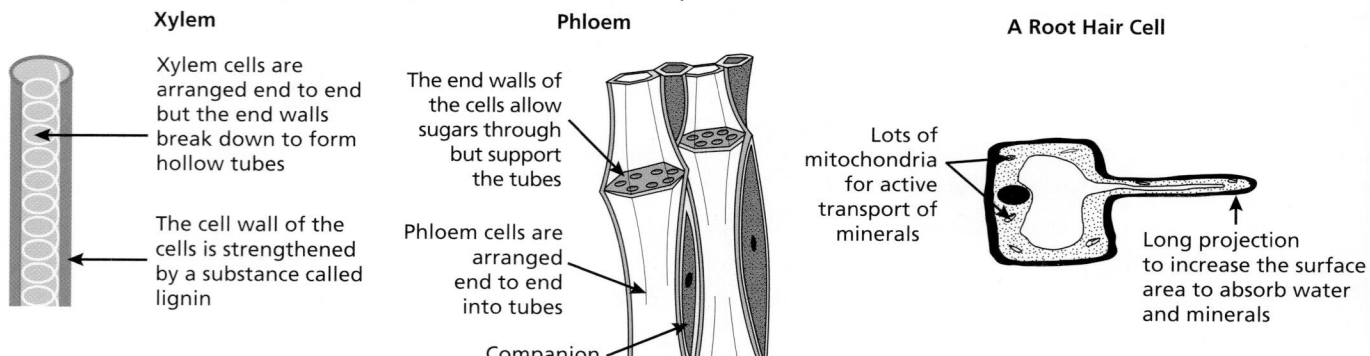

Xylem

Xylem cells are arranged end to end but the end walls break down to form hollow tubes

The cell wall of the cells is strengthened by a substance called lignin

Phloem

The end walls of the cells allow sugars through but support the tubes

Phloem cells are arranged end to end into tubes

Companion cell

A Root Hair Cell

Lots of mitochondria for active transport of minerals

Long projection to increase the surface area to absorb water and minerals

Tissues, Organs and Systems

- In most organisms, cells are arranged into **tissues**.
- A tissue is a group of cells with a similar structure and function, which all work together to do a job, e.g.
  - muscle tissue contracts to produce movement
  - glandular tissue produces substances such as enzymes and hormones
  - epithelial tissue covers organs.
- **Organs** are groups of different tissues, which all work together to perform a specific job.
- Each organ may contain several tissues.
- For example, the stomach is an organ that contains:
  - muscle tissue that contracts to churn the contents
  - glandular tissue to produce digestive juices
  - epithelial tissue to cover the outside and inside of the stomach.
- Organs are organised into **organ systems**, which are groups of organs working together to do a particular job.
- The digestive system is an example of an organ system, in which several organs work together to digest and absorb food.
- Lots of organ systems work together to make an organism.

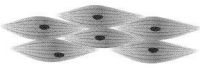

Muscle tissue
Can contract to bring about movement

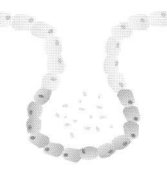

Glandular tissue
Can produce substances such as enzymes and hormones

Epithelial tissue
Covers all parts of the body

The Stomach

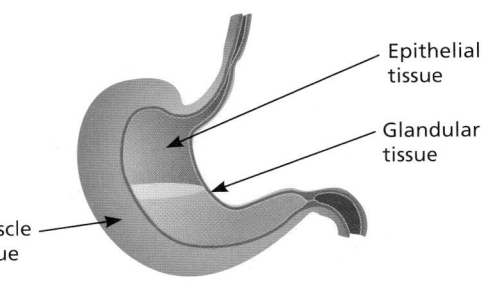

Epithelial tissue

Glandular tissue

Muscle tissue

The Digestive System

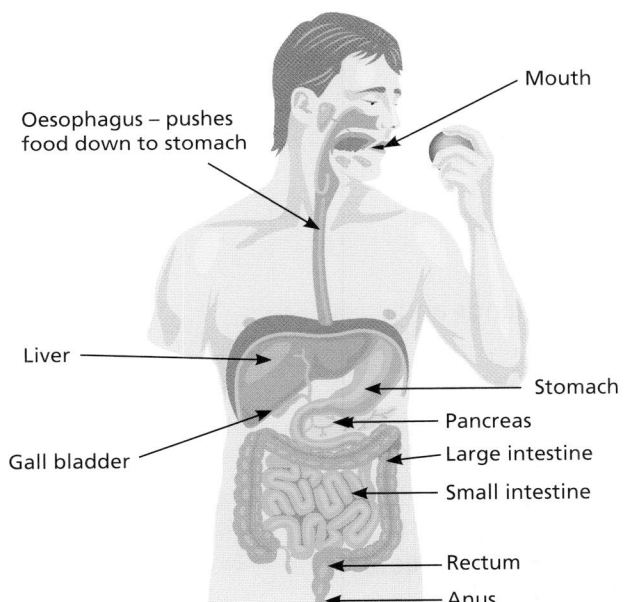

Mouth

Oesophagus – pushes food down to stomach

Liver

Stomach

Pancreas

Large intestine

Small intestine

Gall bladder

Rectum

Anus

Quick Test

1. Why do cells differentiate?
2. Why do sperm cells contain lots of mitochondria?
3. What is a group of cells with a similar structure and function called?
4. What is the function of epithelia?
5. Is the heart a tissue, an organ or an organ system?

Key Words

specialised
tissue
organ
organ system

Digestion

You must be able to:

- Explain how enzymes work
- Describe the role of enzymes in digestion
- Explain how bile can speed up the digestion of lipids.

Enzymes

- Enzymes are biological catalysts – they speed up chemical reactions in living organisms.
- Enzymes have a number of properties:
  - They are all large proteins.
  - There is a space within the protein molecule called the active site.
  - Each enzyme catalyses a specific reaction.
  - They work best at a specific temperature and pH called the optimum.
- The 'lock and key theory' is a model used to explain how enzymes work: the chemical that reacts is called the substrate (key) and it fits into the enzyme's active site (lock).
- High temperature and extremes of pH make enzymes change shape. This is called denaturing.
- The enzyme cannot work once it has been denatured, because the substrate cannot fit into the active site – the lock and key no longer fit together.

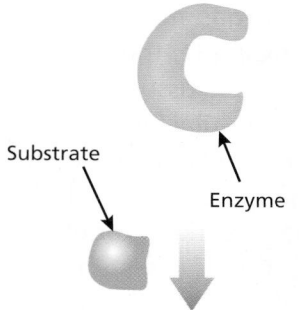

Substrate

Enzyme

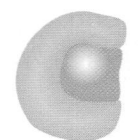

Combined substrate and enzyme. Reaction can take place.

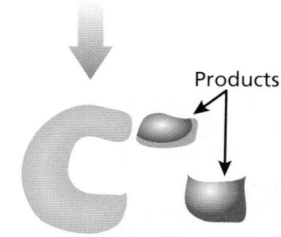

Products

Substrate is broken down and enzyme can be reused.

Enzymes in Digestion

- Digestive enzymes are produced by specialised cells in glands and in the lining of the gut:
  - ❶ The enzymes pass out of the cells into the digestive system.
  - ❷ They come into contact with food molecules.
  - ❸ They catalyse the breakdown of large insoluble food molecules into smaller soluble molecules.
- The digestive enzymes, protease, lipase and carbohydrase, digest proteins, lipids (fats and oils) and carbohydrates to produce smaller molecules that can be easily absorbed into the bloodstream.

> ### Key Point
>
> The 'Lock and Key Theory' is an example of how models are used in science to try and explain observations.

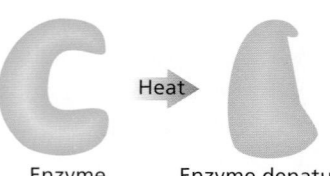

Heat

Enzyme

Enzyme denatured by heat

REQUIRED PRACTICAL	
Use qualitative reagents to test for a range of carbohydrates, lipids and proteins.	
Sample Method	**Considerations, Mistakes and Errors**
1. To test for sugars, e.g. glucose, add Benedict's reagent and heat in a water bath for two minutes. If sugar is present, it will turn red. 2. To test for starch add iodine solution. If starch is present, it will turn blue-black. 3. To test for protein add biuret reagent. If protein is present, it will turn purple.	• Do not boil the mixture for a long time, because any starch present might break down into sugar and test positive. • Refer to 'iodine solution' not 'iodine'. • Sometimes the purple colour is difficult to see. Try holding the test tube in front of a sheet of white paper.

- Amylase:
  - is produced in the salivary glands and the pancreas
  - is a carbohydrase that breaks down starch into sugar (maltose).
- Protease:
  - is produced in the stomach, pancreas and small intestine
  - breaks down proteins into amino acids.
- Lipase:
  - is produced in the pancreas and small intestine
  - breaks down lipids (fats) into fatty acids and glycerol.

Bile and Digestion

- **Bile** is a liquid made in the liver and stored in the gall bladder.
- It is alkaline to neutralise hydrochloric acid from the stomach.
- It also emulsifies fat to form small droplets, increasing the surface area for enzymes to act on.
- The alkaline conditions and large surface area increase the rate at which fat is broken down by lipase.

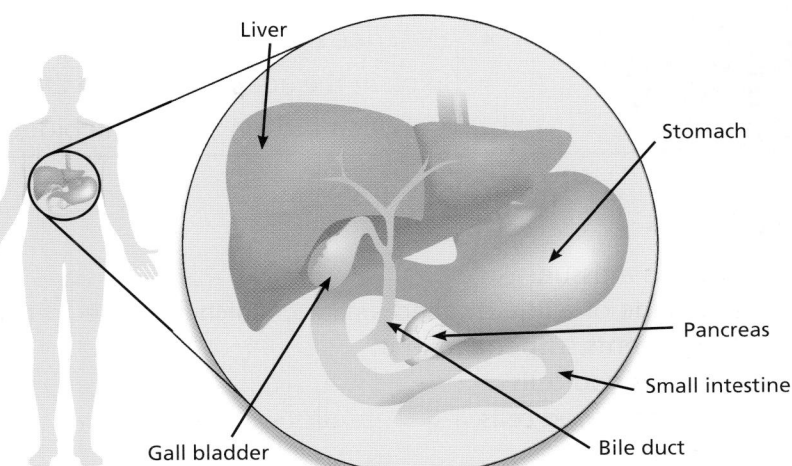

Liver

Stomach

Pancreas

Small intestine

Bile duct

Gall bladder

REQUIRED PRACTICAL	
Investigate the effect of pH on the rate of reaction of amylase enzyme.	
Sample Method 1. Put a test tube containing starch solution and a test tube containing amylase into a water bath at 37°C. 2. After 5 minutes add the amylase solution to the starch. 3. Every 30 seconds take a drop from the mixture and test it for starch using iodine solution. 4. Record how long it takes for the starch to be completely digested. 5. Repeat the experiment at different pH values using different buffer solutions.	**Considerations, Mistakes and Errors** • The solutions need to be left in the water bath for a while to reach the correct temperature before they are mixed. • After mixing, the tube must be kept in the water bath. • A buffer solution must be used to keep the reaction mixture at a certain fixed pH.
Variables • The independent variable is the one deliberately changed – in this case, the pH. • The dependent variable is the one that is measured – in this case, the time taken for the starch to be digested. • The control variables are kept the same – in this case, temperature, concentration and volume of starch and amylase.	**Hazards and Risks** • Care must be taken if a Bunsen burner is used to heat the water bath. • Take care not to spill iodine solution on the skin.

> **Quick Test**
>
> 1. What type of molecule is an enzyme?
> 2. Give **two** factors that affect the rate at which enzymes work.
> 3. Where are protease enzymes produced in the body?
> 4. What type of enzyme breaks down lipids?
> 5. Where is bile produced?

> **Key Words**
>
> catalyst
> active site
> optimum
> lock and key theory
> denature
> protease
> lipase
> carbohydrase
> amylase
> bile

Blood and the Circulation

You must be able to:

- Explain how the components of blood are adapted for their roles
- Explain how the different types of blood vessels are adapted for their functions
- Describe the structure and function of the heart
- Explain the adaptations that allow gaseous exchange to take place between the lungs and the blood.

Blood

- Blood is a tissue.
- It is made of a liquid called **plasma**, which has three different components suspended in it:
  - red blood cells
  - white blood cells
  - platelets.
- Plasma transports various chemical substances around the body, such as the products of digestion, hormones, antibodies, urea and carbon dioxide.
- Red blood cells:
  - contain **haemoglobin**, which binds to oxygen to transport it from the lungs to the tissues and cells, which need it for respiration
  - do not contain a nucleus, so there is more room for haemoglobin
  - are very small, so they can fit through the tiny capillaries
  - are shaped like biconcave discs, giving them a large surface area that oxygen can quickly diffuse across.
- White blood cells:
  - help to protect the body against infection
  - can change shape, so they can squeeze out of the blood vessels into the tissues or surround and engulf microorganisms.
- Platelets are fragments of cells, which collect at wounds and trigger blood clotting.

Blood Vessels

- The blood passes around the body in blood vessels.
- The body contains three different types of blood vessel:

> ### Key Point
>
> In the lungs:
>
> haemoglobin + oxygen → oxyhaemoglobin
>
> In the tissues:
>
> oxyhaemoglobin → haemoglobin + oxygen

Red Blood Cell

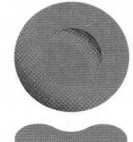

No nucleus, so packed full of haemoglobin to absorb oxygen.

White Blood Cell

Can change shape in order to engulf and destroy invading microorganisms.

Arteries	Veins	Capillaries
• Take blood from your heart to your organs. • Thick walls made from muscle and elastic fibres.	• Take blood from your organs to your heart. • Thinner walls and valves to prevent backflow.	• Allow substances needed by the cells to pass out of the blood. • Allow substances produced by the cells to pass into the blood. • Narrow, thin-walled blood vessels.

The Heart

- The heart pumps blood around the body in a **double circulatory system**.
- Blood passes through the heart twice on each circuit.
- There are four chambers in the heart:
  - the left and right **atria**, which receive blood from veins
  - the left and right **ventricles**, which pump the blood out into arteries.

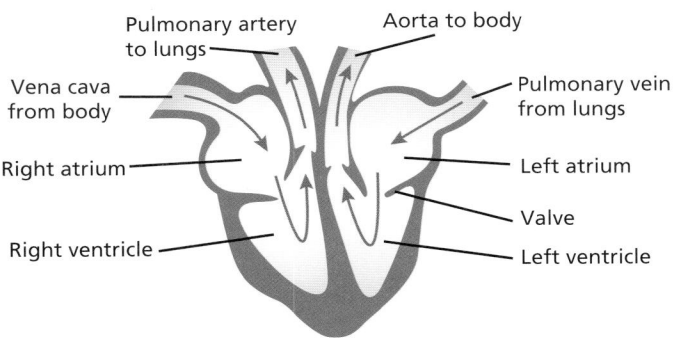

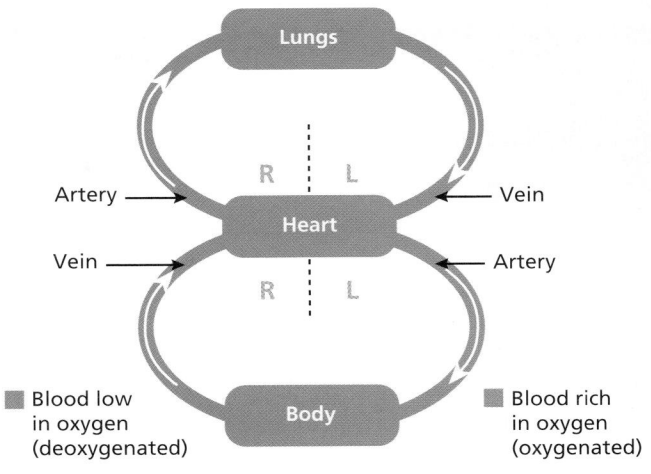

- Blood enters the heart through the atria.
- The atria contract and force blood into the ventricles.
- The ventricles then contract and force blood out of the heart.
- Valves make sure the blood flows in the correct direction.
- The natural resting heart rate is controlled by a group of cells located in the right atrium, which act as a **pacemaker**.
- Artificial pacemakers are electrical devices used to correct irregularities in the heart rate.

Gaseous Exchange

- The heart sends blood to the lungs via the **pulmonary artery**.
- Air obtained by breathing reaches the lungs through the **trachea** (windpipe), which has rings of **cartilage** to prevent it collapsing.
- The trachea divides into two tubes – the **bronchi**.
- The bronchi divide to form **bronchioles**.
- The bronchioles divide until they end in tiny air sacs called **alveoli**.
- There are millions of alveoli and they are adapted to be very efficient at exchanging oxygen and carbon dioxide:
  - They have a large, moist surface area.
  - They have a very rich blood supply.
  - They are very close to the blood capillaries, so the distance for gases to diffuse is small.
- The blood is taken back to the heart through the **pulmonary vein.**

Key Point

The pulmonary artery is unusual because, unlike other arteries, it carries deoxygenated blood. The pulmonary vein carries oxygenated blood.

Key Words

plasma
haemoglobin
double circulatory system
atria
ventricle
pacemaker
pulmonary artery
trachea
cartilage
bronchi
bronchioles
alveoli
pulmonary vein

Quick Test

1. Which component of blood makes it clot?
2. How are red blood cells adapted to carry oxygen?
3. Which type of blood vessel carries blood away from the heart?
4. In which chamber does deoxygenated blood enter the heart?
5. What do the heart and veins contain to prevent backflow of blood?

Non-Communicable Diseases

You must be able to:

- Explain the difference between non-communicable and communicable diseases
- Explain, with examples, what is meant by 'risk factors'
- Describe the causes and treatments for certain heart diseases
- Describe the main causes and types of cancer.

Health and Disease

- Good health is a state of physical and mental wellbeing.
- A disease is caused by part of the body not working properly. This can affect physical and / or mental health.
- Diseases can be divided into two main types: **communicable diseases** and **non-communicable diseases**.
- Non-communicable diseases cannot be spread between organisms, but communicable diseases can.
- There are many examples of how different diseases can interact with each other:
  - Viruses infecting cells can be the trigger for cancers, such as cervical cancer.
  - Diseases of the immune system mean that an individual is more likely to catch infectious diseases, e.g. people with HIV are more likely to get tuberculosis.
  - Immune reactions triggered by a pathogen can cause allergies, such as skin rashes and asthma.
  - If a person is physically ill, this can lead to depression and mental illness.
  - Poor diet, stress and difficult life situations can increase the likelihood of developing certain diseases.
- Non-communicable diseases, such as diabetes, can change a person's life and cost countries large sums of money.
- About 10% of the health budget in Britain is spent on diabetes.

> ### Key Point
>
> A causal mechanism is the process by which a cause brings about an effect.
>
> A causal mechanism has been found that links smoking to lung cancer. It is the action of the chemicals in the tar.

Risk Factors

- Non-communicable diseases are often caused by the interaction of a number of factors.
- These factors are called **risk factors**, because they make it more likely that a person will develop the disease.
- Risk factors can be:
  - aspects of a person's lifestyle, e.g. lack of exercise
  - substances in the person's body or environment, e.g. chemicals from smoking.
- Sometimes there is a clear link between a risk factor and the chance of getting a disease, e.g. obesity and Type 2 diabetes.
- This does not necessarily mean that the risk factor causes the disease.
- Scientists need to look for a **causal mechanism** to prove that a risk factor is involved.
- Some causal mechanisms have been found, linking some diseases and risk factors.

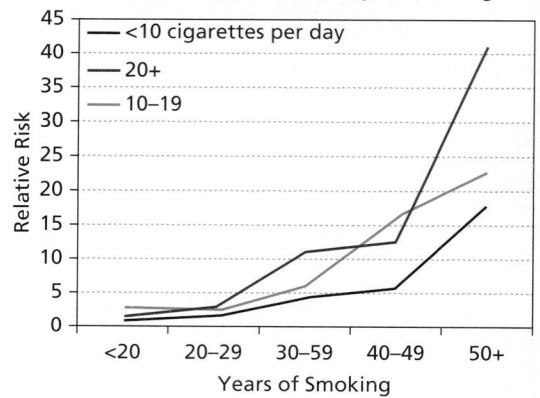

Relative Risk of Lung Cancer in Men According to Duration and Intensity of Smoking

Key: <10 cigarettes per day; 20+; 10–19
Y-axis: Relative Risk (0–45)
X-axis: Years of Smoking (<20, 20–29, 30–59, 40–49, 50+)

Disease	Proven Risk Factors
Cardiovascular disease	Lack of exercise / smoking / high intake of saturated fat
Type 2 diabetes	Obesity
Liver and brain damage	Excessive alcohol intake
Lung diseases, including lung cancer	Smoking
Skin cancer	Ionising radiation, e.g. UV light
Low birth weight in babies	Smoking during pregnancy
Brain damage in babies	Excessive alcohol intake during pregnancy

Diseases of the Heart

- In **coronary heart disease**, layers of fatty material build up inside the coronary arteries and narrow them.
- Treatments for coronary heart disease include:
  - **stents** to keep the coronary arteries open
  - **statins** to reduce blood cholesterol levels and slow down the rate at which fatty materials build up.
- In some people, heart valves may become faulty, developing a leak or preventing the valve from opening fully.
- Faulty valves can be replaced using biological or mechanical valves.
- For cases of heart failure:
  - a donor heart, or heart and lungs, can be transplanted
  - artificial hearts can be used to keep patients alive while waiting for a heart transplant or to allow the heart to recover.

Cancer

- Cancer is a non-communicable disease.
- Scientists have identified lifestyle risk factors for some types of cancer, e.g. smoking, obesity, common viruses and UV exposure.
- There are also genetic risk factors for some cancers, which may run in families, e.g. some genes make the carrier more susceptible to certain types of breast cancer.
- Cancer is caused by uncontrolled cell division. This can form masses of cells called **tumours**.
- There are two main types of tumours:
  - **Benign** tumours do not spread around the body.
  - **Malignant** tumours spread, in the blood, to different parts of the body where they form secondary tumours.

Coronary Heart Disease

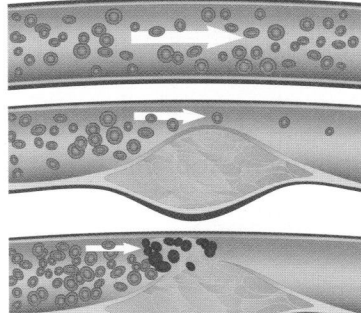

Key Point

The narrowing of coronary arteries reduces the flow of blood, so not enough oxygen can reach the heart muscle.

Key Words

health
disease
communicable disease
non-communicable disease
risk factor
causal mechanism
coronary heart disease
stent
statin
tumour
benign
malignant

Quick Test

1. State whether each of these diseases are **non-communicable** or **communicable**:
   a) Flu   b) Scurvy (lack of vitamin C)   c) Lung cancer
2. Why can spending too long in the sun result in skin cancer?
3. Give **one** risk factor for Type 2 diabetes.
4. Describe how coronary heart disease can cause heart muscle cells to stop contracting.

Transport in Plants

You must be able to:

- Explain how various plant tissues are adapted for their functions
- Describe how water is transported through a plant
- Describe how dissolved food substances are transported in a plant.

Plant Tissues

Tissue	Function
Epidermis	Covers the outer surfaces of the plant for protection.
Palisade mesophyll	The main site of photosynthesis in the leaf.
Spongy mesophyll	Air spaces between the cells allow gases to diffuse through the leaf.
Xylem vessels	Transports water and minerals through the plant, from roots to leaves. Also supports the plant.
Phloem vessels	Transports dissolved food materials through the plant.
Meristem tissue	Found mainly at the tips of the roots and shoots, where it can produce new cells for growth.

- Plant tissues are gathered together to form organs.
- The leaf is a plant organ.
- The structures of tissues in the leaf are related to their functions:

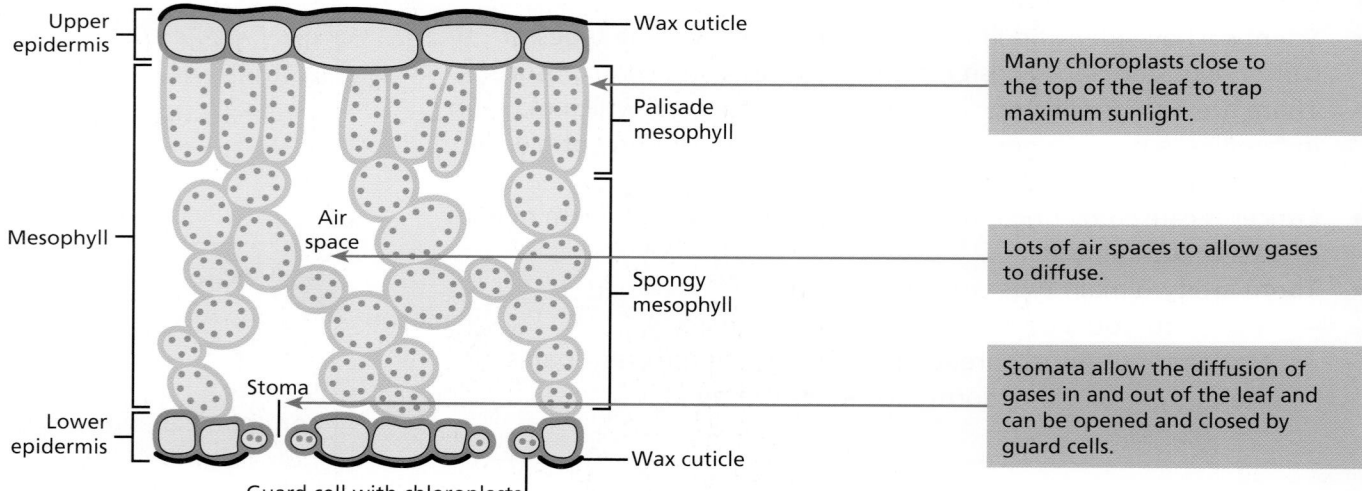

Water Transport

- Water enters the plant from the soil, through the root hair cells, by osmosis (see pages 22–23).
- Root hair, xylem and phloem cells are specialised to transport water, minerals and sugars around the plant (see page 24).

- This water contains dissolved minerals.
- The water and minerals are transported up the xylem vessels, from the roots to the stems and leaves.
- At the leaves, most of the water will evaporate and diffuse out of the **stomata** (small pores).
- The loss of water from the leaves is called **transpiration**.
- It helps to draw water up the xylem vessels from the roots.
- There are many factors that can affect the rate of transpiration:
  - An increase in temperature will increase the rate, as more energy is transferred to the water to allow it to evaporate.
  - Faster air flow will increase the rate, as it will blow away water vapour allowing more to evaporate.
  - Increased light intensity will increase the rate, as it will cause stomata to open.
  - An increase in humidity will decrease the rate as the air contains more water vapour, so the concentration gradient for diffusion is lower.
- In the leaf, the role of guard cells is to open and close stomata.
- At night the stomata are closed. This is because carbon dioxide is not needed for photosynthesis, so closing the stomata reduces water loss.
- When water is plentiful, guard cells take up water and bend. This causes the stomata to open, so gases for photosynthesis are free to move in and out of the stomata along with water from transpiration.
- When water is scarce, losing water makes the stomata change shape and close. This stops the plant from losing more water through transpiration.

Guard Cells

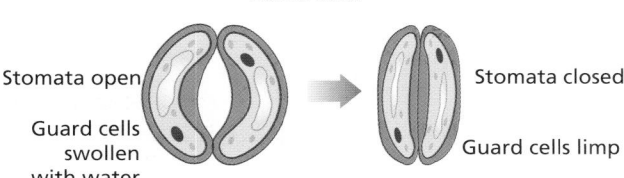

Stomata open · Guard cells swollen with water · Stomata closed · Guard cells limp

- The rate of transpiration from a cut shoot can be estimated by measuring the rate at which the shoot takes up water.
- This is only an estimate because not all of the water taken up by a shoot is lost – a very small percentage is used in the leaf.

Translocation

- Phloem tissue transports dissolved sugars from the leaves to the rest of the plant.
- This movement of food through phloem tissue is called **translocation**.
- Phloem cells are adapted for this function (see page 24).

(see page 24)

Key Point

In most plants, the stomata are mainly found on the bottom surface of the leaf. This means that the sun does not shine directly on them, reducing water loss.

Measuring Water Uptake by a Leafy Shoot Using a Potometer

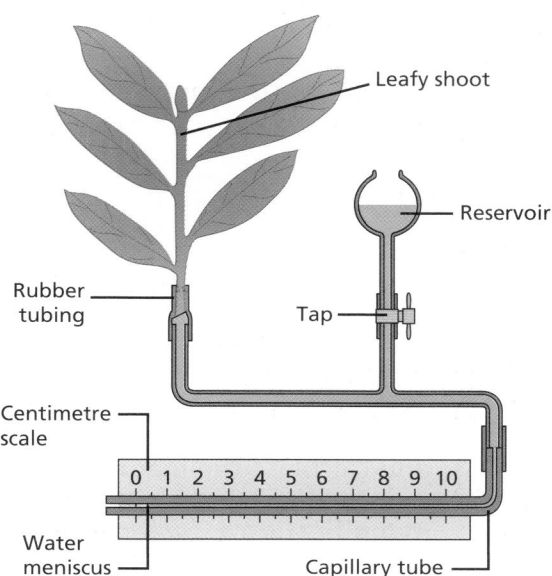

Leafy shoot
Reservoir
Rubber tubing
Tap
Centimetre scale
0 1 2 3 4 5 6 7 8 9 10
Water meniscus
Capillary tube

Key Point

Plants cannot stop transpiration completely. This is because carbon dioxide is needed for photosynthesis, so water will always escape.

Key Words

stomata
transpiration
translocation

Quick Test

1. In plant leaves, which tissue is the main site of photosynthesis?
2. What substances are transported by xylem tissue?
3. Give **two** environmental factors that slow down transpiration.
4. Why do stomata close at night?

Practice Questions

Cell Structure

1. Complete **Table 1** with a tick (✓) or cross (✗) to show if the structures are present or absent in the cells listed.

Table 1

Type of Cell	Nucleus	Cytoplasm	Cell Membrane	Cell Wall
Plant cell	✓	✓	✓	✓
Bacterial cell	✗	✓	✗	✓
Animal cell	✗	✓	✓	✗

[3]

Total Marks _____ / 3

Investigating Cells

1. **Figure 1** shows an image of a palisade cell.

   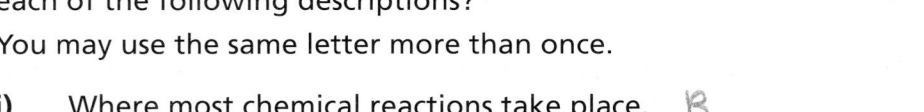

   Figure 1

   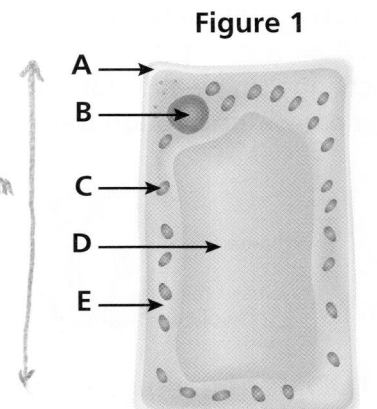

   a) The actual height of the cell is 0.1mm.

   $\frac{5}{} \quad \frac{50mm}{0.1mm} = 500mm$

   Use a ruler to measure the height of the cell in the diagram 5cm and calculate the magnification of the image in **Figure 1**.

   $$\text{magnification} = \frac{\text{size of image}}{\text{size of real object}}$$

   [2]

   b) Which structure, **A**, **B**, **C**, **D** or **E**, in the cell in **Figure 1** matches each of the following descriptions?
   You may use the same letter more than once.

   i)   Where most chemical reactions take place. B   [1]

   ii)   Made of cellulose. E   [1]

   iii) Contains chlorophyll. C   [1]

   iv)   Controls the cell's functions. B   [1]

   v)   Gives the cell rigidity and strength. A   [1]

   vi)   Filled with cell sap. E   [1]

   vii) Where photosynthesis occurs. B   [1]

   viii) Contains chromosomes. B   [1]

   Total Marks _____ / 10

Cell Division

1 A magazine published this article about stem cells:

> In 2006, scientists at Kyoto University managed to turn body cells from rats into stem cells by inserting some genes.
>
> The inserted genes seemed to reprogramme the cells so that they could use all their genes again. This made the cells act like embryonic stem cells.
>
> Doctors are very interested in this discovery. This is because they hope that the technique could be used to produce stem cells in a way that fewer people will object to.

a) What are embryonic stem cells? *Cells that can form new life* **[2]**

b) Explain why stem cells may be useful to doctors and why this discovery might make fewer people object to their use. **[5]**

Total Marks _____ / 7

Transport In and Out of Cells

1 a) Complete the sentences about diffusion.

Diffusion is the spreading of _____ from an area of _____

concentration to an area of _____ concentration.

The greater the difference in concentration, the _____ the rate of diffusion. **[4]**

b) Glucose is reabsorbed into the blood in the kidneys by active transport.

Give **two** ways in which active transport is different from diffusion. **[2]**

2 Katie decides to investigate osmosis. This is the method she uses:

1. Prepare five boiling tubes containing different sugar solutions.
2. Cut five potato chips of equal size.
3. Weigh each potato chip and place one in each tube.
4. Leave for 24 hours.
5. Reweigh the potato chips.

Table 1 shows Katie's results.

Table 1

Concentration of Sugar Solution (mol/dm³)	Mass of Potato Chip Before (g)	Mass of Potato Chip After (g)	Difference in Mass (g)
0.00	1.62	1.74	0.12
0.25	1.72	1.62	−0.10
0.50	1.69	1.62	−0.07
0.75	1.76	1.60	−0.16
1.00	1.74	1.59	−0.15

a) What is the independent variable in Katie's investigation? **[1]**

b) Name a variable that should be kept constant during Katie's investigation. **[1]**

c) In which concentration did the potato gain mass? **[1]**

d) Explain why this potato chip gained mass. **[3]**

e) How could Katie increase the reliability of her results? **[1]**

3 Complete **Table 2** by putting a tick (✓) or a cross (✗) in each of the blank boxes.

Table 2

	Osmosis	Diffusion	Active Transport
Can cause a substance to enter a cell		✓	
Needs energy from respiration	✗		
Can move a substance against a concentration gradient	✗		
Is responsible for oxygen moving into the red blood cells in the lungs			✗

[4]

Total Marks .... / 17

Levels of Organisation

1 Some cells are specialised to carry out a specific function.

Draw **one** line from each description to the correct function of that type of cell.

Description	Function
A cell that is hollow and forms tubes	to contract
A cell that has a flagellum	to transport water
A cell that is full of protein fibres	to carry nerve impulses
A cell that has a long projection with branched endings	to swim

[3]

Total Marks _____ / 3

Digestion

1 **a)** Name the organ in the body that produces bile. [1]

b) Where is bile stored in the human body? [1]

c) Into which part of the digestive system is bile released? [1]

d) Bile is needed to neutralise the acid that was added to food in the stomach.

Why is this important? [1]

2 Enzymes are used in industry and in the home.
For example, enzymes can be used in the manufacture of baby food to help predigest certain foods.

a) What type of enzyme is used in industry to predigest proteins? [1]

b) What are produced through the digestion of proteins by enzymes? [1]

Total Marks _____ / 6

Practice Questions

Blood and the Circulation

1. **Figure 1** shows some of the structures in the thorax.

   a) Name the structures **A**, **B** and **C**. **[3]**

   b) Alveoli are adapted to help gas exchange in the lungs.

      i) Which gas diffuses from the blood into the alveoli? **[1]**

      ii) Which gas diffuses from the alveoli into the blood? **[1]**

Figure 1

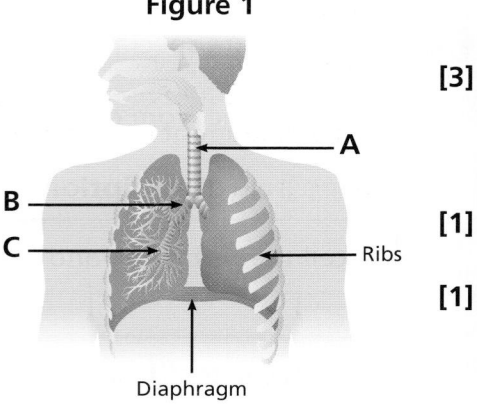

2. The human body has a double circulatory system.

   Complete the sentences.

   Oxygenated blood from the _____ returns to the left atrium of the heart in the

   pulmonary _____. From here, it enters the left ventricle and leaves the heart via

   the aorta to go to the _____. From the body, _____ blood returns via

   the _____ to the right atrium and then leaves the heart in the pulmonary

   artery to go to the lungs. **[5]**

   > **Total Marks** _____ / 10

Non-Communicable Diseases

1. Some non-communicable diseases have proven risk factors linked to them.

   Draw **one** line from each disease to the risk factor linked to it.

Disease	Proven Risk Factors
Type 2 diabetes	smoking
liver damage	excess alcohol intake
lung cancer	obesity
skin cancer	UV light

   [3]

2 Emphysema is a lung disease that irreversibly damages the alveoli in the lungs.

Two men do the same amount of daily exercise.

One man is in good health. The other man has emphysema.

Measurements are taken to show:
- The amount of air that enters their lungs when they inhale.
- The amount of oxygen that enters their blood.

Table 1 shows the results.

Table 1

	Healthy Man	Man with Emphysema
Total Air Flowing Into Lungs (dm³ per minute)	89.5	38.9
Oxygen Entering Blood (dm³ per minute)	2.5	1.2

a) Calculate the percentage difference between the total air flowing into the two men's lungs per minute. [2]

b) Explain how the changes to the lungs caused by emphysema can account for the difference in the oxygen figures. [3]

c) Explain why the man with emphysema will struggle to carry out exercise. [2]

Total Marks _____ / 10

Transport in Plants

1 Circle **three** phrases that could be used to correctly complete the sentence.

absorbing oxygen   giving off water vapour   absorbing carbon dioxide

giving off carbon dioxide   giving off oxygen   absorbing water

During the day, when the stomata are open, the leaf is _____. [3]

2 Plants have two separate transport tissues.

a) Give the name of the tissue responsible for transporting water and mineral ions. [1]

b) Give the name of the tissue responsible for transporting dissolved sugars. [1]

Total Marks _____ / 5

Pathogens and Disease

You must be able to:

- Describe the main types of disease-causing pathogen
- Describe the symptoms and method of spread of measles, HIV, salmonella and gonorrhoea
- Explain the role of the mosquito in the spread of malaria
- Describe the symptoms and method of spread of rose black spot in plants.

Pathogens and Disease

- **Pathogens** are microorganisms that cause infectious (communicable) diseases.
- Pathogens may infect plants or animals.
- They can be spread by:
  - direct contact
  - water or air
  - **vectors** (organisms that carry and pass on the pathogen without getting the disease).
- The spread of infectious diseases can be reduced by:
  - simple hygiene measures, e.g. washing hands and sneezing into a handkerchief
  - destroying vectors
  - isolating infected individuals, so they cannot pass the pathogen on
  - giving people at risk a vaccination (see page 43).

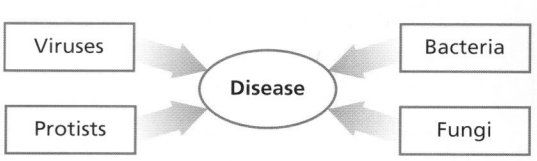

Viral Diseases

- Viruses reproduce rapidly in body cells, causing damage to the cells.
- **Measles** is a disease caused by a virus:
  - The symptoms are fever and a red skin rash.
  - The measles virus is spread by breathing in droplets from sneezes and coughs.
  - Although most people recover well from measles, it can be fatal if there are complications, so most young children are vaccinated against measles.
- **HIV** (human immunodeficiency virus) causes AIDS:
  - It is spread by sexual contact or exchange of body fluids, e.g. it can be transmitted in blood when drug users share needles.
  - At first, HIV causes a flu-like illness.
  - If untreated, the virus enters the lymph nodes and attacks the body's immune cells.
  - Taking antiviral drugs can delay this happening.
  - Late stage HIV, or AIDS, is when the body's immune system is damaged and cannot fight off other infections or cancers.
- **Tobacco mosaic virus (TMV)** occurs in tobacco plants and many other species, including tomatoes:
  - The distinctive 'mosaic' pattern of discolouration reduces the chlorophyll content of leaves.
  - It therefore affects photosynthesis and plant growth.

The Virus that Causes Measles

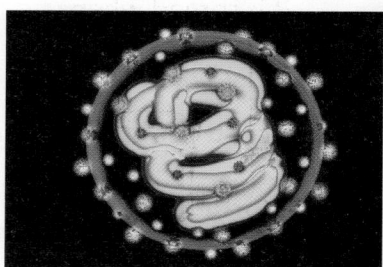

Key Point

It is not the HIV virus that directly kills people with AIDS. It is other infections, such as pneumonia, that a healthy body would usually be able to survive.

Bacterial Diseases

- Bacteria may damage cells directly or produce **toxins** (poisons) that damage tissues.
- **Salmonella** is a type of food poisoning caused by bacteria:
  - The bacteria are ingested in food, which may not have been cooked properly or may not have been prepared in hygienic conditions.
  - The bacteria secrete toxins, which cause fever, abdominal cramps, vomiting and diarrhoea.
  - Chicken and eggs can contain the bacteria, so chickens in the UK are vaccinated against salmonella to control the spread.
- **Gonorrhoea** is a sexually transmitted disease (STD) caused by bacteria:
  - It is spread by sexual contact.
  - The symptoms are a thick, yellow or green discharge from the vagina or penis and pain when urinating.
  - It used to be easily treated with penicillin, but many resistant strains have now appeared.
  - The use of a barrier method of contraception, e.g. a condom can stop the bacteria being passed on.

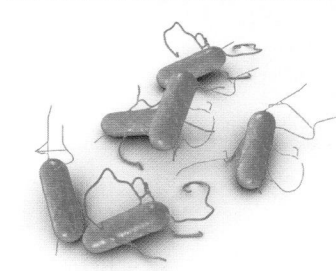

The Bacteria that Cause Salmonella

Protists and Disease

- Protists are single-celled organisms.
- However, unlike bacteria, they are eukaryotic.
- **Malaria** is caused by a protist:
  - The protist uses a particular type of mosquito as a vector.
  - It is passed on to a person when they are bitten by the mosquito.
  - Malaria causes severe fever, which reoccurs and can be fatal.
  - One of the main ways to stop the spread is to stop people being bitten, e.g. by killing the mosquitoes or using mosquito nets.

Mosquitoes Transmit Malaria

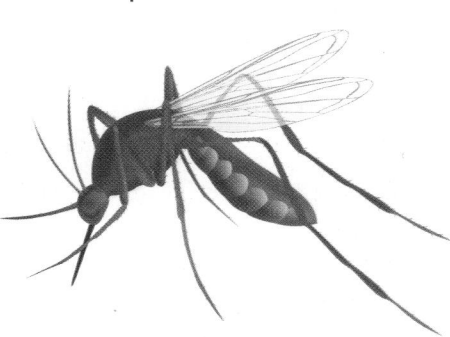

Fungal Diseases

- **Rose black spot** is a fungal disease:
  - It is spread when spores are carried from plant to plant by water or wind.
  - Purple or black spots develop on leaves, which often turn yellow and drop early.
  - The loss of leaves will stunt the growth of the plant because photosynthesis is reduced.
  - It can be treated by using fungicides and removing and destroying the affected leaves.

 Key Point

In malaria, the protist is the pathogen for the disease. The mosquito is acting as a parasite when it feeds on a person.

> **Quick Test**
>
> 1. State one simple precaution that can stop pathogens being spread by droplets in the air.
> 2. Why is it important to keep uncooked meat separate from cooked meat?
> 3. Why does the contraceptive pill **not** prevent the spread of gonorrhoea?
> 4. Why should leaves infected with rose black spot be removed and burned?

 Key Words

pathogen
vector
toxin

Human Defences Against Disease

You must be able to:

- Describe how the body tries to prevent pathogens from entering
- Describe how the immune system reacts if pathogens do enter the body
- Explain the process of immunity and how vaccinations work.

Preventing Entry of Pathogens

- The body has a number of non-specific defences against disease.
- These are defences that work against all pathogens, to try and stop them entering the body.

The Body's Defences

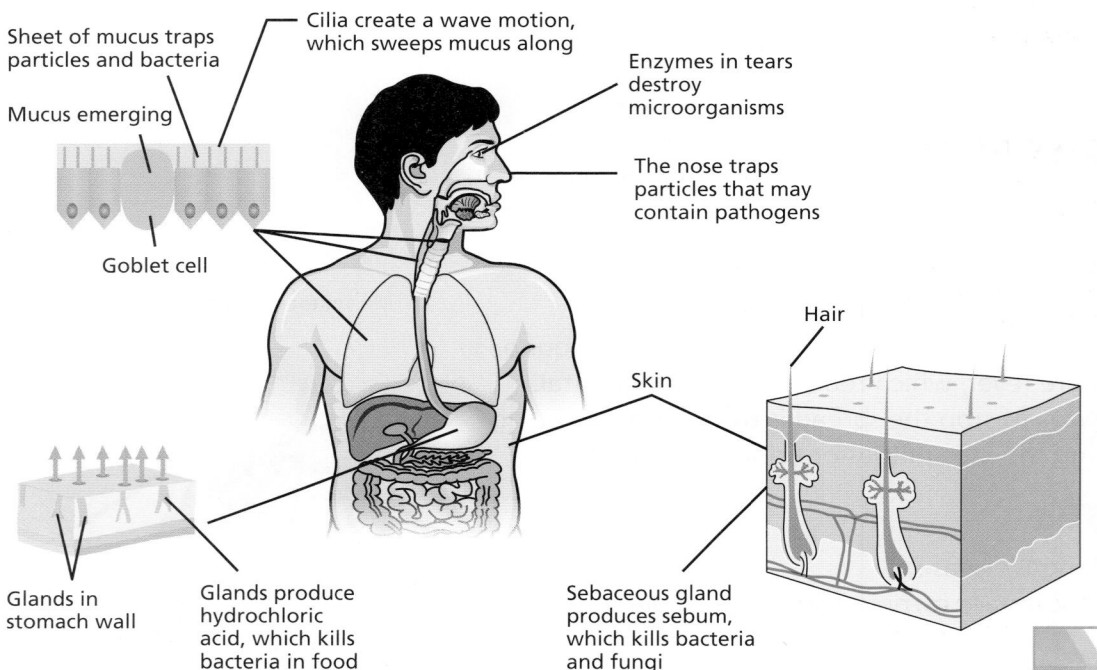

Sheet of mucus traps particles and bacteria

Mucus emerging

Goblet cell

Cilia create a wave motion, which sweeps mucus along

Enzymes in tears destroy microorganisms

The nose traps particles that may contain pathogens

Hair

Skin

Glands in stomach wall

Glands produce hydrochloric acid, which kills bacteria in food

Sebaceous gland produces sebum, which kills bacteria and fungi

The Immune System

- If a pathogen enters the body, the immune system tries to destroy it.
- White blood cells help to defend against pathogens through:
  - phagocytosis, which involves the pathogen being surrounded, engulfed and digested

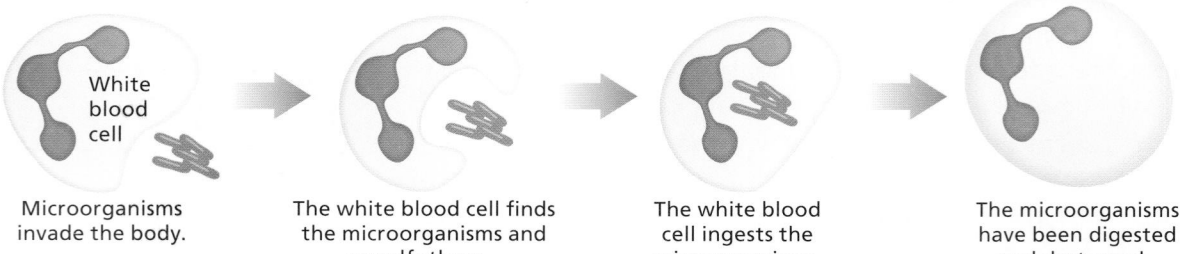

White blood cell

Microorganisms invade the body.

The white blood cell finds the microorganisms and engulfs them.

The white blood cell ingests the microorganisms.

The microorganisms have been digested and destroyed.

– the production of special protein molecules called antibodies, which attach to antigen molecules on the pathogen

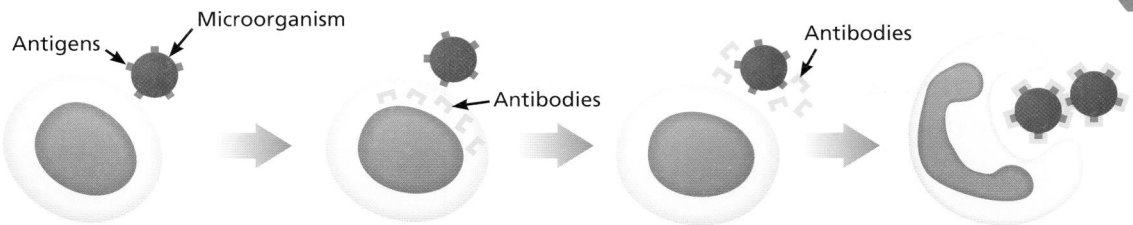

Antigens are markers on the surface of the microorganism.

The white blood cells become sensitised to the antigens and produce antibodies.

The antibodies then lock onto the antigens.

This causes the microorganisms to clump together, so that other white blood cells can digest them.

– the production of antitoxins, which are chemicals that neutralise the poisonous effects of the toxins.

Boosting Immunity

- If the same pathogen re-enters the body, the white blood cells respond more quickly to produce the correct antibodies.
- This quick response prevents the person from getting ill and is called immunity.
- When a person has a vaccination, small quantities of dead or inactive forms of a pathogen are injected into the body.
- Vaccination stimulates the white blood cells to produce antibodies and to develop immunity.

1 A weakened / dead strain of the microorganism is injected. Antigens on the modified microorganism's surface cause the white blood cells to produce specific antibodies.

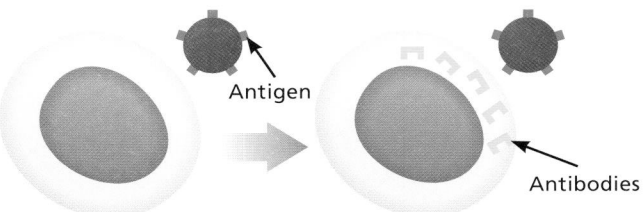

Antigen

Antibodies

2 The white blood cells that are capable of quickly producing the specific antibody remain in the bloodstream.

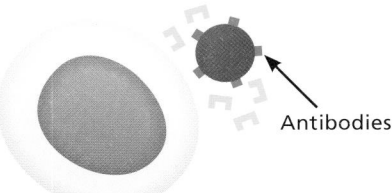

Antibodies

- If a large proportion of the population can be made immune to a pathogen, then the pathogen cannot spread very easily.

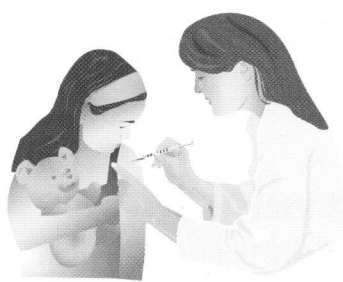

Key Words

non-specific defences
immune system
phagocytosis
antibody
antigen
antitoxin
immunity
vaccination

Quick Test

1. How does the stomach help to kill pathogens?
2. What is phagocytosis?
3. What is the name of the protein molecules made by white blood cells when they detect a pathogen?
4. Why does vaccination use a dead or weakened pathogen?

Treating Diseases

You must be able to:

- Explain how antibiotics have saved lives and why their use is now under threat
- Describe how new drugs are developed.

Antibiotics

- **Antibiotics**, e.g. penicillin, are medicines that kill bacteria inside the body. However, they cannot destroy viruses.
- Doctors will prescribe certain antibiotics for certain diseases.
- The use of antibiotics has greatly reduced deaths from infections.
- However, bacterial strains resistant to antibiotics are increasing (see page 69).
- **MRSA** is a strain of bacteria that is resistant to antibiotics.
- To reduce the rate at which resistant strains of bacteria develop:
  - doctors should **not** prescribe antibiotics:
    - o unless they are really needed
    - o for non-serious infections
    - o for viral infections.
  - patients must complete their course of antibiotics so that all bacteria are killed and none survive to form resistant strains.

> ### Key Point
>
> Antibiotics only kill bacteria. They do not kill viruses.

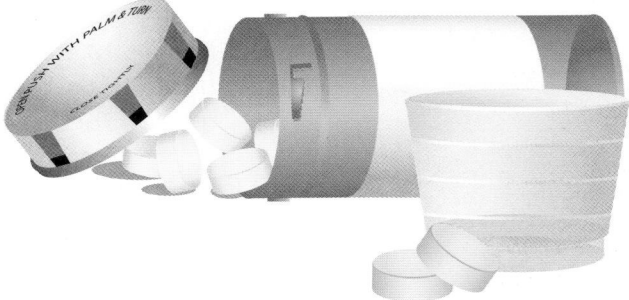

Developing New Drugs

- There is a constant demand to produce new drugs:
  - New painkillers are developed to treat the symptoms of disease but they do not kill the pathogens.
  - Antiviral drugs are needed that will kill viruses without also damaging the body's tissues.
  - New antibiotics are needed as resistant strains of bacteria develop.
- Traditionally drugs were extracted from plants and microorganisms:
  - **Digitalis** is a heart drug that originates from foxgloves.
  - **Aspirin** is a pain killer that originates from willow.
  - **Penicillin** was discovered by Alexander Fleming from the *Penicillium* mould.

Foxgloves

- Now most new drugs are synthesised (made) by chemists in the pharmaceutical industry. However, the starting point may still be a chemical extracted from a plant.
- New medical drugs have to be tested and trialled before being used to make sure they are safe (not toxic).
- If a drug is found to be safe, it is then tested on patients to:
  - see if it works
  - find out the optimum dose.
- These tests on patients are usually **double-blind trials**:
  - some patients are given a **placebo**, which does not contain the drug, and some patients are given the drug
  - patients are allocated randomly to the two groups
  - neither the doctors nor the patients know who has received a placebo and who has received the drug.
- New painkillers are developed to treat the symptoms of disease – they do not kill pathogens.
- New antiviral drugs are needed that will kill viruses without damaging the body's tissues. This is not easy to achieve.

Key Point

The purpose of a double-blind trial is to ensure that it is completely fair. If the patients or doctors knew whether it was the drug or a placebo being used, it might influence the outcome of the test.

Quick Test

1. Why can antibiotics not be used to destroy HIV?
2. What term describes bacterial pathogens that are not affected by antibiotics?
3. Why did people once chew willow bark if they had a headache?
4. Why does the title 'double-blind trial' include the word 'double'?

Key Words

antibiotics
MRSA
digitalis
aspirin
penicillin
double-blind trial
placebo

Photosynthesis

You must be able to:

- Recall the word equation for photosynthesis
- Understand that photosynthesis is an endothermic reaction
- Explain how various factors can change the rate of photosynthesis
- Describe how the products of photosynthesis are used by plants.

Photosynthesis

- The equation for photosynthesis is:

carly didn't want light getting out. for the photo

LEARN

$$\text{carbon dioxide} + \text{water} \xrightarrow{\text{light}} \text{glucose} + \text{oxygen}$$
$$CO_2 \qquad\qquad H_2O \qquad\qquad C_6H_{12}O_6 \qquad O_2$$

- To produce glucose molecules by photosynthesis, energy is required.
- This is because the reactions are endothermic (take heat in).
- The energy needed is supplied by sunlight.
- It is trapped by the green chemical chlorophyll, which is found in chloroplasts.

Factors Affecting Photosynthesis

- There are several factors that may affect the rate of photosynthesis.

HT At any moment, the factor that stops the reaction going any faster is called the limiting factor.

- **Temperature**
  - o As the temperature increases, so does the rate of photosynthesis.
  - o This is because more energy is provided for the reaction.
  - o As the temperature approaches 45°C, the rate of photosynthesis drops to zero because the enzymes controlling photosynthesis have been destroyed.
- **Carbon dioxide concentration**
  - o As the concentration of CO_2 increases, so does the rate of photosynthesis.
  - o This is because CO_2 is needed in the reaction.

  HT After reaching a certain point, an increase in CO_2 has no further effect. CO_2 is no longer the limiting factor.

- **Light intensity**
  - o As light intensity increases, so does the rate of photosynthesis.
  - o This is because more energy is provided for the reaction.

  HT After reaching a certain point, any increase in light has no further effect. It is no longer the limiting factor.

- **Chlorophyll concentration**
  - o This does not vary in the short term but may change if plants are grown in soil without enough minerals to make chlorophyll.

Key Point

During photosynthesis, the energy from sunlight is converted to chemical energy in the form of glucose molecules.

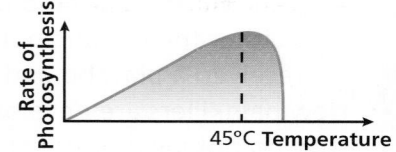

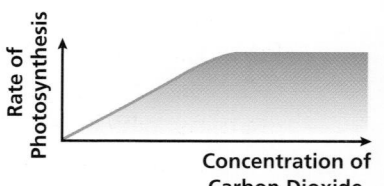

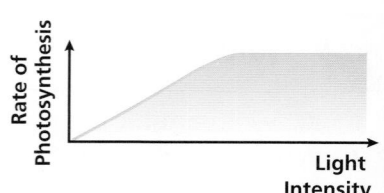

HT By looking at a graph, it is possible to say what the limiting factor is at any point.

HT Greenhouses can be used to increase the rate of photosynthesis. By controlling lighting, temperature and carbon dioxide, farmers can increase the growth rate of their crops.

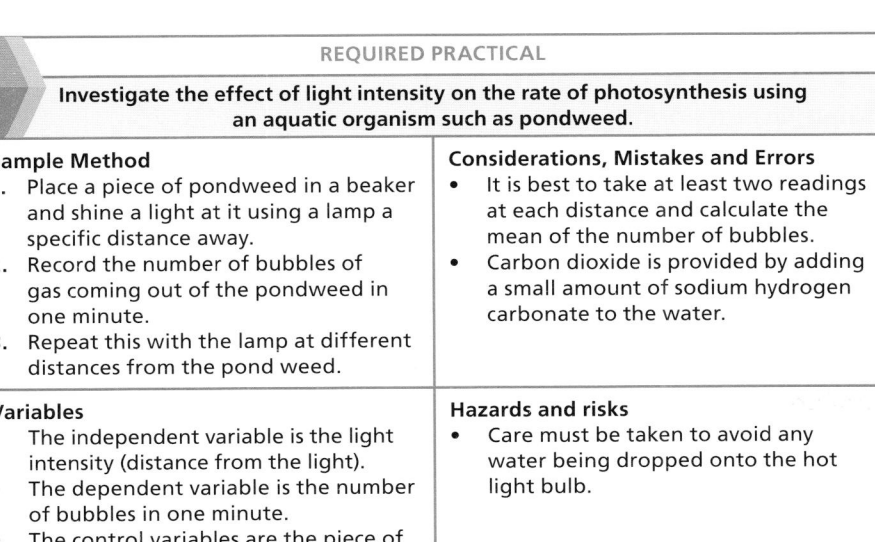

At this point on the graph light intensity is limiting

Here light intensity is not limiting but temperature has become the limiting factor

 Key Point

Farmers have to carefully work out if the extra cost of lighting and heating will be paid for by the extra growth that their crops achieve.

REQUIRED PRACTICAL
Investigate the effect of light intensity on the rate of photosynthesis using an aquatic organism such as pondweed.

Sample Method	**Considerations, Mistakes and Errors**
1. Place a piece of pondweed in a beaker and shine a light at it using a lamp a specific distance away. 2. Record the number of bubbles of gas coming out of the pondweed in one minute. 3. Repeat this with the lamp at different distances from the pond weed.	• It is best to take at least two readings at each distance and calculate the mean of the number of bubbles. • Carbon dioxide is provided by adding a small amount of sodium hydrogen carbonate to the water.
Variables • The independent variable is the light intensity (distance from the light). • The dependent variable is the number of bubbles in one minute. • The control variables are the piece of pondweed, the temperature, and the concentration of carbon dioxide.	**Hazards and risks** • Care must be taken to avoid any water being dropped onto the hot light bulb.

HT When light intensity is studied, doubling the distance between the lamp and the pondweed will reduce the light intensity by a quarter. This is called the **inverse square law**.

Converting Glucose

- The glucose produced in photosynthesis may be used by the plant during respiration to provide energy.
- Glucose may also be changed into other products such as:
  - insoluble starch, which is stored in the stem, leaves or roots
  - fat or oil, which is also stored, e.g. in seeds
  - cellulose, to strengthen cell walls
  - proteins, which are used for growth and for enzymes.
- To produce proteins from glucose, plants also use nitrate ions, which are absorbed from the soil.

 Key Point

Nitrate ions are needed to make proteins because amino acids contain nitrogen, but glucose does not.

Quick Test

1. Name the green pigment essential for photosynthesis.
2. Where do plants obtain the carbon dioxide used in photosynthesis?
3. List **three** factors that may limit the rate of photosynthesis.
4. What do plants need, in addition to glucose, to make proteins?

 Key Words

endothermic reaction
chlorophyll
HT limiting factor
HT inverse square law

Respiration and Exercise

You must be able to:

- Explain why respiration is important in living organisms
- Compare the processes of aerobic and anaerobic respiration
- Explain the changes that occur in respiration during exercise
- Describe respiration as a part of the metabolism of the body.

The Importance of Respiration

- Respiration is an example of an **exothermic reaction**.
- It releases energy from glucose molecules for use by the body.
- Organisms need this energy:
  - for chemical reactions to build larger molecules
  - for movement
  - to keep warm.
- Respiration in cells can be **aerobic** (with oxygen) or **anaerobic** (without oxygen).

sounds like air (oxygen)

Aerobic Respiration

- The equation for aerobic respiration is the same in all organisms:

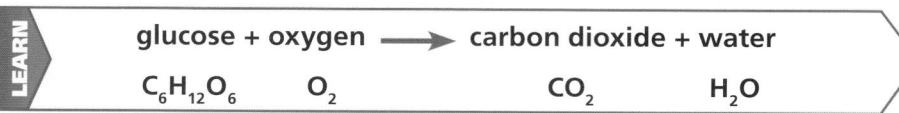

LEARN

glucose + oxygen ⟶ carbon dioxide + water

$C_6H_{12}O_6$ O_2 CO_2 H_2O

Anaerobic Respiration

- In anaerobic respiration, the glucose is not completely broken down.
- This means that it transfers much less energy than aerobic respiration.
- The process of anaerobic respiration is different in animals to the process found in plants and yeast.
- In animals, lactic acid is produced:

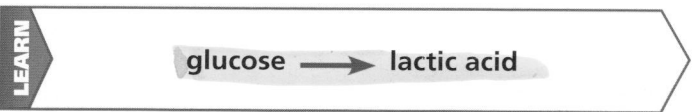

LEARN

glucose ⟶ lactic acid

- In plants and yeast, alcohol (ethanol) and carbon dioxide are produced:

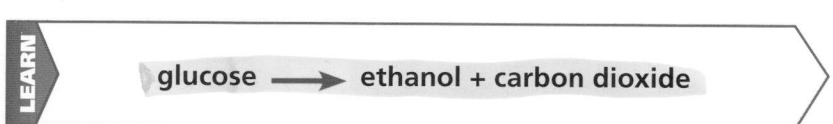

LEARN

glucose ⟶ ethanol + carbon dioxide

- Anaerobic respiration in yeast cells is called **fermentation**.
- It is important in the manufacture of bread and alcoholic drinks.

Exercise and Respiration

- During exercise, the body demands more energy, so the rate of respiration needs to increase.

> ### Key Point
>
> Ethanol is the type of alcohol made by fermentation and is found in alcoholic drinks. The carbon dioxide produced can also be trapped to make the drink fizzy.

- The heart rate, breathing rate and breath volume all increase to supply the muscles with more oxygen and glucose for the increase in aerobic respiration.

Diffusion Between a Capillary and a Working Muscle Cell

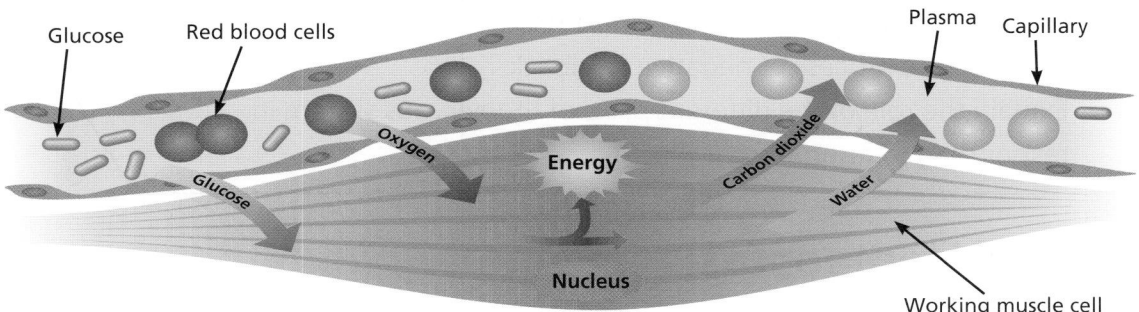

- During periods of vigorous activity, the muscles may not get supplied with enough oxygen, so anaerobic respiration starts to take place in the muscle cells.
- This causes a build-up of lactic acid and creates an oxygen debt.
- The lactic acid causes the muscles to hurt and stops them contracting efficiently. Lactic acid is a poison, so needs to be got rid of quickly.
- Once exercise is finished, the oxygen debt must be 'repaid'.

HT After exercise, blood flowing through the muscles transports the lactic acid to the liver where it is broken down.

HT The oxygen debt is the amount of extra oxygen the body needs after exercise to react with the lactic acid and remove it from the cells.

> **Key Point**
>
> Deep breathing for some time after exercise is used to 'pay back' the oxygen debt.

Metabolism

- Metabolism is the sum of all the chemical reactions in a cell or in the body.
- These reactions are controlled by enzymes and many need a transfer of energy.
- This energy is transferred by respiration and used to make new molecules.
- This includes:
  - the conversion of glucose to starch, glycogen and cellulose
  - the formation of lipid molecules from a molecule of glycerol and three molecules of fatty acids
  - the use of glucose and nitrate ions to form amino acids, which are used to synthesise proteins
  - the breakdown of excess proteins into urea for excretion.

> **Quick Test**
>
> 1. Which type of respiration requires oxygen?
> 2. Why do we need to eat more in cold weather?
> 3. Name the waste product of anaerobic respiration in animals
> 4. HT Where in the body is lactic acid broken down after exercise?
> 5. Why might your muscles hurt if you run a long race?

> **Key Words**
>
> exothermic reaction
> aerobic respiration
> anaerobic respiration
> fermentation
> lactic acid
> oxygen debt
> metabolism

Homeostasis and the Nervous System

You must be able to:

- Explain why homeostasis is so important
- HT Describe the process of negative feedback
- Explain the role of the different parts of the nervous system in responding to a stimulus.

The Importance of Homeostasis

- Homeostasis is the regulation of the internal conditions of a cell or organism in response to internal and external changes.
- Homeostasis is important because it keeps conditions constant for enzyme action and cell functions.
- Homeostasis includes the control of:
  - blood glucose concentration
  - body temperature
  - water and ion levels.
- The control systems may involve:
  - responses using nerves
  - chemical responses using hormones.

> **Key Point**
>
> Homeostasis is the maintenance of a constant internal environment by compensating for changes.

Control Systems

- All control systems include:
  - cells called receptors, which detect stimuli (changes in the environment)
  - coordination centres (such as the brain, spinal cord and pancreas), which receive and process information from receptors
  - effectors (muscles or glands), which bring about responses that restore optimum levels.

HT This type of control mechanism is called negative feedback:

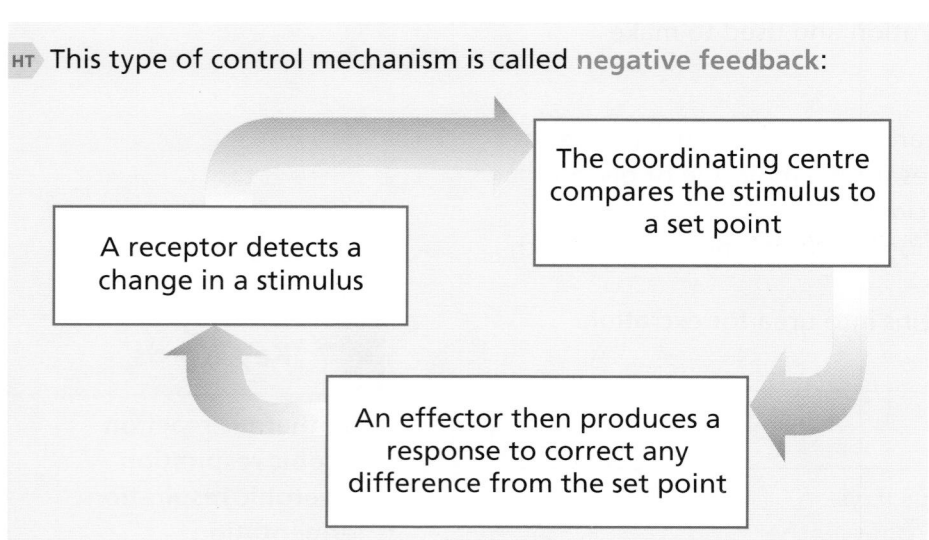

A receptor detects a change in a stimulus

The coordinating centre compares the stimulus to a set point

An effector then produces a response to correct any difference from the set point

The Nervous System

- The nervous system enables humans to react to their surroundings and coordinate their behaviour.
- Information from receptors passes to the **central nervous system (CNS)** (the brain and spinal cord).
- The CNS coordinates the response of effectors, i.e. muscles contracting or glands secreting hormones.
- Reflex actions are automatic and rapid so they can protect the body. They do not involve the conscious part of the brain:
  1. The pain stimulus is detected by receptors.
  2. Impulses from the receptor pass along a sensory neurone to the CNS.
  3. An impulse then passes through a relay neurone.
  4. A motor neurone carries an impulse to the effector.
  5. The effector (usually a muscle) responds, e.g. to withdraw a limb away from the source of pain.
- Neurones are not directly connected to each other.
- They communicate with each other via **synapses** (gaps between neurones).
- When an electrical impulse reaches a synapse, a chemical is released that diffuses across the gap between the two neurones.
- This causes an electrical impulse to be generated in the second neurone.

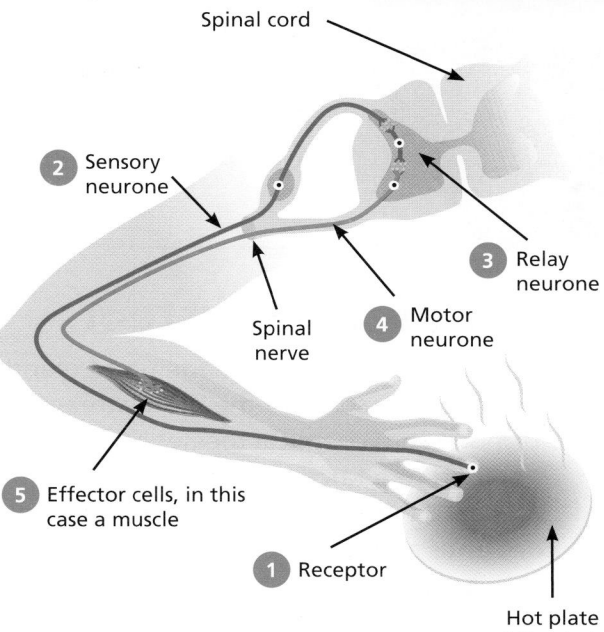

Spinal cord
2 Sensory neurone
3 Relay neurone
Spinal nerve
4 Motor neurone
5 Effector cells, in this case a muscle
1 Receptor
Hot plate (stimulus)

A Synapse

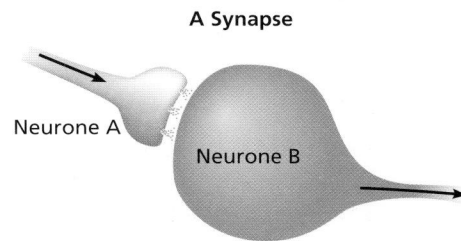

Neurone A
Neurone B

REQUIRED PRACTICAL	
Investigating the effect of a factor on human reaction time.	
Sample Method Reaction time can be investigated by seeing how quickly a dropped ruler can be caught between finger and thumb: 1. The experimenter holds a metre ruler vertically from the end. 2. The subject has their finger and thumb a small distance apart, either side of the ruler, on the 50cm line. 3. The experimenter lets go of the ruler and the subject has to trap it. 4. The distance the ruler travels from the 50cm line is noted. 5. The experiment is repeated on subjects that have just drunk coffee or cola and subjects that have not.	**Considerations, Mistakes and Errors** • It is very difficult to control the variables in this experiment. • To obtain reliable results, large numbers of subjects need to be tested and averages taken.
Variables • The independent variable is whether the subject has taken in caffeine or not. • The dependent variable is the distance that the ruler travels. • The control variables are the age, sex and mass of the subjects.	**Hazards and risks** • There are limited risks with this experiment.

Quick Test

1. Which type of neurone is responsible for sending impulses from the receptors to the CNS?
2. What is the gap between two neurones called?
3. Which **two** structures make up the CNS?

Key Words

homeostasis
receptors
effectors
HT negative feedback
central nervous system (CNS)
synapse

Hormones and Homeostasis

You must be able to:

- Describe the principles of hormonal coordination and control
- Describe the location of the main hormone-producing glands
- Explain how hormones are used to control blood glucose levels.

The Endocrine System

- The **endocrine system** is made up of glands that secrete hormones directly into the bloodstream.
- **Hormones** are chemical messengers that are carried in the blood to a target organ where they produce an effect.
- Compared with effects of the nervous system, the effects of hormones are slower and act for longer.
- The **pituitary gland** in the brain is a 'master gland'.
- It secretes several hormones in response to body conditions.
- Some of these hormones act on other glands to stimulate other hormones to be released and bring about effects.

Main Glands that Produce Hormones in the Human Body

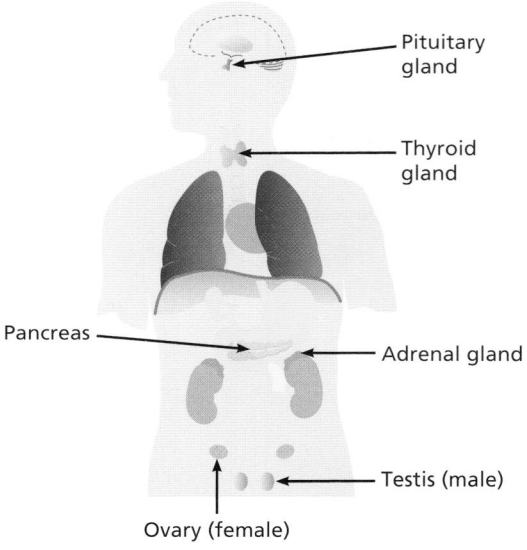

- Pituitary gland
- Thyroid gland
- Pancreas
- Adrenal gland
- Testis (male)
- Ovary (female)

HT **Adrenaline:**
- is produced by the adrenal glands in times of fear or stress
- increases the heart rate, boosting the delivery of oxygen and glucose to the brain and muscles
- prepares the body for 'flight or fight'.

HT **Thyroxine:**
- is produced by the thyroid gland
- increases the metabolic rate
- controls growth and development in young animals
- is controlled by negative feedback.

> ### Key Point
>
> Enzymes from glands like the salivary gland pass into tubes called ducts. Endocrine glands are sometimes called ductless glands, because the hormones pass into the blood.

Control of Blood Glucose

- Blood glucose concentration is monitored and controlled by the pancreas.
- If the blood glucose concentration is too high:
  - the pancreas releases more of the hormone insulin
  - insulin causes glucose to move from the blood into the cells
  - in liver and muscle cells, excess glucose is converted to glycogen for storage.

HT If the blood glucose concentration is too low:
  - the pancreas releases glucagon
  - glucagon stimulates glycogen to be converted into glucose and released into the blood.

HT This is an example of negative feedback (see page 50).

- **Type 1 diabetes** is a disorder that:
  - is caused by the pancreas failing to produce sufficient insulin
  - results in uncontrolled high blood glucose levels
  - is normally treated with insulin injections.
- **Type 2 diabetes** is a disorder that:
  - is caused by the body cells no longer responding to insulin
  - has obesity as a risk factor
  - is treated with a carbohydrate-controlled diet and regular exercise.
- The graph shows the effect of injecting insulin on blood glucose levels in people with Type 1 and Type 2 diabetes:
  - In people with Type 1 diabetes, the insulin lowers the blood glucose level by stimulating the liver to covert it into glycogen.
  - In people with Type 2 diabetes, there is little effect because the cells do not respond to insulin.

> ### Key Point
>
> Type 2 diabetes **cannot** be treated with insulin. Type 1 diabetes **cannot** be treated by controlling the diet.

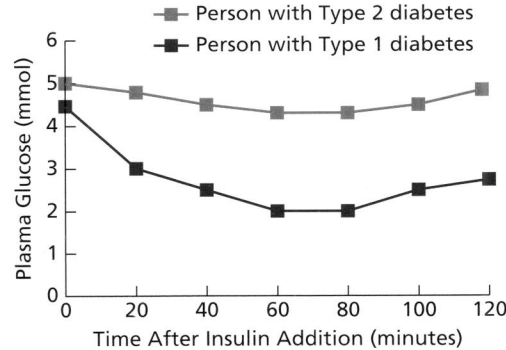

> ### Quick Test
>
> 1. Where is the thyroid gland?
> 2. Where is insulin made?
> 3. What effect does insulin have on liver cells?
> 4. How is Type 1 diabetes treated?

> ### Key Words
>
> endocrine system
> hormone
> pituitary gland
> **HT** adrenaline
> **HT** thyroxine
> **Type 1 diabetes**
> **Type 2 diabetes**

Hormones and Reproduction

You must be able to:

- Describe the roles of the main sex hormones in the body
- Explain how hormones control the menstrual cycle
- Describe the different methods of contraception
- HT Explain how hormones can be used to treat infertility.

The Sex Hormones

- Hormones play many roles in controlling human reproduction.
- During puberty, the sex hormones cause secondary sexual characteristics to develop.
- Oestrogen, from the ovaries, is the main female sex hormone.
- In females, at puberty, eggs begin to mature and be released. This is called ovulation.
- Testosterone is the main male sex hormone. It is produced by the testes and stimulates sperm production.
- After puberty, men produce sperm continuously, but women have a monthly cycle of events called the menstrual cycle.
- Several other hormones are involved in a woman's menstrual cycle.

Control of the Menstrual Cycle

- There are four hormones involved in control of the menstrual cycle:

Hormone	Secreted by	Function in the Menstrual Cycle
Follicle stimulating hormone (FSH)	Pituitary gland	• Causes eggs to mature in the ovaries in the first part of the cycle. HT Stimulates the ovaries to produce oestrogen.
Oestrogen	Ovaries	HT Inhibits FSH release. HT Stimulates LH release. • Makes the lining of the uterus grow again after menstruation.
Luteinising hormone (LH)	Pituitary gland	• Stimulates the release of the egg from the ovary (ovulation).
Progesterone	Empty follicle in the ovaries	• Maintains the lining of the uterus during the second half of the cycle. HT Inhibits both FSH and LH release.

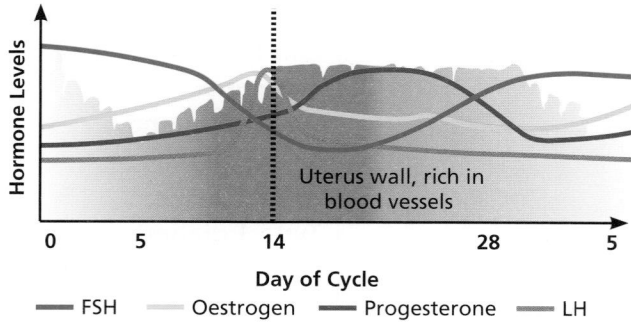

Hormone Levels — Day of Cycle: 0, 5, 14, 28, 5

Uterus wall, rich in blood vessels

— FSH   — Oestrogen   — Progesterone   — LH

Key Point

An egg is released by ovulation approximately every 28 days, but this timescale can vary considerably between different women.

Reducing Fertility

- Fertility can be reduced by a variety of methods of contraception.
- Hormonal methods include:
  - oral contraceptives (the combined pill) that contain oestrogen and progesterone, which inhibit FSH production so that no eggs are released
  - an injection, implant or skin patch of slow release progesterone to stop the release of eggs for a number of months or years.
- Non-hormonal methods include:
  - barrier methods, such as condoms and diaphragms, that prevent the sperm from reaching an egg
  - intrauterine devices, which prevent embryos from implanting in the uterus
  - spermicidal creams, which kill or disable sperm
  - not having intercourse when an egg may be in the oviduct
  - surgical methods of male and female sterilisation, such as cutting the sperm ducts or tying the fallopian tubes.

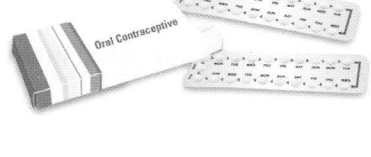

HT Increasing Fertility

- Doctors may give FSH and LH in a fertility drug to a woman if her own level of FSH is too low to stimulate eggs to mature.
- **In vitro fertilisation** (IVF) treatment involves:
  - giving a woman FSH and LH to stimulate the growth of many eggs
  - collecting the eggs from the woman
  - fertilising the eggs with sperm from the father in the laboratory
  - inserting one or two embryos into the woman's uterus (womb).
- Fertility treatment gives a woman the chance to have a baby. However:
  - it is emotionally and physically stressful
  - the success rates are not high
  - it can lead to multiple births, which are a risk to both the babies and the mother.

IVF Under a Microscope

Injected with father's sperm

Holds the egg in place

Unfertilised human egg cell

Key Point

Both contraception and fertility treatment are technological applications of science that raise ethical issues. People have many different views on these treatments.

Key Words

oestrogen
ovulation
testosterone
menstrual cycle
follicle stimulating hormone (FSH)
luteinising hormone (LH)
progesterone
contraception
HT fertility drug
HT in vitro fertilisation

Quick Test

1. What is the name of the male sex hormone?
2. What organ secretes FSH and LH?
3. Which hormone causes the release of an egg from the ovaries?
4. HT Where does fertilisation take place in IVF?

Review Questions

Cell Structure

1 Use words from the box to complete the sentences.

cell wall	cytoplasm	in a nucleus	plasmids	flagella
free within the cell	guard cell	cytotoxin		

A bacterial cell consists of _____ surrounded by a cell membrane.

Outside the cell membrane is a _____ .

The main chromosome in bacteria is found _____ .

There are also small loops of DNA called _____ . **[4]**

Total Marks _____ / 4

Investigating Cells

1 Fiona and Isaac are studying the diagram of a single-celled organism, called *Euglena*, shown in **Figure 1**.

Figure 1

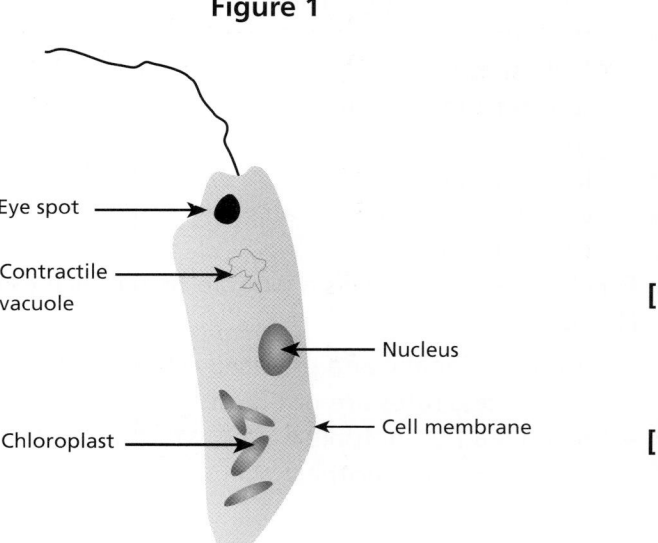

Eye spot

Contractile vacuole

Nucleus

Chloroplast

Cell membrane

a) Fiona finds out that *Euglena* is 0.02mm wide.

How wide is this in micrometres (μm)? **[1]**

b) Fiona thinks *Euglena* is a plant cell.

Why might she think this? **[1]**

c) Isaac says that *Euglena* cannot be a plant cell.

Suggest **one** reason why he might say this. **[1]**

d) *Euglena* reproduces by splitting into two in a process similar to mitosis.

What must happen in the nucleus of *Euglena* before it divides? **[1]**

Total Marks _____ / 4

Cell Division

Figure 1

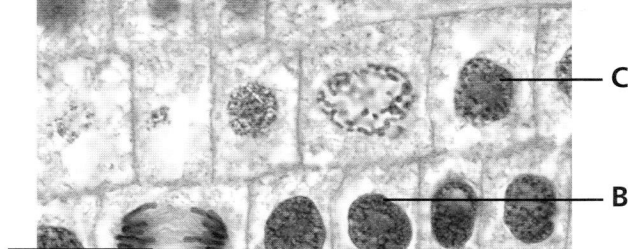

1 Karen wants to study cells dividing by mitosis. She uses cells from the tip of a plant root. She stains the cells with a dye and looks at them under a light microscope.

Figure 1 shows the photograph that she takes.

a) Cell **C** is actually 0.03mm wide.

Use a ruler to measure the width of the cell in Karen's photograph and calculate the magnification of his photograph.

$$\text{magnification} = \frac{\text{size of image}}{\text{size of real object}}$$ **[2]**

b) Which cell, **A**, **B** or **C** is about to divide? You must explain your answer. **[2]**

Total Marks _____ / 4

Transport In and Out of Cells

1 A student carries out an investigation into the effect of different concentrations of sugar solution on the mass of potato chips.

He weighs five chips separately and places each one into a different concentration of sugar solution. After one hour he removes the chips and then reweighs them. **Table 1** shows the results.

Table 1

Concentration of Sugar Solution (mol/dm³)	Mass of Chip at Start (g)	Mass of Chip After 1 hour (g)	Change in Mass (g)	Percentage Change in Mass
0.0	2.6	2.8	0.2	7.7
0.2	2.5	2.5	0.0	0.0
0.4	2.6	2.2	−0.4	−15.4
0.6	2.7	2.1	−0.6	

a) Work out the percentage change in mass for the potato chip in 0.6mol/dm³ sugar solution. **[2]**

b) Explain the result of the potato chip in 0.2mol/dm³ sugar solution. **[3]**

c) Before the student reweighs each chip he rolls it on some tissue paper.

Explain why he does this. **[2]**

Total Marks _____ / 7

Review Questions

Levels of Organisation

1 Organs are made up of a number of tissues.

Draw **one** line to join each type of tissue to its function.

Tissue	Function
glandular	can carry electrical impulses
nervous	can produce enzymes and hormones
muscular	a lining / covering tissue
epithelial	can contract to bring about movement

[3]

Total Marks .... / 3

Digestion

1 Different enzymes act on specific nutrients.

Draw **one** line to match each enzyme to the nutrient that it works on and draw **one** line to match each nutrient with its smaller subunit.

Enzyme	Nutrient	Subunit
protease	fats	glycerol and fatty acids
amylase	proteins	amino acids
lipase	starch	maltose

[3]

Total Marks .... / 3

Blood and the Circulation

1 **Figure 1** shows a specialised cell.

a) What is the name of this cell? [1]

b) Why does this cell not have a nucleus? [1]

Figure 1

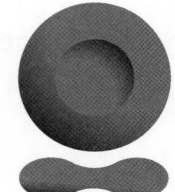

2 a) Circle the correct options to complete the following sentences.

   i)   The human circulation system is a **single / double / triple** system.   [1]

   ii) Blood passes through the heart **once / twice / three times / four times** on each circuit.   [1]

   iii) The blood vessels that carry blood away from the heart are called **arteries / veins /
        capillaries**.   [1]

   b) Complete the following sentences.

   Plasma transports carbon dioxide from the .... to the

   .... and it transports .... from the

   small intestine to the organs.   [3]

   > **Total Marks** .... / 8

Non-Communicable Diseases

1 Statins are drugs prescribed to lower cholesterol levels in the blood.

   a) What effect may high cholesterol levels have on blood vessels?   [2]

   b) Describe why these changes to blood vessels can be dangerous.   [4]

   > **Total Marks** .... / 6

Transport in Plants

1 Use words from the box to complete the sentences.

gaps	epidermis	chloroplasts	phloem	water	mesophyll	gases

Like animals, plants have different tissues.

The tissue found covering the outer layers of the leaf is called the ....

The .... tissue carries out photosynthesis.

The cells of this tissue have .... between them to allow easy passage

of gases.   [3]

   > **Total Marks** .... / 3

Practice Questions

Pathogens and Disease

1 Draw **one** line from each type of pathogen to the disease that it causes.

bacterium		rose black spot
virus		malaria
protist		salmonella
fungus		measles

[3]

Total Marks _____ / 3

Human Defences Against Disease

1 Complete the sentences about the vaccination of children against various diseases.

Vaccines contain _____ or _____ pathogens.

These stimulate the white blood cells to produce _____ .

This results in the children becoming _____ to the disease.   **[4]**

Total Marks _____ / 4

Treating Diseases

1 *Staphylococcus aureus* is a bacterium found on human skin. It can sometimes cause skin infections. Most strains of this bacterium are sensitive to antibiotics and so these infections are easily cured.

   a) i) What is an antibiotic?   **[2]**

   ii) Name one type of pathogen that antibiotics do not affect.   **[1]**

   b) A local newspaper reported that:

> Some strains of the bacterium *Staphylococcus aureus* are resistant to an antibiotic called **methicillin. They may also be resistant to many other commonly prescribed antibiotics. This is called multiple resistance.**
>
> **It is difficult to get rid of these 'superbugs' when patients are infected with them.**
>
> **To try and reduce the risk of 'superbugs' appearing, we have to take care when using antibiotics.**

i) What is the abbreviation used for multiple resistant strains of *Staphylococcus aureus* that are resistant to many antibiotics including methicillin? **[1]**

ii) List the precautions that are being taken with antibiotics to try and stop resistant strains developing. **[3]**

> Total Marks _____ / 7

Photosynthesis

1 a) Complete the **word** equation for photosynthesis.

   carbon dioxide + _____ ⟶ glucose + _____ **[2]**

   b) Apart from the chemicals given in the equation above, what **two** other resources are required for photosynthesis? **[2]**

   c) Plants need nitrates to produce proteins. Where do plants obtain nitrates from?

   Tick **one** box.

   Soil ☐   Oxygen ☐   Leaves ☐   Photosynthesis ☐ **[1]**

> Total Marks _____ / 5

Respiration and Exercise

1 Two students want to find out who is the fittest.
They carry out a simple investigation by performing star jumps for three minutes.
They record their pulse rate before the activity and every minute afterwards.

Table 1 shows the results.

Table 1

Time (mins)	Pulse Rate (bpm)	
	Student A	Student B
Before activity	68	72
1 minute after	116	160
2 minutes after	120	175
3 minutes after	116	168
4 minutes after	72	148
5 minutes after	66	92
6 minutes after	68	76

The results for Student A are plotted on the graph in **Figure 1** below.

Figure 1

a) Add the data for Student B to the graph. [2]

b) Suggest which student is fitter. You must give a reason for your answer. [2]

c) Both students experienced fatigue in their muscles.

Explain why this happens. [3]

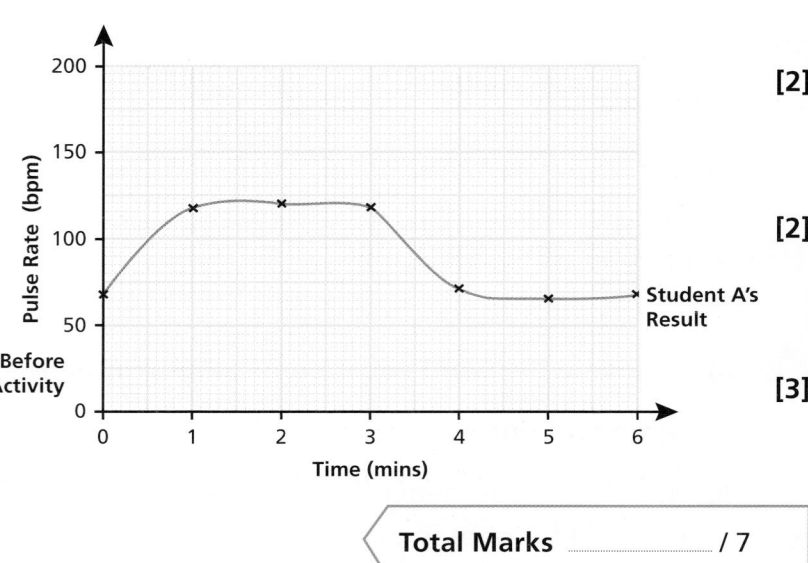

Total Marks .... / 7

Homeostasis and the Nervous System

1 The nervous system allows organisms to react to their surroundings.

a) Put the following words in the correct order to show the pathway for receiving and responding to information. [3]

| relay neurone response stimulus receptor sensory neurone |
| effector motor neurone |

b) Where in the pathway would you find a synapse? [2]

2 **Figure 1** shows a neurone.

Figure 1

a) Which letter on **Figure 1** shows each of these structures?

i) Cell membrane [1]

ii) Nucleus [1]

iii) Effector [1]

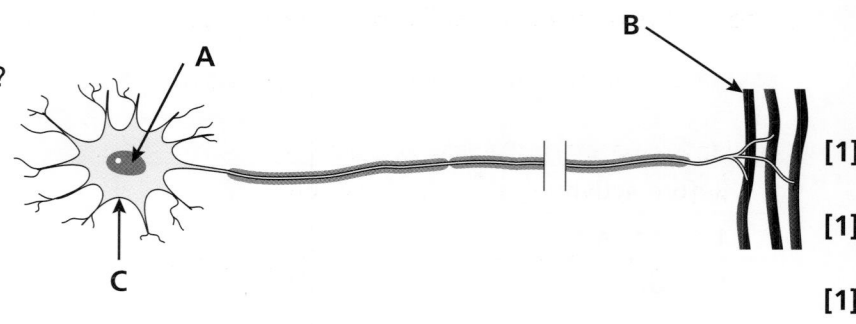

b) What type of neurone is shown in **Figure 1**? [1]

Total Marks .... / 9

Hormones and Homeostasis

1 A doctor gave a patient some glucose to eat.

He then measured the glucose and insulin levels in the patient's blood over a one-hour period.

The graph in **Figure 1** shows the doctor's results.

Figure 1

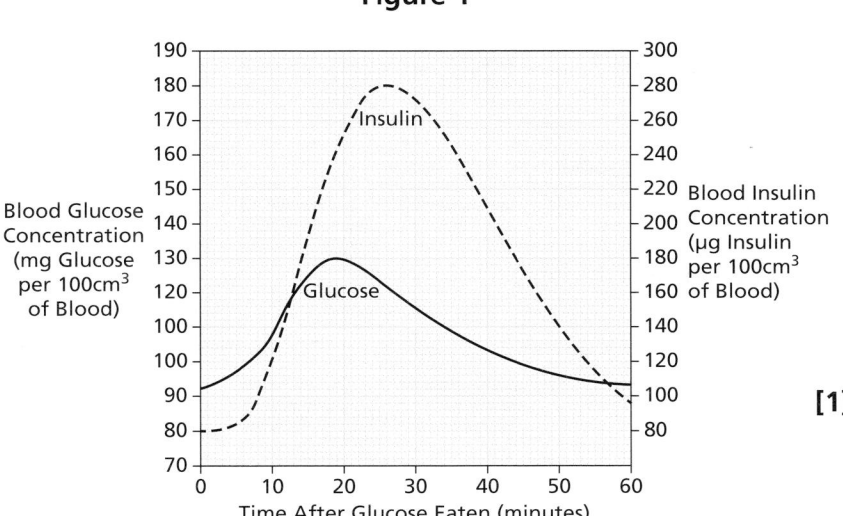

a) What was the level of glucose in the patient's blood before they ate the glucose? **[1]**

b) What was the maximum insulin level in the patient's blood during the one-hour period? **[1]**

c) When did the maximum level of insulin occur? **[1]**

d) Explain how the insulin in the patient's blood allows the glucose level to return to normal. **[3]**

e) The doctor performed this investigation to show that the patient did not have Type 1 diabetes.

Describe how the graph would differ if the patient did have Type 1 diabetes. **[2]**

Total Marks _____ / 8

Hormones and Reproduction

1 The first contraceptive pills contained large amounts of oestrogen.

a) Where in the body is oestrogen produced? **[1]**

b) Nowadays, contraceptive pills contain a much lower dose of oestrogen, which is combined with progesterone. Some birth control pills contain progesterone only.

Suggest **one** reason for the reduction of oestrogen levels in birth control pills. **[1]**

Total Marks _____ / 2

Sexual and Asexual Reproduction

You must be able to:

- Describe some examples of asexual reproduction in different organisms
- Explain why sexual reproduction involves meiosis
- Describe how the genetic material is arranged in a cell.

Asexual Reproduction

- **Asexual reproduction** involves:
  - only one parent
  - no fusion of gametes, so no mixing of genetic information
  - the production of genetically identical offspring (clones)
  - mitosis (see pages 20–21).
- Many plants reproduce asexually and in different ways, e.g.
  - strawberry plants send out long shoots called **runners**, which touch the ground and grow a new plant
  - daffodils produce lots of smaller bulbs, which can grow into new plants.

> ### Key Point
>
> Gardeners use asexual reproduction to produce large numbers of identical plants.

Daffodils

Strawberry Plant

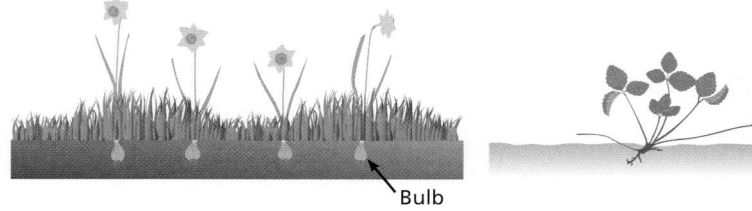

Bulb

Runner

- Many fungi reproduce asexually by spores.
- Malarial protists reproduce asexually when they are in the human host.

Sexual Reproduction and Meiosis

- Sexual reproduction involves the fusion (joining) of male and female gametes:
  - sperm and egg cells in animals
  - pollen and egg cells in flowering plants.
- This leads to a mix of genetic information, which produces variation in the offspring.
- The formation of gametes involves **meiosis**.

Meiosis

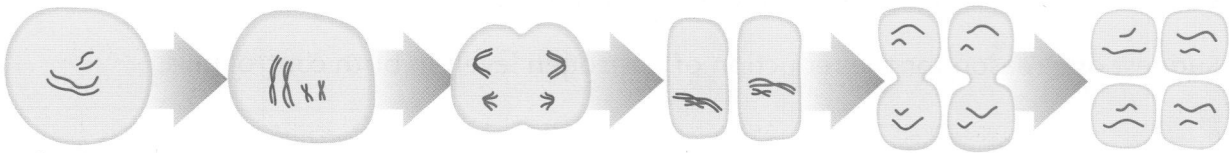

| Cell with two pairs of chromosomes (diploid cell). | Each chromosome replicates itself. | Chromosomes part company and move to opposite poles. | Cell divides for the first time. | Copies now separate and the second cell division takes place. | Four haploid cells (gametes), each with half the number of chromosomes of the parent cell. |

- When a cell divides by meiosis:
  - copies of the genetic information are made
  - the cell divides twice to form four gametes, each with a single set of **chromosomes**
  - all gametes are genetically different from each other.
- Meiosis is important because it halves the number of chromosomes in gametes.
- This means that fertilisation can restore the full number of chromosomes.

sperm		egg		fertilised egg cell
23 chromosomes	**+**	23 chromosomes	**=**	46 chromosomes (23 pairs) – half from mother (egg) and half from father (sperm)

The Genome

- The genetic material in the nucleus of a cell is made of a chemical called **DNA**.
- The DNA is contained in structures called chromosomes.
- A **gene** is a small section of DNA on a chromosome.
- Each gene codes for a particular sequence of amino acids, to make a specific protein.
- The **genome** of an organism is the entire genetic material of that organism.
- The whole human genome has now been studied and this may have some important uses in the future, e.g.
  - doctors can search for genes linked to different types of disorder
  - it can help scientists to understand the cause of inherited disorders and how to treat them
  - scientists can investigate how humans may have changed over time.

Key Point

A person now has the choice to have their genome tested to see how likely it is that they may get certain disorders. This may be a difficult decision to take.

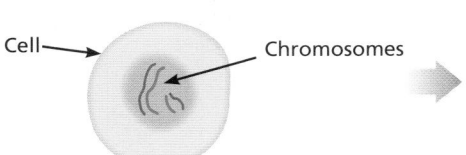
Cell — Chromosomes

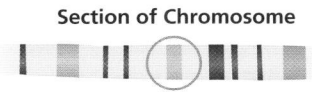
Section of Chromosome

Section of DNA

Quick Test

1. What are strawberry runners?
2. How many gametes are made when one cell divides by meiosis?
3. The body cells of chickens have 78 chromosomes. How many chromosomes are in each gamete?
4. What are chromosomes made from?

Key Words

asexual reproduction
gamete
runners
meiosis
chromosomes
DNA
gene
genome

Patterns of Inheritance

You must be able to:

- Explain the different terms used to describe genetic inheritance
- Predict the outcome of genetic crosses using genetic diagrams
- Describe examples of human genetic disorders
- Explain how sex is determined in humans.

Genetic Inheritance

- Some characteristics are controlled by a single gene, e.g. fur colour in mice and red-green colour blindness in humans.
- Each gene may have different forms called **alleles**, e.g. the gene for the attachment of earlobes has two alleles – attached or free.
- An individual always has two alleles for each gene:
  - One allele comes from the mother.
  - One allele comes from the father.
- The combination of alleles present in a gene is called the **genotype**, e.g. bb.
- How the alleles are expressed (what characteristic appears) is called the **phenotype**, e.g. blue eyes.
- Alleles can either be **dominant** or **recessive**.
- If the two alleles present are the same, the person is **homozygous** for that gene, e.g. BB or bb.
- If the alleles are different, they are **heterozygous**, e.g. Bb.

> ### Key Point
>
> A dominant allele is always expressed, even if only one copy is present. A recessive allele is only expressed if two copies are present, i.e. no dominant allele is present.

Genetic Crosses

- Most characteristics are controlled by several genes working together.
- If only one gene is involved, it is called **monohybrid inheritance**.
- Genetic diagrams or **Punnett squares** can be used to predict the outcome of a monohybrid cross.
- These diagrams use: capital letters for dominant alleles and lower case letters for recessive alleles.
- For example, for earlobes:
  - the allele for a free lobe is dominant, so **E** can be used
  - the allele for an unattached lobe is recessive, so **e** can be used.
- These Punnett squares show the possible outcomes of three crosses:

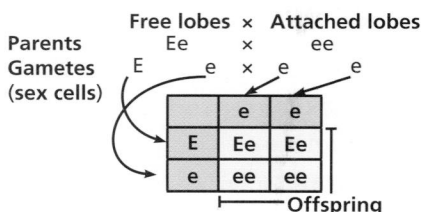

Each offspring will have a 1 in 2 chance of having attached lobes (because the dominant allele is present in half the crosses).

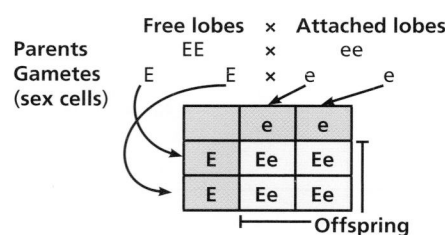

Each offspring will have free lobes (because the dominant allele is present in each cross).

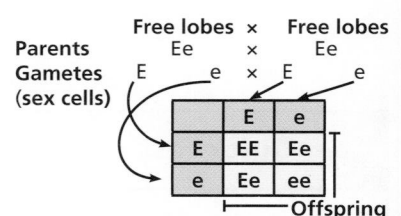

Each offspring has a 3 in 4 chance of having free lobes (because the dominant allele is present in three out of four crosses).

Inherited Disorders

- Some human disorders are inherited and are caused by the inheritance of certain alleles:
  - **Polydactyly** (having extra fingers or toes) is caused by a dominant allele.
  - **Cystic fibrosis** (a disorder of cell membranes) is caused by a recessive allele.

Sex Determination

- Only one pair out of the 23 pairs of chromosomes in the human body carries the genes that determine sex.
- These are called the **sex chromosomes**.
- In females, the two sex chromosomes are identical and are called X chromosomes (XX).
- Males inherit an X chromosome and a much shorter chromosome, called a Y chromosome (XY).
- As with all chromosomes, offspring inherit:
  - one sex chromosome from the mother (X)
  - one sex chromosome from the father (X or Y).

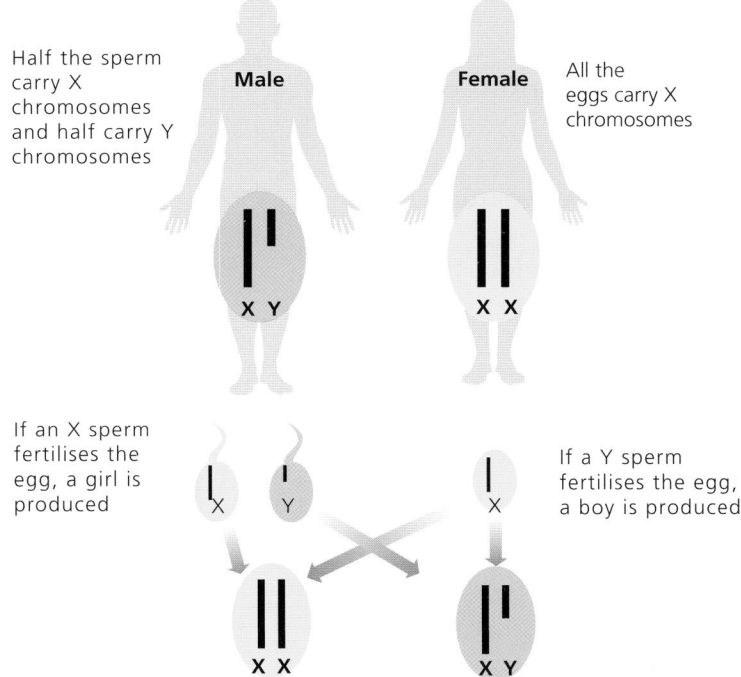

Half the sperm carry X chromosomes and half carry Y chromosomes

Male

Female

All the eggs carry X chromosomes

X Y

X X

If an X sperm fertilises the egg, a girl is produced

If a Y sperm fertilises the egg, a boy is produced

X

Y

X

X X

X Y

> **Key Words**
>
> allele
> genotype
> phenotype
> dominant
> recessive
> homozygous
> heterozygous
> monohybrid inheritance
> Punnett square
> polydactyly
> cystic fibrosis
> sex chromosomes

> **Quick Test**
>
> 1. What are the different forms of a gene called?
> 2. What do we call a combination of one dominant and one recessive allele?
> 3. Is cystic fibrosis caused by a **recessive** or **dominant** allele?
> 4. What sex chromosomes are present in a male liver cell?

Variation and Evolution

You must be able to:

- Describe the main sources of variation between individuals
- Explain Darwin's theory of natural selection
- Describe some of the evidence for evolution.

Variation

- In a population, differences in the characteristics of individuals are called **variation**.
- This variation may be due to differences in:
  - the genes that individuals have inherited (genetics)
  - the conditions in which individuals have developed (environment)
  - a combination of both genetic and environmental causes.
- Sexual reproduction produces different combinations of alleles and, therefore, variation.
- Only mutations create new alleles.

Evolution

- **Evolution** is the gradual change in the inherited characteristics of a population over time.
- This may lead to the formation of a new species.
- Many people have put forward theories to explain how evolution may occur.
- The theory that most scientists support is called **natural selection**, which was put forward by Charles Darwin.
- It states that all species have evolved from simple life forms that first developed more than three billion years ago.
- Natural selection occurs because:
  - Within a particular species, more individuals are born than can survive.
  - Due to the differences in their genes, the individuals may show variation in their phenotypes.
  - Individuals with the characteristics best suited to the environment are more likely to survive.
  - These individuals breed and the genes that enabled them to survive are passed on to their offspring.
- Different populations of the same species can evolve differently.
- Differences in their environments may mean that different characteristics are favoured by natural selection.
- New species are formed when two populations of one species become so different that they can no longer interbreed to produce fertile offspring.

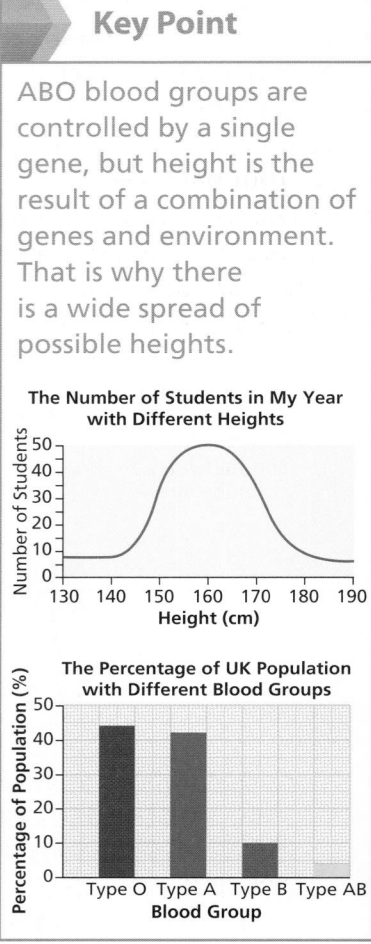

Key Point

ABO blood groups are controlled by a single gene, but height is the result of a combination of genes and environment. That is why there is a wide spread of possible heights.

The Number of Students in My Year with Different Heights

The Percentage of UK Population with Different Blood Groups

Evidence for Evolution

- When Darwin's theory of evolution by natural selection was first published, the mechanism of inheritance and variation was not known.
- It has now been shown that characteristics are passed on from one generation to the next by genes.
- There is also some evidence for evolution provided by **fossils**.
- Fossils are the remains of organisms from hundreds of thousands of years ago that are found in rocks.
- Fossils may be formed in various ways:
  - from the hard parts of animals that do not decay easily
  - from parts of organisms that have not decayed, because one or more of the conditions needed for decay was absent
  - when parts of the organisms are replaced by other materials as they decay
  - as preserved traces of organisms, e.g. footprints, burrows and root pathways.
- Scientists have used fossils to look at how organisms have gradually changed over long periods of time.
- Although fossils have been useful to scientists there are problems.
- There are gaps in the fossil record, because:
  - Many early forms of life were soft-bodied, which means that they have left very few traces behind.
  - What traces there were may have been destroyed by geological activity.

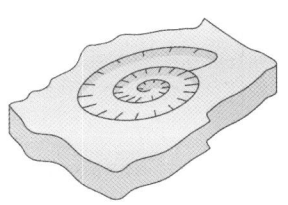 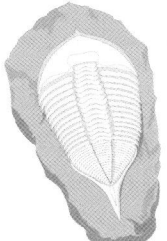

> ### Key Point
>
> When scientists describe natural selection now, they can talk about alleles being passed on, which will cause changes to the phenotypes in a population.

- The development of antibiotic-resistant strains of bacteria can be explained using the theory of natural selection:
  - Bacteria can evolve rapidly because they reproduce at a fast rate.
  - When they reproduce, mutations occur.
  - Some mutated bacteria might be resistant to antibiotics and are not killed.
  - These bacteria survive and reproduce, so a resistant strain develops (see page 44).
- There is still much debate today among scientists over the theory of evolution and the origins of life.

Quick Test

1. Who developed the theory of evolution by natural selection?
2. Use the theory of evolution by natural selection to explain how some parasites found in livestock may have become resistant to the drugs traditionally used to kill them.
3. Give **one** reason why there are gaps in the fossil record.

Key Words

variation
evolution
natural selection
fossils

Manipulating Genes

You must be able to:

- Describe the process of selective breeding
- Explain how genetic engineering can be used to change organisms' characteristics.

Selective Breeding

- Humans have been using selective breeding, or artificial selection, for thousands of years to produce:
  - food crops from wild plants
  - domesticated animals from wild animals.
- It is the process by which humans breed plants and animals with particular, desirable genetic characteristics.
- Selective breeding involves several steps:
  1. Choose parents that best show the desired characteristic.
  2. Breed them together.
  3. From the offspring, again choose those with the desired characteristic and breed.
  4. Continue over many generations.
- The type of characteristic that could be selected includes:
  - disease resistance in food crops
  - animals that produce more meat or milk
  - domestic dogs with a gentle nature
  - large or unusual flowers.
- However, selective breeding can lead to 'inbreeding', where some breeds are particularly prone to disease or inherited defects.

> **Key Point**
>
> Owners have to be very careful when mating pedigree dogs to make sure that they are not too closely related.

Example of Selective Breeding

| Choose the spottiest two to breed… | … and then the spottiest of their offspring… | … to eventually get Dalmatians. |

Genetic Engineering

- Genetic engineering is a more recent way of bringing about changes in organisms.
- It involves changing the characteristics of an organism by introducing a gene from another organism.

> **Key Point**
>
> Genetic engineering is a good example of a new technology that could be very useful. However, there are ethical issues to consider.

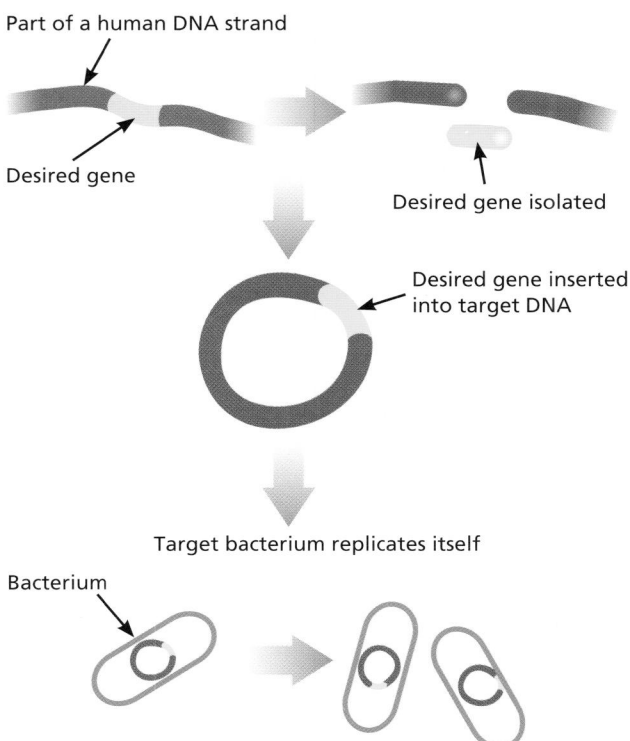

Part of a human DNA strand

Desired gene

Desired gene isolated

Desired gene inserted
into target DNA

Target bacterium replicates itself

Bacterium

HT In genetic engineering:

① Enzymes are used to isolate the required gene.

② This gene is inserted into a vector, e.g. a bacterial plasmid or virus.

③ The vector is used to insert the gene into the required cells.

HT If the genes are put into the cells of animals or plants at the egg or embryo stage, then all cells in the organism will get the new gene.

- Plant crops have been genetically engineered to:
  - be resistant to diseases, insects or herbicide attack
  - produce bigger, better fruits.
- Crops that have had their genes modified in this way are called genetically modified (GM) crops.
- Some people are concerned about GM crops and the possible long-term effects on populations of wild flowers and insects and on human health (if consumed).
- Fungi or bacterial cells have been genetically engineered to produce useful substances, e.g. human insulin to treat Type 1 diabetes.
- In the future, it may be possible to use genetic modification to cure or prevent some inherited diseases in humans.

Quick Test

1. State **one** characteristic that farmers might selectively breed cows for?
2. HT What is used to cut genes from chromosomes in genetic engineering?
3. Describe **one** concern about GM crops.

Key Words

selective breeding
genetic engineering
genetically modified (GM)

Classification

You must be able to:

- Explain why classification systems have changed over time
- Describe why organisms may become extinct
- Describe how evolutionary trees are constructed.

Principles of Classification

- Traditionally, living things have been classified into groups based on their structure and characteristics.
- One of the main systems used was developed by Carl Linnaeus.
- Linnaeus classified living things into:
  kingdom → phylum → class → order → family → genus → species
- Organisms are named by the **binomial system**, i.e. they have two parts to their Latin name:
  - The first part is their **genus**.
  - The second part is their **species**.
- New models of classification were proposed because:
  - microscopes improved, so scientists learnt more about cells
  - biochemical processes became better understood.
- Due to evidence, e.g. from genetic studies, there is now a **three-domain system** developed by Carl Woese.
- In this system organisms are divided into:
  - archaea (primitive bacteria, usually living in extreme environments)
  - bacteria (true bacteria)
  - eukaryota (including protists, fungi, plants and animals).

> ### Key Point
>
> The scientific name for lion is *Panthera leo* and for tiger it is *Panthera tigris*. This shows that they are in the same genus but different species. The cheetah is called *Acinonyx jubatus* – it is in a different genus and species.
>
>

Extinction

- Throughout the history of life on Earth, different organisms have been formed by evolution and some organisms have become **extinct**.
- Extinction may be caused by:
  - changes to the environment over geological (long periods of) time
  - new predators
  - new diseases
  - new, more successful competitors
  - a single catastrophic event, e.g. massive volcanic eruptions or collisions with asteroids.
- For example, the great auk is now extinct due to over-hunting.

Great Auk

Evolutionary Trees

- Evolutionary trees are a method used by scientists to show how they think organisms are related.
- They use current classification data for living organisms and fossil data for extinct organisms.
- Below is an evolutionary tree to show how scientists think that primates are related.
- It shows, for example, that modern humans (Homo sapiens) are most closely related to chimps and bonobos. They shared a common ancestor just over 5 million years ago.
- It is based on fossil records and also new techniques, such as studies of the animals' DNA.

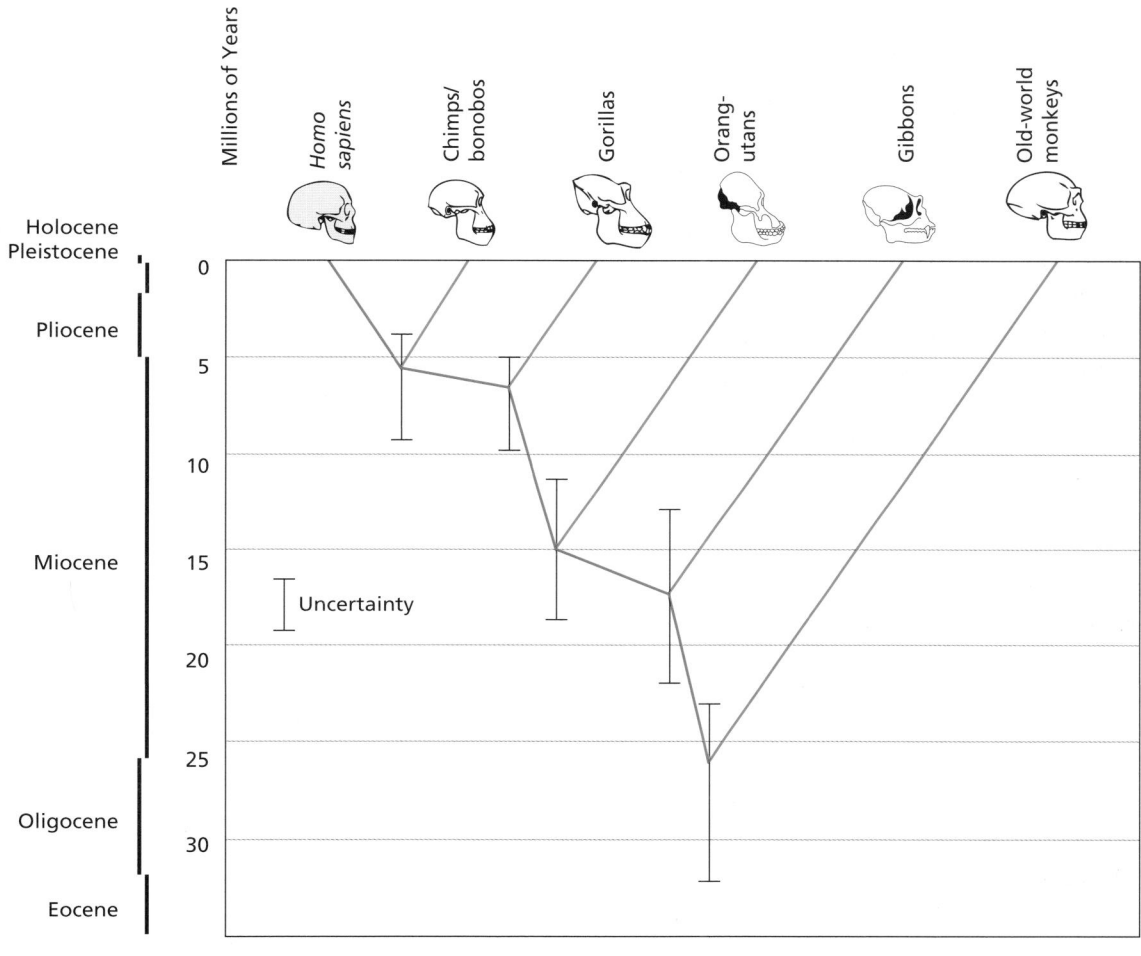

Ecosystems

You must be able to:

- Describe different levels of organisation in an ecosystem
- Explain why different organisms in an ecosystem show interdependence
- Describe the factors that determine where organisms can live
- Describe some of the techniques used by scientists to study ecosystems.

Relationships Between Organisms

- An ecosystem is all the organisms living in a habitat *and* the non-living parts of the habitat.
- There are different levels of organisation in an ecosystem:
  - individual organisms
  - populations – groups of individuals of the same species
  - communities – made up of many populations living together.
- To survive and reproduce, organisms require certain resources from their habitat and the other living organisms there.
- Trying to get enough of these resources results in competition.
- Plants in a community or habitat often compete with each other for light, water, space and mineral ions from the soil.
- Animals often compete with each other for food, mates and territory.
- As well as competing with each other, species also rely on each other for food, shelter, pollination, seed dispersal, etc. This is called interdependence.
- Because of interdependence, removing one species from a habitat can affect the whole community.
- In a stable community, all the species and environmental factors are in balance so that population sizes stay fairly constant.
- Tropical rainforests and ancient oak woodlands are examples of stable communities.

Adaptations

- Factors that can affect communities can be abiotic (non-living) or biotic (living).
- Abiotic factors include:
  - light intensity
  - temperature
  - moisture levels
  - soil pH and mineral content
  - wind intensity and direction
  - carbon dioxide levels for plants
  - oxygen levels for aquatic animals.
- Biotic factors include:
  - availability of food
  - new predators arriving
  - new pathogens / diseases
  - one species outcompeting another.

> **Key Point**
>
> All ecosystems should be self-supporting, but they do need energy. Energy is usually transferred into the ecosystem as light energy for photosynthesis in green plants, which are the start of the food chain.

- Organisms have **adaptations** (features) that enable them to survive in the conditions in which they normally live.
- These adaptations may be structural, behavioural or functional.
- Some organisms live in environments that are very extreme, e.g. with high temperature, pressure or salt concentration. These organisms are called **extremophiles**.
- Bacteria living in deep-sea vents are extremophiles.

Studying Ecosystems

- A group of organisms of one species living in a habitat is called a **population**.
- Scientists often want to estimate the size of a population.
- This might involve sampling using a square frame called a **quadrat**.

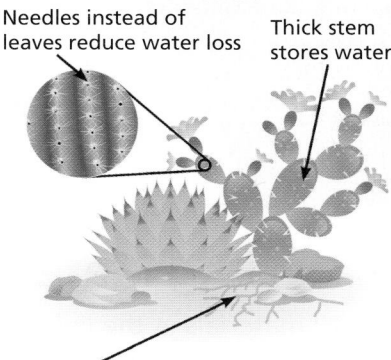

Cactus

Needles instead of leaves reduce water loss

Thick stem stores water

Extensive root system to take in water

REQUIRED PRACTICAL	
Measure the population size of a common species in a habitat.	
Sample Method 1. Place a quadrat on the ground at random. 2. Count the number of individual plants of one species in the quadrat. 3. Repeat this process a number of times and work out the mean number of plants. 4. Work out the mean number of plants in 1m². 5. Measure the area of the whole habitat and multiply the number of plants in 1m² by the whole area.	**Considerations, Mistakes and Errors** • The main consideration in the experiment is making sure that the quadrats are placed at random. Using random numbers to act as coordinates can help with this. • The more samples that are taken, then the more accurate the estimate should be.
Variables • The dependent variable is the number of plants in the quadrat.	**Hazards and Risks** • Care should be taken to wash hands after ecology work in a habitat. • Care should be taken throwing quadrats – throw low to the ground, not up in the air.

> ### Key Point
>
> Bacteria that live at high temperatures have enzymes that are very resistant to denaturing.

- To see how plants are spread or distributed in a habitat:
  1. Stretch a long tape, called a **transect line**, across the area.
  2. Place a quadrat down at regular intervals along the line.
  3. Count the plants in the quadrat each time.

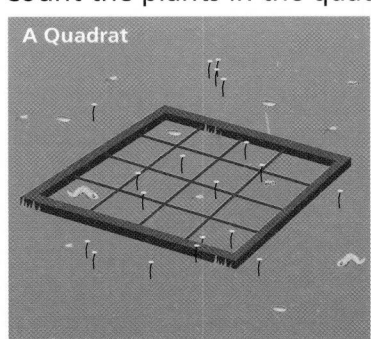

A Quadrat

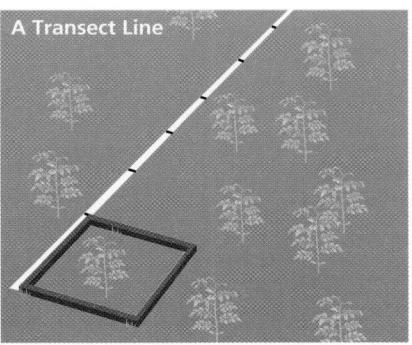

A Transect Line

> ### Quick Test
>
> 1. Name **one** stable community.
> 2. What resources do plants compete for?
> 3. What are extremophiles?
> 4. What is a quadrat?
> 5. How can an estimate of a population be made more accurate?

> ### Key Words
>
> ecosystem <br> competition <br> interdependence <br> abiotic <br> biotic <br> adaptation <br> extremophile <br> population <br> quadrat <br> transect line

Cycles and Feeding Relationships

You must be able to:

- Explain how carbon and water are recycled in nature
- Explain how feeding relationships can be shown by food chains and predator–prey graphs.

Recycling Materials

- All materials in the living world need to be recycled so that they can be used again in future organisms.
- The **carbon cycle** describes how carbon is recycled in nature.
- It relies on decomposers to return carbon to the atmosphere as carbon dioxide through respiration.
- The main process in the carbon cycle that removes carbon dioxide from the air is photosynthesis.
- Any action that reduces photosynthesis could lead to an increase in carbon dioxide levels in the air.

> **Key Point**
>
> The carbon cycle returns carbon from organisms to the atmosphere as carbon dioxide to be used by plants in photosynthesis.

The Carbon Cycle

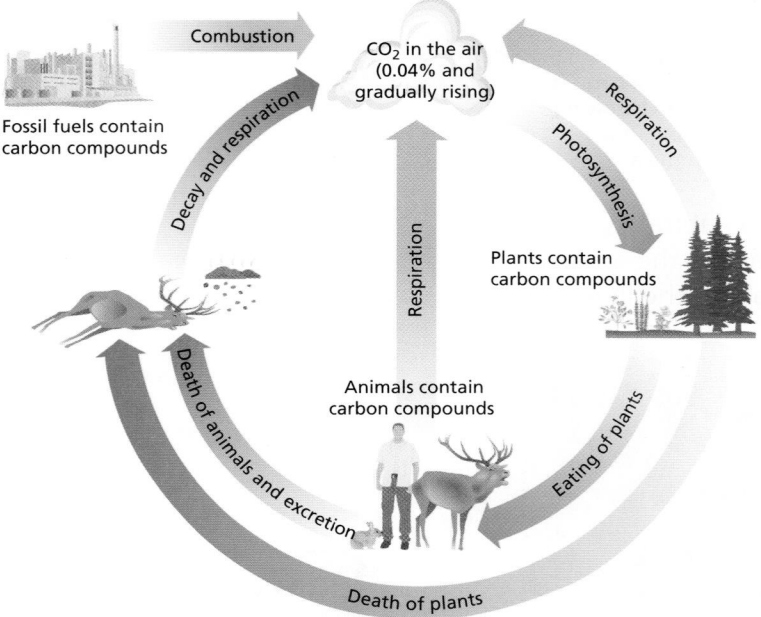

- The **water cycle** describes how fresh water circulates between living organisms, rivers and the sea.
- Transpiration from plants is responsible for returning much of the water to the air.
- Cutting down large areas of forest can disturb the water cycle.

> **Key Point**
>
> The water cycle provides fresh water for plants and animals on land before draining into the seas. Water is continuously evaporated and precipitated.

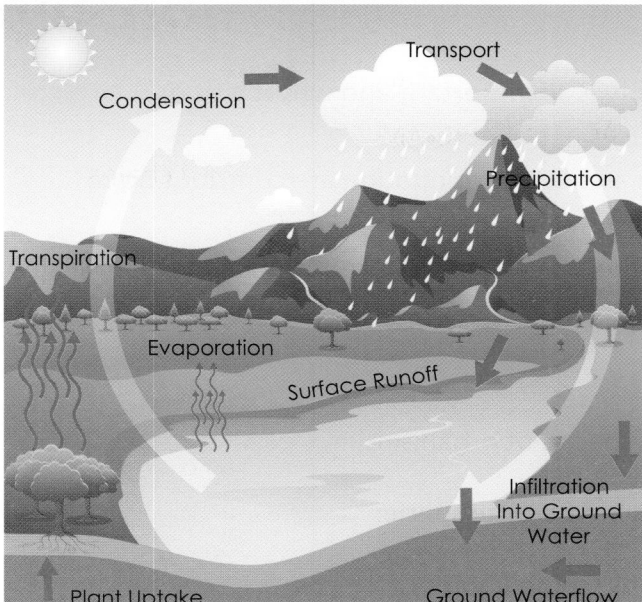

The Water Cycle

Feeding Relationships

- Feeding relationships in a community can be shown in **food chains**.
- All food chains begin with a **producer**, which synthesises (makes) molecules.
- The producer is usually a green plant, which makes glucose molecules by photosynthesis.
- Producers are eaten by primary consumers, which may be eaten by secondary consumers, which in turn may be eaten by tertiary consumers.
- Consumers that eat other animals are **predators** and those that are eaten are **prey**.
- Top consumers are **apex predators**. They are carnivores with no predators.
- In a stable community, the numbers of predators and prey rise and fall in cycles. This can be shown in a predator–prey graph.

 Key Point

In the predator–prey graph, both lines follow the same pattern but the changes in predator numbers happen just after the changes in prey.

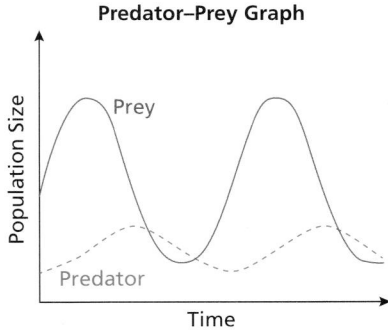

Predator–Prey Graph

 Key Words

carbon cycle
water cycle
food chain
producer
predator
prey
apex predator

Quick Test

1. How can carbon from carbon dioxide in the air get back into compounds in living organisms?
2. In the water cycle, what is the process called that is the reverse of precipitation?
3. Why do predator numbers fall soon after a drop in prey numbers?

Disrupting Ecosystems

You must be able to:

- Explain why biodiversity is so important and why it is at risk
- Describe the main causes of pollution
- Explain how pollution and over-exploitation are contributing to global warming
- Describe some of the steps that are being taken to maintain biodiversity.

Biodiversity

- **Biodiversity** is the variety of all the different species of organisms on Earth.
- A high biodiversity helps ecosystems to be stable because species depend on each other for food and shelter.
- The future of humans, as a species, relies on us maintaining a good level of biodiversity.
- Many human activities (see below) are responsible for reducing biodiversity, so action is now being taken to try to stop this reduction.
- Factors that put biodiversity at risk, and affect the distribution of species in an ecosystem, include changes in:
  - availability of water
  - temperature
  - atmospheric gases.
- These changes may be due to:
  - changes in the seasons
  - geographic activity, e.g. volcanoes or storms
  - human interaction.

Pollution

- **Pollution** kills plants and animals, which can reduce biodiversity.
- The human population is increasing rapidly and, in many areas, there is also an increase in the standard of living.
- This means that more resources are used and more waste is produced.
- Unless waste and chemical materials are properly handled, more pollution will be caused.
- Pollution can occur:
  - in water, from sewage, fertilisers or toxic chemicals
  - in air, from gases, e.g. sulfur dioxide, which dissolves in moisture in the atmosphere to produce **acid rain**
  - on land, from landfill and toxic chemicals, e.g. pesticides and herbicides, which may be washed from land into water.

Overexploitation

- Humans can also put biodiversity at risk by taking too many resources out of the environment.
- Building, quarrying, farming and dumping waste can all reduce the amount of land available for other animals and plants.
- Producing garden compost destroys peat bogs, reducing the area of this habitat and the variety of different organisms that live there.

Key Point

Many garden centres now sell 'peat-free' compost to try and reduce the destruction of peat bogs.

- The decay or burning of the peat releases carbon dioxide (a greenhouse gas) into the atmosphere.
- Cutting down trees and the destruction of forests is called **deforestation**.
- In tropical areas deforestation has occurred to:
  - provide land for cattle and rice fields to provide more food
  - grow crops from which biofuels can be produced.
- **Global warming** is a gradual increase in the temperature of the Earth.
- Many scientists think that it is being caused by changes in various gases, caused by pollution and deforestation.
- These gases include carbon dioxide and methane.
- There are a number of biological consequences of global warming:
  - loss of habitat, when low-lying areas are flooded by rising sea levels
  - changes in the distribution of species in areas where temperature or rainfall has changed
  - changes to the migration patterns of animals.

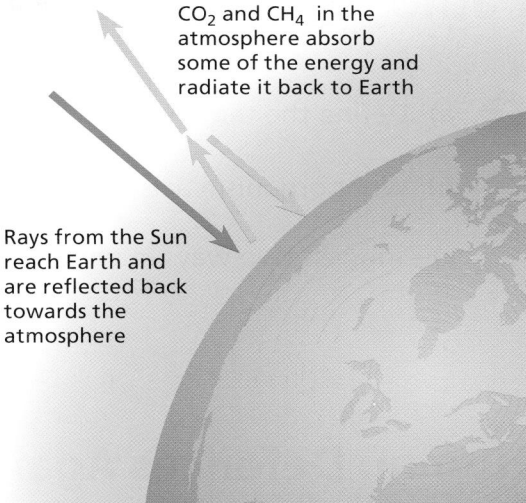

CO_2 and CH_4 in the atmosphere absorb some of the energy and radiate it back to Earth

Rays from the Sun reach Earth and are reflected back towards the atmosphere

Conserving Biodiversity

- Scientists and governments have taken steps to reduce pollution and over-exploitation to help maintain biodiversity. These include:
  - setting up breeding programmes for endangered species
  - protecting rare habitats, e.g. coral reefs, mangroves and heathland
  - encouraging farmers to keep margins and hedgerows in fields
  - reducing deforestation and carbon dioxide emissions
  - recycling resources rather than dumping waste in landfill.

> **Key Point**
>
> Any increase in carbon dioxide levels will cause photosynthesis rates to increase. However, there is a limit to how well plants will be able to reduce the effect of global warming, especially if huge areas of rainforests are cut down.

Quick Test

1. Give **one** reason why the human population is producing more waste.
2. What can acidic pollutant gases cause?
3. Give **one** reason why deforestation is occurring.
4. What effect does deforestation have on the amount of carbon dioxide in the atmosphere and why?

Key Words

biodiversity
pollution
acid rain
deforestation
global warming

Review Questions

Pathogens and Disease

1 **a)** Define the term 'vector'. [1]

b) What organism acts as a vector for the protist that causes malaria? [1]

c) Suggest **two** ways in which people can protect themselves from this vector. [2]

Total Marks _____ / 4

Human Defences Against Disease

1 **a)** Use words from the box to complete the sentences about vaccination.

Each word may be used once, more than once or not at all.

| antibodies | antibiotics | antiseptics | dead | live | red | toxins | white |

During vaccination, _____ or weakened pathogens are injected into the body.

This causes _____ blood cells to make _____.

Later, when _____ pathogens enter the body, they are destroyed quickly. [4]

b) Vaccinations can provide immunity.

Give **one** other way by which a person can become immune to a pathogen. [1]

2 In March 2009, a nine-year-old girl was found to be infected with a new strain of the H1N1 swine flu virus. Over the next year many more people were found to have the swine flu virus.

The graph in **Figure 1** shows the number of reported cases of swine flu in the first 10 days of May 2010.

a) How many cases of swine flu had been reported by 5th May? [1]

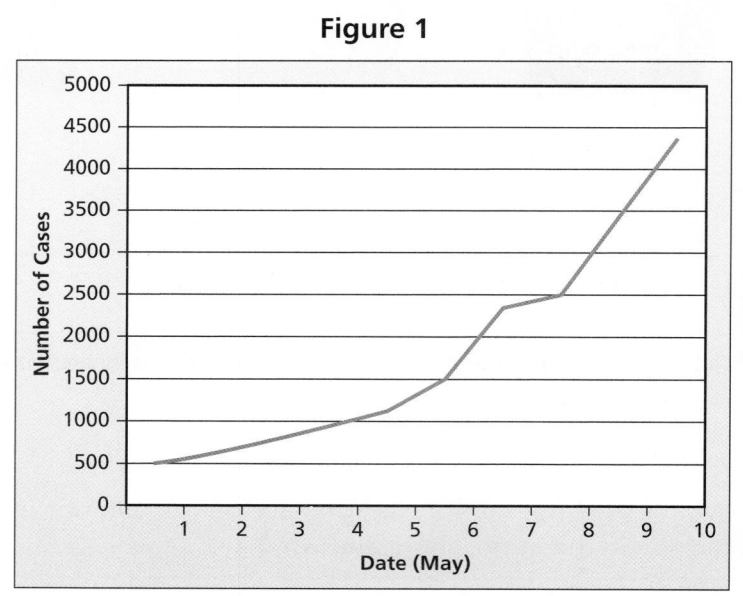

Figure 1

b) Which period showed the largest increase in the number of reported cases? [1]

c) Suggest why the spread of disease was so rapid. [2]

d) Why is it difficult to kill viruses inside the body? [2]

Total Marks _____ / 11

Treating Diseases

1 Many people in the world have a disease called arthritis.
Their joints become very painful.

Here is an article about a new arthritis drug.

To Use or Not to Use?

Arthritis is a very painful condition.

The problem with many drugs used to treat arthritis is that they have side effects.

One new drug was recently developed and tested on animals like mice and rats with no side effects. However, after a long-term study on human patients, side effects were noticed.

In this study, the drug was compared with a placebo. After 18 months of taking the drug, the risk of a patient having a heart attack was 15 out of 1000 compared with 7.5 out of 1000 for patients taking the placebo.

A decision has to be made about whether to use the drug even though it increases the risk of heart disease.

---

a) It is important that all drugs are tested on humans, and not just animals, before they are widely used.

Why is this? [1]

b) The long-term study used a placebo.

What is a placebo and why is it used? [3]

c) Some new drugs for arthritis are still allowed to be used even though they carry a slight risk.

Why do you think that this is? [2]

Total Marks _____ / 6

Review Questions

Photosynthesis

1 Rebecca investigates how quickly pondweed produces oxygen by photosynthesis.

Figure 1 shows the apparatus that she uses.

Rebecca adds different masses of sodium hydrogen carbonate to the water.

This gives the pondweed different concentrations of carbon dioxide to use for photosynthesis.

She counts the number of bubbles given off each minute.

Table 1 shows the results.

Figure 1

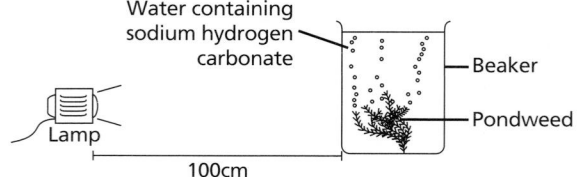

Water containing sodium hydrogen carbonate — Beaker

Pondweed

Lamp

100cm

Table 1

Mass of Sodium Hydrogen Carbonate Added in Grams	0.00	0.10	0.30	0.50	0.65	0.80
Number of Bubbles Given Off in 1 Minute	5	20	36	45	47	47

a) Describe the pattern shown by Rebecca's results. **[2]**

b) Write down **two** ways in which Rebecca made sure that her results were valid. **[2]**

c) The pondweed produces five bubbles of oxygen per minute without any sodium hydrogen carbonate being added to the beaker.

Why is the pondweed still able to produce some oxygen with no extra carbon dioxide? **[1]**

> Total Marks _____ / 5

Respiration and Exercise

1 **Figure 1** shows one method of making wine.

a) Name the gas that bubbles up through the mixture. **[1]**

b) i) What type of respiration occurs during winemaking? **[1]**

ii) What substance gradually builds up in the wine as a result of this kind of respiration? **[1]**

c) Explain why sugar is used during winemaking. **[2]**

Figure 1

Water

Bubbles of gas

Mixture of sugar, yeast and fruit juice

> Total Marks _____ / 5

Homeostasis and the Nervous System

1 Reflex actions are designed to prevent the body from being harmed.

a) What type of neurone carries a signal to the spinal cord in a reflex action? [1]

b) What is the junction between two neurones called? [1]

Total Marks _____ / 2

Hormones and Homeostasis

1 a) What does insulin cause glucose to be converted into? [1]

b) In which organ is this product mainly stored? [1]

c) HT What hormone does the body produce that converts this product back into glucose, and where is it produced? [1]

d) Which statement describes one of the causes of Type 1 diabetes?
Tick **one** box.

The pancreas does not produce insulin. ☐

The kidneys do not remove glucose from the blood. ☐

The liver does not respond to insulin. ☐

The liver does not produce insulin. ☐ [1]

Total Marks _____ / 4

Hormones and Reproduction

1 The female menstrual cycle is controlled by hormones.

a) Which hormone is secreted from the pituitary gland and causes eggs to mature in the ovaries? [1]

b) Which hormone is released from the ovaries and causes the uterus lining to thicken? [1]

c) Explain how hormones given to women can:

i) HT Increase fertility   ii) Reduce fertility [4]

Total Marks _____ / 6

Sexual and Asexual Reproduction

1 a) Use the words in the box to label the cell in **Figure 1**.

| chromosomes | nucleus | cytoplasm | cell membrane |

Figure 1

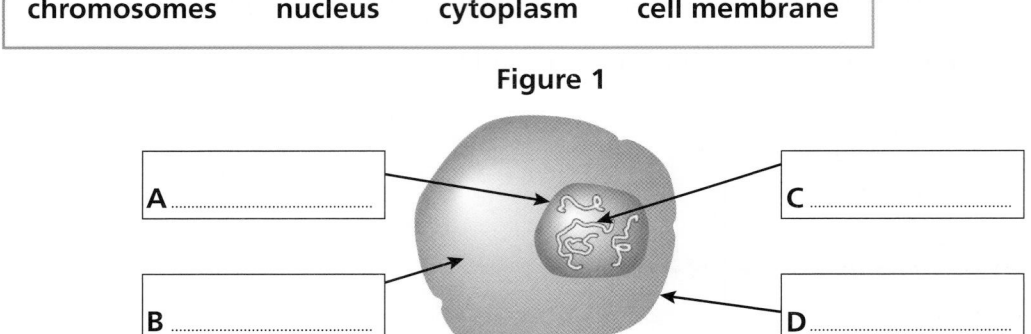

A ....

C ....

B ....

D ....   [4]

b) Choose the correct word to complete each sentence.

Sexual reproduction is the **division / separation / fusion** of the male and female gametes.

The resulting offspring will contain **DNA / cells / enzymes** from both parents.

This gives rise to **fertilisation / differentiation / variation**.   [3]

Total Marks .... / 7

Patterns of Inheritance

1 a) Circle the correct pair of human sex chromosomes.

| XY and YY | XX and XY | XX and YY | XF and XM |

[1]

Figure 1

b) **Figure 1** shows two sets of sex chromosomes.

Label:
- The female sex chromosomes.
- The male sex chromosomes.

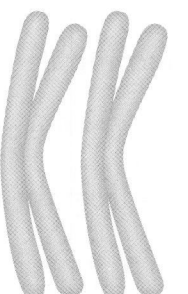
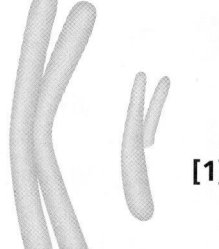

[1]

2 Draw **one** line from each definition to the correct genetic term.

Definition	**Term**
both alleles are the same	phenotype
two different alleles	heterozygous
what the organism looks like	homozygous
an allele that is always expressed if present	dominant

[3]

3 Fruit flies are often used in genetic crosses.
There are two types of wings on fruit flies: short or normal.
Answer the following questions.

Use the letter **N** to represent normal wings and **n** to represent short wings.

a) What is the phenotype for a fly that has the homozygous dominant genotype?   [1]

b) A heterozygous male fly mates with a homozygous recessive female.

Complete the diagram in **Figure 2** for this genetic cross.   [3]

c) What is the percentage chance of the parents having an offspring with short wings?   [2]

Figure 2

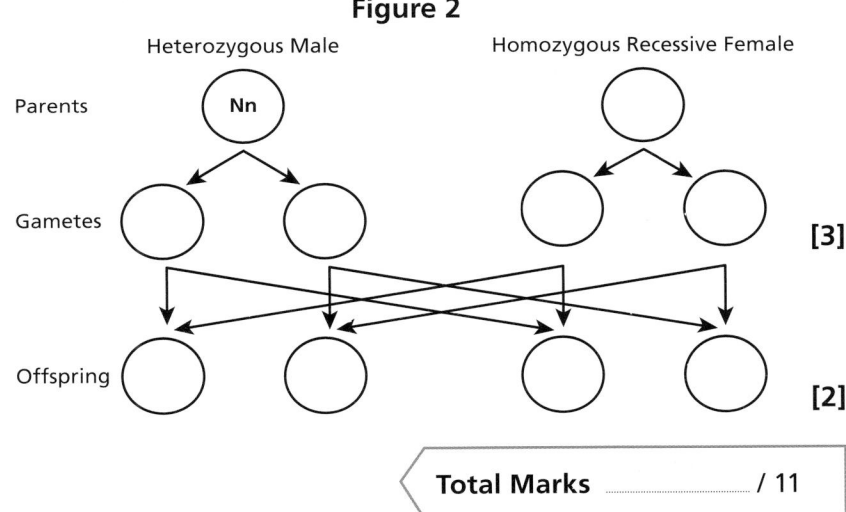

Heterozygous Male   Homozygous Recessive Female

Parents   Nn

Gametes

Offspring

Total Marks _____ / 11

Variation and Evolution

1 It is thought that many years ago members of the giraffe family had short necks.
Giraffes now have longer necks, which allow them to reach food higher in the trees.

Use the theory of evolution by natural selection to explain how modern giraffes with longer necks may have evolved.   [3]

Total Marks _____ / 3

Practice Questions

Manipulating Genes

1 HT **Figure 1** shows the **first** stage in the process of insulin production using genetic engineering.

Figure 1

a) What do scientists use to 'cut out' the insulin gene from the chromosome? [1]

b) The 'cut' gene is then inserted into a bacterium.

Why are bacteria good host cells for the 'cut' insulin gene? [2]

Total Marks _____ / 3

Classification

1 a) Describe **one** way in which fossils are formed. [1]

b) Explain why fossils can be quite hard to find. [1]

c) Many fossils are of animals that are extinct.

Give **three** factors that could contribute to the extinction of a species. [3]

d) Give an example of a **species** that is now extinct. [1]

Total Marks _____ / 6

Ecosystems

1 A number of animals live in the Sahara desert.

a) Suggest **two** major problems that animals living in the desert have to deal with. [2]

b) The cactus is a plant that is adapted to survive in desert environments.

Suggest how the following adaptations help the cactus to survive:

i) The cactus has a thick stem. [1]

ii) The cactus has spines instead of leaves. [1]

Total Marks _____ / 4

Cycles and Feeding Relationships

1 **Figure 1** shows some parts of the carbon cycle.

Figure 1

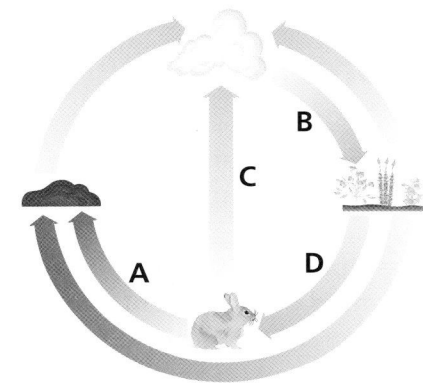

a) Choose the letter from **Figure 1** that corresponds with each process.

i) Egestion

iii) Photosynthesis   **[2]**

ii) Feeding

iv) Respiration   **[2]**

b) Name **one** other process, not shown on the diagram, which releases carbon dioxide into the air.   **[1]**

Total Marks .... / 5

Disrupting Ecosystems

1 Circle the correct options to complete the sentences.

When deforestation occurs in **tropical / arctic / desert** regions, it has a devastating impact on the environment.

The loss of **trees / animals / insects** means that less photosynthesis takes place, so less **oxygen / nitrogen / carbon dioxide** is removed from the atmosphere.

It also leads to a reduction in **variation / biodiversity / mutation**, because some species may become **devolved / damaged / extinct** as **habitats / land / farms** are destroyed.   **[6]**

Total Marks .... / 6

Atoms, Elements, Compounds and Mixtures

You must be able to:

- Define the terms: atom, element, mixture and compound
- Write word equations for reactions
- Describe and explain how mixtures can be separated by physical processes.

Atoms, Elements and Compounds

- All substances are made of **atoms**.
- An atom is the smallest part of an **element** that can exist.
- An element is a substance that contains only one sort of atom.
- There are about 100 different elements.
- Elements are displayed in the periodic table.
- The atoms of each element are represented by a different chemical symbol, e.g. sodium = Na, carbon = C and iron = Fe.
- Most substances are **compounds**.
- A compound contains atoms of two or more elements, which are chemically combined in fixed proportions.
- Compounds are represented by a combination of numbers and chemical symbols called a 'chemical formula'.
- Scientists use chemical formulae to show:
  - the different elements in a compound
  - how many atoms of each element one molecule of the compound contains.
- For example:
  - water, H_2O, contains 2 hydrogen (H) atoms and 1 oxygen (O) atom
  - sulfuric acid, H_2SO_4, contains 2 hydrogen (H) atoms, 1 sulfur (S) atom and 4 oxygen (O) atoms.
- Compounds can only be separated into their component elements by chemical reactions or electrolysis.

> **Key Point**
>
> Elements are made of only one type of atom. They are displayed in the periodic table and can be represented by one or two letters called symbols.

Equations

- You can sum up what has happened during a chemical reaction by writing a word equation or balanced symbol **equation**.
- The **reactants** (the substances that react) are on the left-hand side of the equation.
- The **products** (the new substances that are formed) are on the right-hand side of the equation.
- The total mass of the products of a chemical reaction is always equal to the total mass of the reactants. This is because no atoms are lost or made.
- The products of a chemical reaction are made from exactly the same atoms as the reactants.
- For example, when magnesium is burned in oxygen, magnesium oxide is produced. This can be written in a word equation:

magnesium + oxygen ⟶ magnesium oxide

Separating Mixtures

- **Mixtures** consist of two or more elements or compounds, which are not chemically combined.
- The components of a mixture retain their own properties, e.g. in a mixture of iron and sulfur, the iron is still magnetic and the sulfur is still yellow.
- Mixtures can be separated by physical processes – these processes do not involve chemical reactions.
- **Filtration** is used to separate soluble solids from insoluble solids, e.g. a mixture of salt (soluble) and sand (insoluble) can be separated by dissolving the salt in water and then filtering the mixture.
- **Crystallisation** is used to obtain a soluble solid from a solution, e.g. salt crystals can be obtained from a solution of salty water:
  1. The mixture is gently warmed.
  2. The water evaporates leaving crystals of pure salt.
- **Simple distillation** is used to obtain a solvent from a solution.

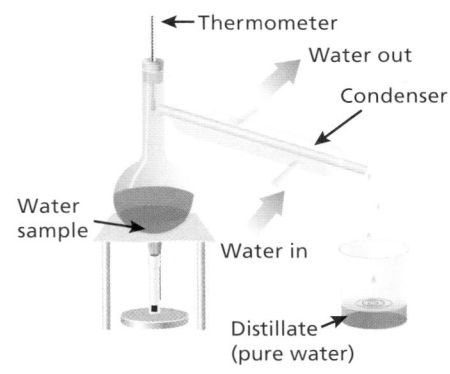

Thermometer
Water out
Condenser
Water sample
Water in
Distillate (pure water)

REQUIRED PRACTICAL	
Analysis and purification of water samples from different sources.	
Sample Method 1. Use a pH probe or suitable indicator to analyse the pH of the sample. 2. Set up the equipment as shown. 3. Heat a set volume to 100°C so that the water changes from liquid to gas. 4. The water collects in the condenser and changes state from gas to liquid. Collect this pure water in a beaker. 5. When all the water from the sample has evaporated, measure the mass of solid that remains to find the amount of dissolved solids present in the sample.	**Hazards and Risks** • There is a risk of the experimenter burning themselves on hot equipment, so care must be taken during and after the heating process.

- **Fractional distillation** is used to separate mixtures in which the components have different boiling points, e.g. oxygen and nitrogen can be obtained from liquid air by fractional distillation because they have different boiling points.

- **Chromatography** is used to separate the different soluble, coloured components of a mixture, e.g. the different colours added to a fizzy drink can be separated by chromatography.

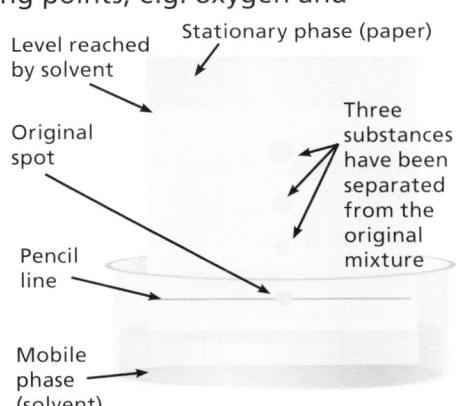

Level reached by solvent
Stationary phase (paper)
Original spot
Three substances have been separated from the original mixture
Pencil line
Mobile phase (solvent)

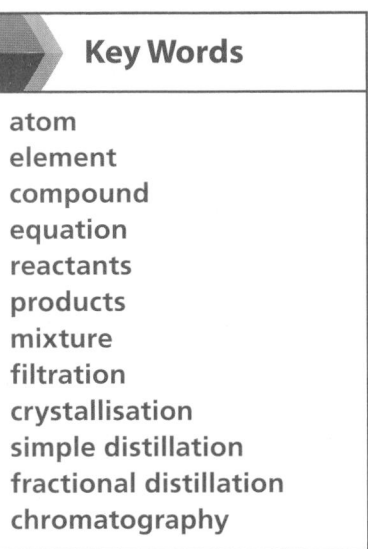

Key Words

atom
element
compound
equation
reactants
products
mixture
filtration
crystallisation
simple distillation
fractional distillation
chromatography

Quick Test

1. Define the term 'atom'.
2. How is an element different to a compound?
3. What does the formula $CaCO_3$ tell you about the compound calcium carbonate?
4. What process can be used to extract pure water from salt water?

Atoms and the Periodic Table

You must be able to:

- Explain how the scientific model of the atom has changed over time
- Recall the relative electrical charge and mass of subatomic particles
- Use the atomic number and mass number to work out the number of protons, neutrons and electrons in an atom or ion
- Describe what an isotope is
- Work out the electron configuration of the first 20 elements.

Scientific Models of the Atom

- In early models, atoms were thought to be tiny spheres that could not be divided into simpler particles.
- In 1898, Thomson discovered electrons and the representation of the atom had to be changed.
- Overall, an atom is neutral, i.e. it has no charge.
- Thomson thought atoms contained tiny, negative **electrons** surrounded by a sea of positive charge. This was the 'plum-pudding' model.
- Later, Geiger and Marsden carried out an experiment in which they bombarded a thin sheet of gold with alpha particles.
- Although most of the positively charged alpha particles passed straight through the atoms, a tiny number were deflected back towards the source.
- Rutherford looked at these results and concluded that the positive charge in an atom must be concentrated in a very small area.
- This area was named the 'nucleus' and the resulting model became known as the 'nuclear' model of the atom.
- Bohr deduced that electrons must orbit the nucleus at specific distances, otherwise they would spiral inwards.
- Later experiments showed that the nucleus is made of smaller particles:
  - some of which have a positive charge and are called **protons**
  - some of which have no charge and are called **neutrons**.

Subatomic Particles

- Atoms are very, very small and typically have an atomic radius of about 0.1nm or 1×10^{-10}m.
- Atoms contain three types of subatomic particle:

Subatomic Particle	Relative Mass	Relative Charge
proton	1	+1
neutron	1	0
electron	very small	−1

- Almost all of the mass of an atom is in the nucleus.
- However, the radius of the nucleus is less than $\frac{1}{10\,000}$ of the atomic radius of the atom, so most of an atom is empty space.
- Atoms have no overall charge because they contain an equal number of protons and electrons.
- All atoms of a particular element have the same number of protons.

> ### Key Point
>
> Scientists look at the evidence available and use it to put together a model of what appears to be happening. As new evidence emerges, they re-evaluate the model. If the model no longer works, they change it.

The Geiger and Marsden Experiment

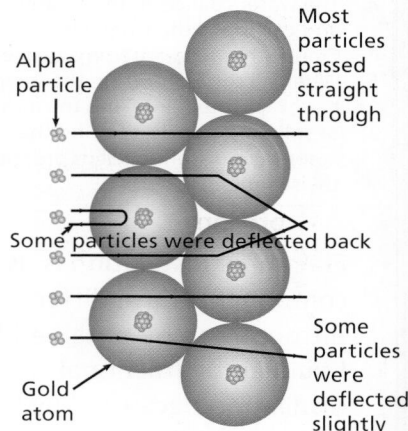

Alpha particle · Most particles passed straight through · Some particles were deflected back · Gold atom · Some particles were deflected slightly

The Nuclear Model

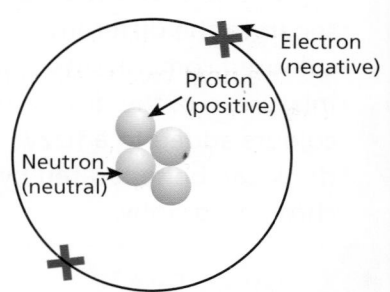

Electron (negative) · Proton (positive) · Neutron (neutral)

> ### Key Point
>
> A nanometer (nm) is one billionth of a metre, i.e. $\frac{1}{1\,000\,000\,000}$ or 1×10^{-9}.

- Atoms of different elements have different numbers of protons.
- The number of protons in an atom is called its **atomic number**.
- The sum of the protons *and* neutrons in an atom is its **mass number**.
- In the modern periodic table, elements are arranged in order of increasing atomic number.

 LEARN

number of neutrons = mass number – atomic number

How many protons, electrons and neutrons are there in $^{23}_{11}Na$? ← 23 is the mass number and 11 is the atomic number.

11 protons
11 electrons
12 neutrons (23 – 11)

Isotopes and Ions

- **Isotopes** of an element have the same number of protons but a different number of neutrons, i.e. they have the same atomic number but a different mass number.
- For example, chlorine has two isotopes:

$^{35}_{17}Cl$ ←
17 protons
17 electrons
18 neutrons (35 – 17)

$^{37}_{17}Cl$ ←
17 protons
17 electrons
20 neutrons (37 – 17)

- Atoms can gain or lose electrons to become **ions**:
    - Metal atoms lose electrons to form positive ions.
    - Non-metal atoms gain electrons to form negative ions.

Fluorine is a non-metal. It forms negative ions when it gains an electron.

$^{23}_{11}Na^+$ ← Sodium is a metal. It forms positive ions when it loses an electron.

11 protons
10 electrons (11 – 1)
12 neutrons

$^{19}_{9}F^-$ ←

9 protons
10 electrons (9 + 1)
10 neutrons

Electron Configuration

- The electrons in an atom occupy the lowest available shell or energy level.
- For the first 20 elements:
    - the first shell can only hold a maximum of two electrons
    - the next two shells can each hold a maximum of eight electrons.
- The **electron configuration** of an atom shows how the electrons are arranged around the nucleus in shells.

Sodium 2,8,1

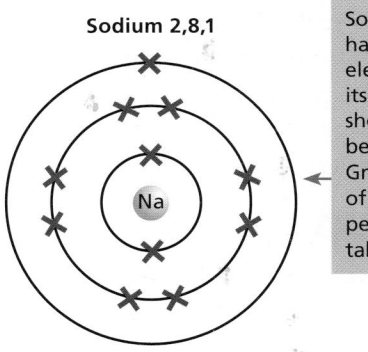

Sodium has one electron in its outer shell, so it belongs to Group 1 of the periodic table.

Quick Test

1. Describe the plum-pudding model of the atom.
2. An atom of potassium has an atomic number of 19 and a mass number of 39. State the number of protons, neutrons and electrons in this atom.
3. An ion of potassium-39 has a +1 charge. State the number of protons, neutrons and electrons in this ion.

Key Words

electron
proton
neutron
atomic number
mass number
isotope
ion
electron configuration

The Periodic Table

You must be able to:

- Describe the key stages in the development of the periodic table
- Describe and explain the properties of Group 0, Group 1 and Group 7 elements.

The Development of the Periodic Table

- When John Newlands tried to put together a periodic table in 1864, only 63 elements were known. Many were still undiscovered.
- Newlands arranged the known elements in order of atomic weight.
- He noticed periodicity (repeated patterns), although the missing elements caused problems.
- However, strictly following the order of atomic weight created issues – it meant some of the elements were in the wrong place.
- Dimitri Mendeleev realised that some elements had yet to be discovered. When he created his table, in 1869, he left gaps to allow for their discovery. He also reordered some elements.
- Each element was placed in a vertical column or 'group' with elements that had similar properties.
- Mendeleev used his periodic table to predict the existence and properties of new elements.
- When the subatomic particles were later discovered, it revealed that Mendeleev had organised the elements in order of increasing atomic number (number of protons).

Group 0

- The elements in Group 0 are known as the noble gases.
- Noble gas atoms have a full outer shell of electrons.
- This means they have a very stable electron configuration, making them very unreactive non-metals.

Group 1

- The elements in Group 1 are known as the alkali metals. They:
  - have one electron in their outermost shell
  - have low melting and boiling points that decrease down the group
  - become more reactive down the group.
- This is because the outer electron gets further away from the influence of the nucleus, so it can be lost more easily.
- Alkali metals are stored under oil because they react very vigorously with oxygen and water, including moisture in the air.
- When alkali metals react with water, a metal hydroxide is formed and hydrogen gas is given off.

$$\text{potassium} + \text{water} \longrightarrow \text{potassium hydroxide} + \text{hydrogen}$$
$$2K(s) + 2H_2O(l) \longrightarrow 2KOH(aq) + H_2(g)$$

Key Point

Elements that have the same number of electrons in their outer shell have similar properties.

Key Point

In the periodic table, elements with the same number of electrons in their outer shell are in the same group, e.g. Group 1 elements all have one electron in their outer shell.

Key Point

Group 1 metals react with oxygen to form metal oxides. For example, sodium reacts with oxygen to form sodium oxide.

Key Point

The reactivity of Group 1 metals increases down the group as the outer electron is lost more easily.

- Group 1 metals have a low density – lithium, sodium and potassium are less dense than water, so float on top of it.

Alkali Metals Reacting with Water

Li   Na   K

- When a metal hydroxide (e.g. potassium hydroxide) is dissolved in water, an alkaline solution is produced.
- Alkali metals react with non-metals to form ionic compounds.
- When this happens, the metal atom loses one electron to form a metal ion with a positive charge (+1).

$$\text{sodium + chlorine} \longrightarrow \text{sodium chloride}$$
$$2Na(s) + Cl_2(g) \longrightarrow 2NaCl(s)$$

Sodium chloride is a white solid that dissolves in water to form a colourless solution.

Group 7

- The Group 7 elements are non-metals and are known as the halogens. They have seven electrons in their outermost shell.
- Reactivity decreases down the group because the outer shell gets further away from the nucleus, so it is less easy to gain an electron.
- Halogens react with metals to produce ionic salts.
- When this happens, the halogen atom gains one electron to form a halide ion with a negative charge (−1).

$$\text{chlorine + potassium} \longrightarrow \text{potassium chloride}$$
$$Cl_2(g) + 2K(s) \longrightarrow 2KCl(s)$$

Key Point

The reactivity of the Group 7 non-metals decreases down the group as it becomes less easy to gain an electron.

- A more reactive halogen will **displace** a less reactive halogen from an aqueous solution of its salt. For example:
  - chlorine will displace bromine from potassium bromide and iodine from potassium iodide
  - bromine will displace iodine from potassium iodide.

$$\text{chlorine + potassium bromide} \longrightarrow \text{potassium chloride + bromine}$$
$$Cl_2(g) + 2KBr(aq) \longrightarrow 2KCl(aq) + Br_2(l)$$

Key Words

Mendeleev
noble gases
alkali metals
halogens
displace

Quick Test

1. Name the Russian chemist who designed the periodic table.
2. Why are Group 0 elements unreactive?
3. Why are the alkali metals stored in oil?

States of Matter

You must be able to:

- Recall the meaning of the state symbols in equations
- Describe how the particles move in solids, liquids and gases
- Use the particle model to explain how the particles are arranged in the three states of matter
- HT Describe the limitations of the particle model.

Three States of Matter

- Everything is made of matter.
- There are three states of matter: solid, liquid and gas.
- These three states of matter are described by a simple model called the 'particle model'.
- In this model, the particles are represented by small solid spheres.
- The model can be used to explain how the particles are arranged and how they move in solids, liquids and gases.
- In solids, the particles:
  - have a regular arrangement
  - are very close together
  - vibrate about fixed positions.
- In liquids, the particles:
  - have a random arrangement
  - are close together
  - flow around each other.
- In gases, the particles:
  - have a random arrangement
  - are much further apart
  - move very quickly in all directions.

Solid

Liquid

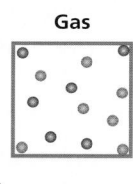

Gas

> ### HT Key Point
>
> This particle model does have some limitations. It does not take into account:
>
> - the forces between the particles
> - the volume (although small) of the particles
> - the space between particles.

Changing States

- When a substance changes state, e.g. from solid to liquid:
  - the particles themselves stay the same
  - the way the particles are arranged changes
  - the way the particles move changes.
- A pure substance will:
  - melt and freeze at one specific temperature – the **melting point**
  - boil and condense at one specific temperature – the **boiling point**.
- The amount of energy required for a substance to change state depends on the amount of energy required to overcome the forces of attraction between the particles.
- The stronger the forces of attraction:
  - the greater the amount of energy needed to overcome them
  - the higher the melting point and boiling point will be.
- Substances that have high melting points due to strong bonds include ionic compounds, metals and giant covalent structures.

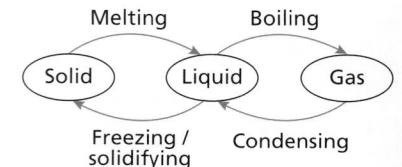

- In substances that contain small molecules:
  - the bonds within the molecules are strong covalent bonds
  - the forces of attraction between the molecules are much weaker
  - only a little energy is needed to overcome the forces between the molecules, so the melting and boiling points are relatively low.

Identifying the State of a Substance

- The melting point and boiling point of a substance can be used to identify its state at a given temperature.

  The table below shows the melting points and boiling points of some Group 7 elements.

  What is the state of each element at 25°C (room temperature)?

Element	Melting Point (°C)	Boiling Point (°C)
fluorine	−220	−188
chlorine	−102	−34
bromine	−7	59
iodine	114	184

  Fluorine and chlorine are gases, bromine is a liquid and iodine is a solid.

25°C is above the boiling points of fluorine and chlorine, so they will be gases.

25°C is above the melting point but below the boiling point of bromine, so it will be a liquid.

25°C is below the melting point of iodine, so it will be a solid.

State Symbols

- Chemical equations are used to sum up what happens in reactions.
- State symbols show the state of each substance involved.

State Symbol	State of Substance
(s)	solid
(l)	liquid
(g)	gas
(aq)	aqueous (dissolved in water)

- For example, when solid magnesium ribbon is added to an aqueous solution of hydrochloric acid:
  - a chemical reaction takes place
  - a solution of magnesium chloride is produced
  - hydrogen gas is produced.
- This can be summed up in a symbol equation:

$$Mg(s) + 2HCl(aq) \longrightarrow MgCl_2(aq) + H_2(g)$$

Quick Test

1. What does the state symbol (l) indicate?
2. What does the state symbol (aq) indicate?
3. How do the particles in a gas move?
4. Why do ionic compounds have high melting points?
5. HT State three limitations of the particle model.

Key Words

matter
particle
melting point
boiling point
aqueous

Ionic Compounds

You must be able to:

- Describe what an ionic bond is
- Explain how ionic bonding involves the transfer of electrons from metal atoms to non-metal atoms to form ions
- Relate the properties of ionic compounds to their structures.

Chemical Bonds

- There are three types of strong chemical bonds:
  - ionic bonds
  - covalent bonds
  - metallic bonds.
- Atoms that have gained or lost electrons are called **ions**.
- Ionic bonds occur between positive and negative ions.

Ionic Bonding

- Ions are formed when atoms gain or lose electrons, giving them an overall charge.
- Ions have a complete outer shell of electrons (the same electron configuration as a noble gas).
- Ionic bonding involves a transfer of electrons from metal atoms to non-metal atoms.
- The metal atoms lose electrons to become positively charged ions.
- The non-metal atoms gain electrons to become negatively charged ions.
- The ionic bond is a strong **electrostatic** force of attraction between the positive metal ion and the negative non-metal ion.

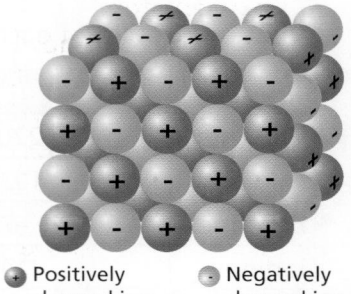

Positively charged ion   Negatively charged ion

Key Point

The arrangement of electrons in an atom can be described in terms of shells or energy levels. Electron configuration diagrams are a good example of using diagrams to represent information.

Sodium forms an ionic compound with chlorine.

Describe what happens when two atoms of sodium react with one molecule of chlorine.

Give your answer in terms of electron transfer.

- Sodium belongs to Group 1 of the periodic table. It has one electron in its outer shell.
- Chlorine belongs to Group 7 of the periodic table. It has seven electrons in its outer shell.
- One chlorine molecule contains two chlorine atoms.
- Each sodium atom transfers one electron to one of the chlorine atoms.
- All four atoms now have eight electrons in their outer shell.
- The atoms become ions, Na^+ and Cl^-.
- The compound formed is sodium chloride, NaCl.

$$2Na + Cl_2 \rightarrow 2NaCl$$

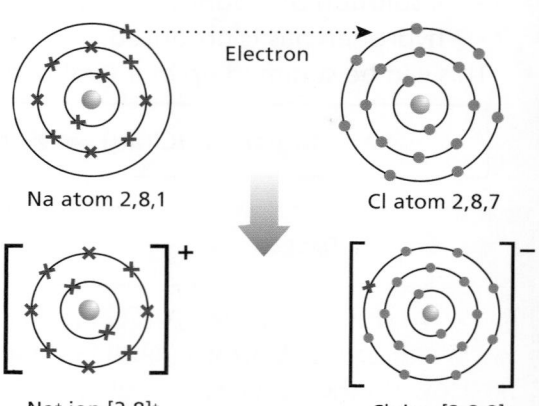

Na atom 2,8,1   Electron   Cl atom 2,8,7

Na^+ ion $[2,8]^+$ Cl^- ion $[2,8,8]^-$

When magnesium is burned, it forms an ionic compound with oxygen.

Describe what happens when two atoms of magnesium react with one molecule of oxygen. Give your answer in terms of electron transfer.

- Magnesium is in Group 2 of the periodic table. It has two electrons in its outer shell.
- Oxygen is in Group 6 of the periodic table. It has six electrons in its outer shell.
- One oxygen molecule contains two oxygen atoms.
- Each magnesium atom loses two electrons to an oxygen atom.
- All four atoms now have eight electrons in their outer shell.
- The atoms become ions, Mg^{2+} and O^{2-}.
- The compound formed is magnesium oxide, MgO.

$$2Mg + O_2 \rightarrow 2MgO$$

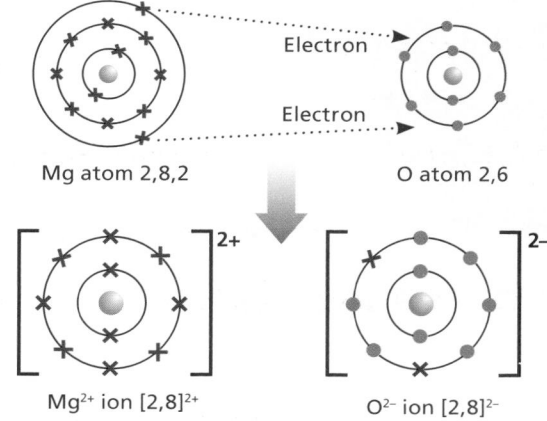

Mg atom 2,8,2   O atom 2,6

Mg^{2+} ion $[2,8]^{2+}$ O^{2-} ion $[2,8]^{2-}$

Properties of Ionic Compounds

- Ionic compounds are giant structures of ions.
- They are held together by strong forces of attraction (electrostatic forces) that act in all directions between oppositely charged ions, i.e. ionic compounds are held together by strong ionic bonds.
- Ionic compounds:
  - have high melting and boiling points
  - do *not* conduct electricity when solid, because the ions cannot move
  - do conduct electricity when **molten** or in solution, because the charged ions are free to move about and carry their charge.

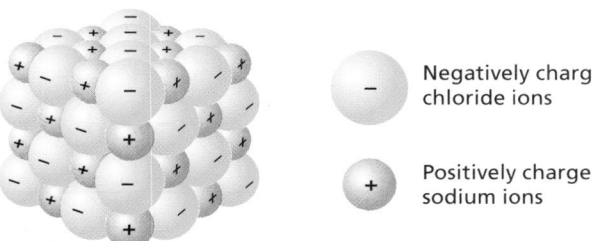

− Negatively charged chloride ions

+ Positively charged sodium ions

Covalent Compounds

You must be able to:

- Describe a covalent bond
- Describe the structure of small molecules
- Explain the properties of small molecules
- Describe the giant covalent structures: diamond, graphite and silicon dioxide
- Explain the properties of giant covalent structures.

Covalent Bonding

- A covalent bond is a shared pair of electrons between atoms.
- Covalent bonds occur in:
  - non-metallic elements, e.g. oxygen, O_2
  - compounds of non-metals, e.g. sulfur dioxide, SO_2.
- For example, a chlorine atom has seven electrons in its outer shell. In order to bond with another chlorine atom:
  - an electron from each atom is shared
  - this gives each chlorine atom eight electrons in the outer shell
  - each atom now has a complete outer shell.
- Covalent bonds in molecules can be shown using dot and cross diagrams.
- Covalent bonds are very strong.
- Some covalently bonded substances consist of small molecules:

A Chlorine Molecule (One Covalent Bond)

Two chlorine atoms

A chlorine molecule (made up of two chlorine atoms)

Outer shells overlap

A Molecule of Ammonia (Three Covalent Bonds)

H N H
H

Molecule	Water H_2O	Chlorine Cl_2	Hydrogen H_2	Hydrogen chloride, HCl	Methane CH_4	Oxygen O_2
Method 1	H O H	Cl Cl	H H	H Cl	H C H (with H top and bottom)	O O
Method 2	H–O–H	Cl–Cl	H–H	H–Cl	H–C–H (with H top and bottom)	O=O (a double bond)

- Others have giant covalent structures, e.g. diamond and silicon dioxide.

Small Molecules

- Small molecules contain a relatively small number of non-metal atoms joined together by covalent bonds.
- The molecules have no overall electrical charge, so they cannot conduct electricity.
- Substances that consist of small molecules usually have low melting and boiling points.
- This is because they have weak intermolecular forces (forces of attraction between the molecules).
- These intermolecular forces are very weak Compared to the strength of the covalent bonds in the molecules themselves.

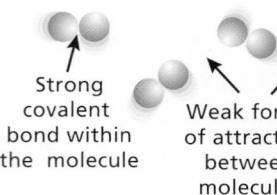

Strong covalent bond within the molecule

Weak forces of attraction between molecules

- The larger the molecules are, the stronger the intermolecular forces between the molecules become.
- This means that larger molecules have higher melting and boiling points.
- Going down Group 7 of the periodic table, the molecules get larger and their melting and boiling points increase.
- This is demonstrated by their states at room temperature:
  - Fluorine and chlorine are gases.
  - Bromine is a liquid.
  - Iodine is a solid.

Giant Covalent Structures

- All the atoms in giant covalent structures are linked by strong covalent bonds.
- These bonds must be broken for the substance to melt or boil.
- This means that giant covalent structures have very high melting and boiling points.
- **Diamond** is a form of carbon:
  - It has a giant, rigid covalent structure (lattice).
  - Each carbon atom forms four strong covalent bonds with other carbon atoms.
  - All the strong covalent bonds mean that it is a very hard substance with a very high melting point.
  - There are no charged particles, so it does not conduct electricity.
- **Graphite** is another form of carbon:
  - It also has a giant covalent structure and a very high melting point.
  - Each carbon atom forms three covalent bonds with other carbon atoms.
  - This results in a layered, hexagonal structure.
  - The layers are held together by weak intermolecular forces.
  - This means that the layers can slide past each other, making graphite soft and slippery.
  - One electron from each carbon atom in graphite is **delocalised**.
  - These delocalised electrons allow graphite to conduct heat and electricity.
- **Silicon dioxide** (or **silica**, SiO_2) has a lattice structure similar to diamond:
  - Each oxygen atom is joined to two silicon atoms.
  - Each silicon atom is joined to four oxygen atoms.

Diamond

Covalent bond between carbon atoms

○ Carbon atom

Graphite

Covalent bond between carbon atoms

○ Carbon atom   Weak bond between layers

Quick Test

1. What is a covalent bond?
2. Why does hydrogen chloride (HCl) have a low boiling point?
3. How does the size of a small molecule affect the strength of the intermolecular forces between molecules?
4. Describe the structure of diamond.
5. Explain how graphite can conduct electricity.

Key Words

covalent bond
intermolecular
diamond
graphite
delocalised
silicon dioxide (silica)

Metals and Special Materials

You must be able to:

- Recall the structure and uses of graphene and fullerenes
- Understand the bonding within and between polymer molecules
- Describe when and why metallic bonding occurs
- Explain the properties of pure metals and alloys.

Graphene

- **Graphene** is a form of carbon. It is a single layer of graphite (see page 99).
- The atoms are arranged in a hexagonal structure, just one atom thick.
- Graphene has some special properties.
- It is very strong, a good thermal and electrical conductor and nearly **transparent**.

Fullerenes

- Carbon can also form molecules known as **fullerenes**, which contain different numbers of carbon atoms.
- Fullerene molecules have hollow shapes, including tubes, balls and cages.
- The first fullerene to be discovered was buckminsterfullerene, C_{60}:
  - It consists of 60 carbon atoms.
  - The atoms are joined together in a series of hexagons and pentagons.
  - It is the most symmetrical and, therefore, most stable fullerene.
- Carbon nanotubes are cylindrical fullerenes with some very useful properties.
- Fullerenes can be used:
  - to deliver drugs in the body
  - in lubricants
  - as catalysts
  - for reinforcing materials, e.g. the frames of tennis rackets, so that they are strong but still lightweight.

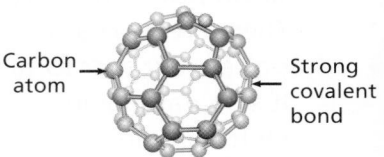

Structure of Buckminsterfullerene

Carbon atom — Strong covalent bond

Structure of a Nanotube

Carbon atom — Strong covalent bond

Polymers

- Polymers consist of very large molecules.
- Plastics are synthetic (man-made) polymers.
- The atoms within the polymer molecules are held together by strong covalent bonds.
- The intermolecular forces between the large polymer molecules are also quite strong.
- This means that polymers are solid at room temperature.
- Poly(ethene), commonly known as polythene, is produced when lots of ethene molecules are joined together in an addition polymerisation reaction.
- It is cheap and strong and is used to make plastic bottles and bags.

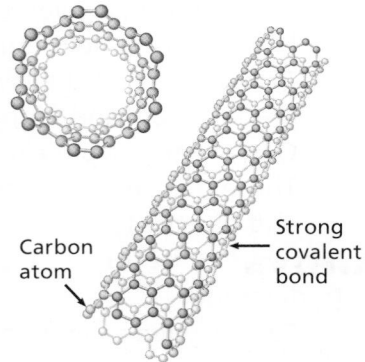

Addition Polymerisation

Metallic Bonding

- Metallic bonding occurs in:
  - metallic elements, such as iron and copper
  - alloys, such as stainless steel.
- Metals have a giant structure in which electrons in the outer shell are **delocalised** (not bound to one atom).
- This produces a regular arrangement (lattice) of positive ions held together by **electrostatic** attraction to the delocalised electrons.
- A **metallic bond** is the attraction between the positive ions and the delocalised negatively charged electrons.

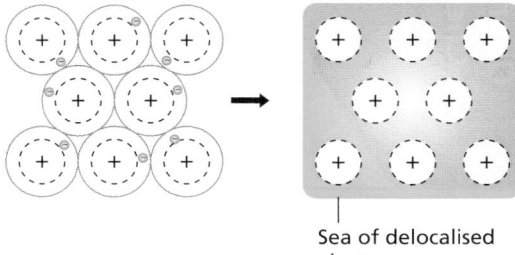

Sea of delocalised electrons

Properties of Metals

- The properties of metals make them very useful.
- Metallic bonds are very strong and most metals have high melting and boiling points. This means that they are useful structural materials.
- The delocalised electrons can move around freely and transfer energy. This makes metals good thermal and electrical conductors.
- The particles in **pure** metals have a regular arrangement.
- The layers are able to slide over each other quite easily, which means that metals can be bent and shaped.
- Traditionally, copper is used to make water pipes because:
  - it is an unreactive metal, so it does not react with water
  - it can be easily shaped.

Metal	Uses	Property
Aluminium	High-voltage power cables, furniture, drinks cans, foil food wrap	Corrosion resistant, **ductile**, **malleable**, good conductivity, low density
Copper	Electrical wiring, water pipes, saucepans	Ductile, malleable, good conductivity
Gold	Jewellery, electrical junctions	Ductile, shiny, good conductivity

> ### Key Point
>
> A metallic bond is the attraction between positive ions and delocalised electrons.

Alloys

- Most metal objects are made from **alloys** – mixtures that contain a metal and at least one other element.
- Pure metals are too soft for many uses.
- In alloys, the added element disturbs the regular arrangement of the metal atoms so the layers do not slide over each other so easily.
- This means alloys are usually stronger and harder than pure metals.

> ### Key Words
>
> graphene
> transparent
> fullerene
> delocalised
> electrostatic
> metallic bond
> pure
> ductile
> malleable
> alloy

> ### Quick Test
>
> 1. What is special about the structure of graphene?
> 2. Which is the most stable fullerene?
> 3. Describe a metallic bond.
> 4. Why is copper a good material for water pipes?

Conservation of Mass

You must be able to:

- Understand the law of conservation of mass
- Work out the relative formula mass of substances
- Understand why reactions that involve gases may appear to show a change in mass
- HT Write balanced half equations and ionic equations.

The Conservation of Mass

- In a chemical reaction, the total mass of the products is equal to the total mass of the reactants.
- This idea is called the **conservation of mass**.
- Mass is conserved (kept the same) because no atoms are lost or made.
- Chemical symbol equations must always be balanced to show this, i.e. there must be the same number of atoms of each element on both sides of the equation.
- For example, when solid iron reacts with copper(II) sulfate solution, a reaction takes place, producing solid copper and iron(II) sulfate solution:

$$Fe(s) + CuSO_4(aq) \longrightarrow Cu(s) + FeSO_4(aq)$$

HT A **half equation** can be used to show what happens to one reactant in a chemical reaction, with electrons written as e^-.

HT The balanced symbol equation for the reaction between iron and copper(II) sulfate can be split into two half equations:

- The iron atoms lose two electrons to form Fe^{2+} ions.

$$Fe(s) \longrightarrow Fe^{2+}(aq) + 2e^-$$

- The Cu^{2+} ions gain two electrons to form copper atoms.

$$Cu^{2+}(aq) + 2e^- \longrightarrow Cu(s)$$

HT **Ionic equations** can be used to simplify complicated equations.

HT They just show the species that are involved in the reaction.

HT The spectator ions (ions not involved in the reaction) are not included.

HT For example, when silver nitrate solution is added to sodium chloride solution, a white precipitate of silver chloride is produced:

$$AgNO_3(aq) + NaCl(aq) \longrightarrow AgCl(s) + NaNO_3(aq)$$

HT In this reaction, the nitrate ions and the sodium ions are spectator ions, so the ionic equation is:

$$Ag^+(aq) + Cl^-(aq) \longrightarrow AgCl(s)$$

HT In chemistry, the term '**species**' refers to the different atoms, molecules or ions that are involved in a reaction.

Key Point

The total mass of the products of a chemical reaction is always equal to the total mass of the reactants. This is because no atoms are lost or made. The products are made from exactly the same atoms as the reactants.

Relative Formula Mass

- The relative formula mass (M_r) of a compound is the sum of the relative atomic masses (A_r) of all the atoms in the numbers shown in the formula. It does not have a unit.
- The relative atomic masses of the atoms are shown in the periodic table.

> What is the relative formula mass of carbon dioxide, CO_2?
> **Relative formula mass = (12 × 1) + (16 × 2)**
> **= 44** ←

CO_2 contains 1 carbon atom with a relative atomic mass of 12 and 2 oxygen atoms with a relative atomic mass of 16.

> Calculate the relative formula mass of calcium nitrate, $Ca(NO_3)_2$.
> **Relative formula mass = 40 + (14 × 2) + (16 × 6)**
> **= 164** ←

Remember that everything inside a set of brackets is multiplied by the number outside the brackets, so $Ca(NO_3)_2$ contains 1 calcium, 2 nitrogen and 6 oxygen atoms.

- Due to conservation of mass, the sum of the relative formula masses of all the reactants is always equal to the sum of the relative formula masses of all the products.

Apparent Changes in Mass

- Some reactions appear to involve a change in mass.
- This happens when reactions are carried out in a non-closed system and include a gas that can enter or leave.
- For example, when magnesium is burned in air to produce magnesium oxide, the mass of the solid increases.
- This is because when the magnesium is burned, it combines with oxygen from the air and the oxygen has mass:

$$2Mg(s) + O_2(g) \longrightarrow 2MgO(s)$$

- If the mass of oxygen is included, the total mass of all the reactants is equal to the total mass of all the products.
- When calcium carbonate is heated, it decomposes to form calcium oxide and carbon dioxide:

$$CaCO_3(s) \longrightarrow CaO(s) + CO_2(g)$$

- The mass of the solid decreases because one of the products is a gas, which escapes into the air.
- If the mass of carbon dioxide is included, the total mass of all the reactants is equal to the total mass of all the products.

 Key Point

The relative atomic mass is an average value that takes account of the abundance of the isotopes of an element. 25% of chlorine atoms have a mass of 37. 75% of chlorine atoms have a mass of 35. The relative atomic mass of chlorine

$$= (\frac{25}{100} \times 37) + (\frac{75}{100} \times 35)$$
$$= 35.5$$

 Key Point

Some reactions appear to involve a change in mass. This happens when reactions are carried out in a non-closed system, so gases can enter or leave.

 Quick Test

1. State the law of conservation of mass.
2. Why must a symbol equation balance?
3. Calculate the relative formula mass of water, H_2O.
4. Calculate the relative formula mass of calcium carbonate, $CaCO_3$.
5. When iron is burned, iron oxide is produced and the mass of the solid increases. Why does the mass of the solid increase during this reaction?

 Key Words

conservation of mass
HT half equation
HT ionic equation
HT species
relative formula mass (M_r)
relative atomic mass (A_r)

Amount of Substance

You must be able to:

HT Recall the number of particles in one mole of any substance

HT Calculate the amount of a substance in moles

HT Calculate the mass of reactants or products from balanced equations

HT Calculate the balancing numbers in equations from the masses of the reactants and the products by using moles

• Calculate the mass of solute in a given volume of solution of known concentration.

HT Amount of Substance

• A **mole (mol)** is a measure of the number of particles (atoms, ions or molecules) contained in a substance.
• One mole of any substance (element or compound) contains the same number of particles – six hundred thousand billion billion or 6.02×10^{23}.
• This value is known as the **Avogadro constant**.
• The mass of one mole of a substance is its relative atomic mass or relative formula mass in grams.

> One mole of sodium atoms contains 6.02×10^{23} atoms.
>
> The relative atomic mass of sodium is 23.0.
>
> One mole of sodium atoms has a mass of 23.0g.

HT Key Point

One mole of any substance (element or compound) will always contain the same number of particles – six hundred thousand billion billion or 6.02×10^{23}. This value is known as the Avogadro constant.

HT Calculating the Amount of Substance

• You can calculate the amount of substance (number of moles) in a given mass of a substance using the formula:

> **LEARN** HT
>
> $$\text{amount (mol)} = \frac{\text{mass of substance (g)}}{\text{atomic (or formula) mass (g/mol)}}$$

> Calculate the number of moles of carbon dioxide in 33g of the compound.
>
> $$\text{amount} = \frac{\text{mass of substance}}{\text{formula mass}}$$
> $$= \frac{33}{44}$$
> $$= 0.75\text{mol}$$

Formula mass of CO_2 = 12 + (16 × 2)
= 44

HT Balanced Equations

• Balanced equations:
  - show the number of moles of each product and reactant
  - can be used to calculate the mass of the reactants and products.
• The numbers needed to balance an equation can be calculated from the masses of the reactants and the products using moles.

Aluminium oxide can be reduced to produce aluminium:

$$Al_2O_3 \rightarrow 2Al + 1\tfrac{1}{2}O_2$$

Calculate the mass of aluminium oxide needed to produce 540g of aluminium.

$$\text{amount of aluminium} = \frac{540}{27} = 20\,mol$$

$$\text{amount of aluminium oxide required} = \frac{20}{2} = 10\,mol$$

$$\text{formula mass of aluminium oxide} = (27 \times 2) + (16 \times 3)$$
$$= 102$$

$$\text{mass of aluminium oxide needed} = 10 \times 102 = 1020g$$

> The equation shows that one mole of aluminium oxide produces two moles of aluminium.

> amount of aluminium = $\dfrac{mass}{atomic\ mass}$
> Relative atomic masses (A_r): Al = 27 and O = 16.

> The equation shows that one mole of aluminium oxide is needed to produce two moles of aluminium, so divide by two.

> mass of aluminium oxide needed = amount (mol) × formula mass

- The numbers needed to balance an equation can be calculated from the masses of the reactants and the products using moles.

In a chemical reaction, 72g of magnesium was reacted with exactly 48g of oxygen molecules to produce 120g of magnesium oxide.

Use the number of moles of reactants and products to write a balanced equation for the reaction.

$$\text{amount of Mg} = \frac{72}{24} = 3\,mol$$

$$\text{amount of O}_2 = \frac{48}{32} = 1.5\,mol$$

$$\text{amount of MgO} = \frac{120}{40} = 3\,mol$$

$$3Mg + 1.5O_2 \rightarrow 3MgO$$

$$2Mg + O_2 \rightarrow 2MgO$$

> Use the masses of the reactants to calculate the number of moles present.

> Divide the number of moles of each substance by the smallest number (1.5) to give the simplest whole number ratio.

> This shows that 2 moles of magnesium react with 1 mole of oxygen molecules to produce 2 moles of magnesium oxide.

HT Limiting Reactants

- Sometimes when two chemicals react together, one chemical is completely used up during the reaction.
- When one chemical is used up, it stops the reaction going any further. It is called the limiting reactant.
- The other chemical, which is not used up, is said to be in excess.

Concentration of Solutions

- Many chemical reactions involve solutions.
- The concentration of a solution may be given by the mass of solute per given volume of solution.
- For example, if 2.00g of solute was dissolved to form 1.00dm³ of solution, the concentration of the solution would be 2.00g/dm³.

Quick Test

1. **HT** 69g of sodium reacts with chlorine to produce sodium chloride:
   $$2Na + Cl_2 \rightarrow 2NaCl$$
   a) Calculate the number of moles of sodium present.
   b) Calculate the number of moles of chlorine (Cl_2) that would be required to react exactly with the sodium.
   c) Calculate the mass of chlorine that would be required to react exactly with the sodium.

Key Words

HT mole (mol)
HT Avogadro constant
HT limiting reactant
concentration

Review Questions

Sexual and Asexual Reproduction

1 Mitosis is the division of body cells to make new cells.

a) When is mitosis **not** used for cell division?
Tick **one** box.

Asexual reproduction ☐   Repair ☐

Gamete production ☐   Growth ☐   **[1]**

b) Complete the sentences about mitosis.

A copy of each _____ is made before a cell divides.

The new cell has the same _____ information as the _____ cell.

Meiosis takes place in the testes, and produces sperm containing 23 _____.   **[4]**

Total Marks _____ / 5

Patterns of Inheritance

1 Rita grows a species of plant that can either have red flowers or white flowers.
She decides to cross a red flowered plant (Rr) with a white flowered plant (rr).

a) Draw a diagram to show this genetic cross.
Use **R** to represent the dominant allele and **r** to represent the recessive allele.   **[4]**

b) Rita grows 24 plants using the seeds from this cross.

Predict the number of red flowered and white flowered plants that are produced.   **[2]**

c) Why is it unlikely that the actual numbers of each type of plant will exactly match this prediction?   **[2]**

Total Marks _____ / 8

Variation and Evolution

1 Variation can be due to inherited factors, environmental factors or a combination of both.
Cathy and Drew are sister and brother. **Table 1** shows how they are different.

Table 1

Cathy	Drew	Inherited (I), Environmental (E) or Both (B)
tongue roller	non roller	
not colour blind	colour blind	
1.5m tall	1.6m tall	
speaks French	does not speak French	

Complete **Table 1** to show if each difference is caused by inherited factors **(I)**, environmental factors **(E)** or a combination of both **(B)**. **[4]**

Total Marks _____ / 4

Manipulating Genes

1 Farmers have been using selective breeding for thousands of years to produce crops and animals with desirable characteristics.

a) Suggest **two** characteristics that might be desirable in food crops. **[2]**

b) Suggest **two** characteristics that might be desirable in dairy cows. **[2]**

c) Describe the main stages in the process of selective breeding. **[4]**

d) Developments in biotechnology mean that genes can now be inserted into crop plants to give them desirable characteristics.

What is the term used to describe crops that have had their genes altered in this way? **[1]**

Total Marks _____ / 9

Classification

1 Scientists frequently study the distribution of the snail, *Cepaea nemoralis*. The snail has a shell that can be brown or yellow.

a) i) What genus does the snail belong to? **[1]**

ii) How could scientists prove that the different coloured snails were all the same species? **[2]**

b) Scientists believe that shell colour affects the body temperature of the snails.
Snails with dark shells warm up faster than those with light shells.
In cold areas, this would be advantageous to the dark-coloured snail.

The average annual temperature in Scotland is 2°C lower than in England.

Use the theory of natural selection to explain why populations of the snail in Scotland contain a higher percentage of dark-coloured snails. **[3]**

> Total Marks _____ / 6

Ecosystems

1 A class of students was asked to estimate the number of daisies on the school football pitch.
The pitch is 60m by 90m.
They decided to use quadrats that were 1m².

a) Which is the best way of using quadrats in this investigation?
Tick **one** box.

Place all the quadrats where there are lots of plants. ☐

Place all the quadrats randomly in the field. ☐

Place all the quadrats where daisies do not grow. ☐ **[1]**

b) Each student collected data by using 10 quadrats.

Table 1 shows the results of one student, Shaun.

Table 1

Quadrat	1	2	3	4	5	6	7	8	9	10
Number of Daisies	5	2	1	0	4	5	2	0	6	3

Calculate the mean number of daisies per quadrat counted by Shaun.
You must show your working. **[2]**

c) Another student, Mandeep, calculated a mean of 2.3 daisies per quadrat from her results.

    i) Use Mandeep's results to estimate the total number of daisies in the whole pitch. You must show your working. **[2]**

    ii) The centre circle has a diameter of 10 metres.

       How many daisies are likely to be in the centre circle? (Use $\pi = 3.14$) **[3]**

> Total Marks .... / 8

Cycles and Feeding Relationships

1 Underline the correct words to complete the sentences about the carbon cycle.

Plants and **animals / algae** remove carbon dioxide from the air.

When plants die, they are broken down by **consumers / decomposers / producers**.

Bacteria and fungi are examples of **consumers / decomposers / producers**. **[3]**

> Total Marks .... / 3

Disrupting Ecosystems

1 a) Complete the sentences.

Some gases in the atmosphere prevent .... from escaping into space.

This is called the .... effect.

Two gases that contribute to this are .... and .... **[4]**

b) Which of the following are possible negative effects of global warming?
Tick **two** boxes.

Climate change	☐	A rise in sea level	☐
Erosion of buildings	☐	Increase in available land	☐
Deforestation	☐		**[2]**

> Total Marks .... / 6

Practice Questions

Atoms, Elements, Compounds and Mixtures

1 What are the substances that react together in a chemical reaction called?
Tick **one** box.

Products ☐   Mixtures ☐   Reactants ☑   Ions ☐   **[1]**

2 Why is the total mass of the reactants in a chemical reaction always equal to the total mass
of the products? *atoms are not made or destroyed during* **[1]**
reactions.

3 Define the term 'mixture'.   *+ can be seperated easily* = **[1]**
two or more substances that are not chemically

4 Suggest a technique that could be used to separate the components of a food colouring *bonded*
used in cupcake icing. **[1]**

> **Total Marks** _____ / 4

Atoms and the Periodic Table

1 Complete **Table 1** to show the relative mass of
the different subatomic particles.

Table 1

Subatomic Particle	Relative Mass
proton	*1*
	1
electron	*very small*

[3]

2 An ion of potassium is represented as $^{39}_{19}K^+$.

Give the number of protons, electrons and neutrons in this ion of potassium. **[3]**
protons: 19   neutrons : 20   electrons : 19

3 What is the radius of a typical atom?
Tick **one** box.

1×10^{-12}m ☐ 1×10^{-10}m ☐ 1×10^{-8}m ☐ 1×10^{-3}m ☑   **[1]**

> **Total Marks** _____ / 7

The Periodic Table

1 **a)** Which group in the periodic table do the alkali metals belong to? **[1]**

b) How many electrons do the alkali metals have in their outermost shell? **[1]**

c) Explain, in terms of electrons, why the alkali metals get more reactive as you go down the group. [2]

d) Why do the alkali metals have to be stored under oil? [2]

> Total Marks _____ / 6

States of Matter

1 What do the state symbols (g) and (aq) mean? [2]

2 HT Give **three** limitations of the particle model. [3]

3 Calcium reacts with hydrochloric acid to produce calcium chloride and hydrogen:

$Ca(s) + 2HCl(aq) \rightarrow CaCl_2(aq) + H_2(g)$

What is the state of the:

a) Calcium?　　　　　b) Hydrochloric acid?　　　　　c) Hydrogen? [3]

> Total Marks _____ / 8

Ionic Compounds

1 **Table 1** shows the charge on some metal ions and non-metal ions.

a) What is an ion? [1]

an atom that has + or − electrons to have a full outer shell

b) In terms of electron transfer, explain why chloride ions have a 1− charge. [2]

it is group 7 and so needs one more electrons to full outer shell

c) In terms of electron transfer, explain why magnesium ions have a 2+ charge. [2]

magnesium is in group 2 + so needs to minus two to have a full outer shell.

Table 1

Metal Ions	Non-Metal Ions
Sodium, Na^+	Chloride, Cl^-
Magnesium, Mg^{2+}	Oxide, O^{2-}
Potassium, K^+	Fluoride, F^-
Calcium, Ca^{2+}	Sulfide, S^{2-}

d) Use the information in the table to determine the formula of:

i) Potassium chloride　　ii) Magnesium sulfide　　iii) Calcium oxide. [3]

e) Magnesium oxide has a very high melting point and can be used to line furnaces.

Explain why the compound magnesium oxide has a high melting point. [2]

> Total Marks _____ / 10

Practice Questions

Covalent Compounds

1 Ammonia, NH_3, and water, H_2O, are both small molecules that contain covalent bonds.

a) Define the term 'covalent bond'. [1]

b) **Figure 1** shows the outer electrons in a nitrogen atom and in a hydrogen atom.

Complete **Figure 2** below to show the electron arrangement in an ammonia molecule. [1]

Figure 1   **Figure 2**

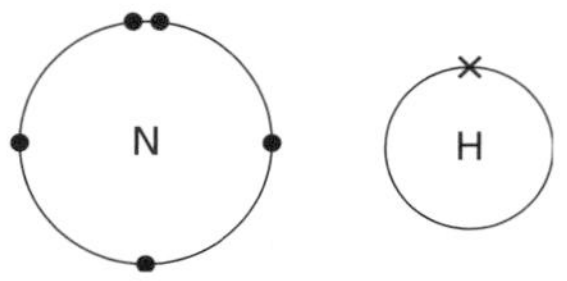

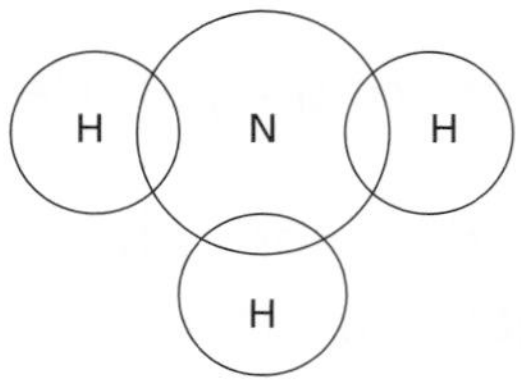

c) Explain why ammonia and water do **not** conduct electricity. [1]

Total Marks _____ / 3

Metals and Special Materials

1 a) What is the chemical symbol for gold? [1]

b) Name the type of bonding that occurs in gold. [1]

c) Pure gold is too soft for many uses. Why is pure gold soft? [2]

d) Gold is often made into an alloy. What is an 'alloy'? [1]

e) Gold is used to make components for computers because it is a very good electrical conductor. Why is this? [2]

Total Marks _____ / 7

Conservation of Mass

1 A student adds a piece of magnesium ribbon to a flask of dilute hydrochloric acid:

$$Mg(s) + 2HCl(aq) \rightarrow MgCl_2(aq) + H_2(g)$$

Why does the mass of the reaction flask go down? [2]

2 What is the relative formula mass of $Ca(NO_3)_2$?

Relative atomic masses (A_r): Ca = 40, N = 14, O = 16

Tick **one** box.

164 ☐   164g ☐   150 ☐   150g ☐   **[1]**

3 HT Magnesium is more reactive than copper.

Magnesium displaces copper from a solution of copper sulfate:

$$Mg + CuSO_4 \rightarrow MgSO_4 + Cu$$

Complete the two half equations for this reaction.

a) $Mg \rightarrow Mg^{2+} +$ ....   **b)** $Cu^{2+} +$ .... $\rightarrow$ ....   **[3]**

Total Marks .... / 6

Amount of Substance

1 HT What unit do chemists use to measure the amount of substance?

Tick **one** box.

Grams ☐   Moles ☐   Kilograms ☐   Tonnes ☐   **[1]**

2 HT The Avogadro constant has a value of 6.02×10^{23}.

a) How many atoms are present in 7g of lithium?   **[1]**

b) How many atoms are present in 24g of carbon?   **[1]**

3 HT Calculate the number of moles in each of these substances:

a) 19g of fluorine, F_2 **[2]**

b) 22g of carbon dioxide, CO_2 **[2]**

c) 17g of hydroxide, OH^- ions   **[2]**

4 HT Complete combustion of carbon produces carbon dioxide, CO_2: $C + O_2 \rightarrow CO_2$

1.8g of carbon was completely burned in oxygen.

Relative atomic masses (A_r): C = 12, O = 16

a) How many moles of carbon were burned?   **[2]**

b) Calculate the mass of carbon dioxide, CO_2, produced in this reaction.   **[2]**

Total Marks .... / 13

Reactivity of Metals

You must be able to:

- Recall that when metals form metal oxides, the metals are oxidised
- Recall that metals can be placed into a reactivity series
- Recall that a more reactive metal can displace a less reactive metal from a solution of its salt
- Explain how metals less reactive than carbon can be extracted from their oxides
- **HT** Explain oxidation and reduction in terms of electron transfer.

Oxidation and Reduction

- In **oxidation** reactions, a substance often gains oxygen.
- In **reduction** reactions, a substance often loses oxygen.
- Oxidation and reduction always occur together.
- Metals react with oxygen to form metal oxides.
- For example, when magnesium is burned in air it reacts with oxygen to form magnesium oxide.
- The magnesium gains oxygen in the reaction, so it is oxidised:

> **magnesium + oxygen ⟶ magnesium oxide**
>
> $2Mg + O_2 \longrightarrow 2MgO$

- Metal oxides can be reduced by removing oxygen.
- For example, when lead(IV) oxide is heated with carbon:
  - the lead(IV) oxide loses oxygen so it is reduced
  - the carbon gains oxygen so it is oxidised.

> **lead(IV) oxide + carbon ⟶ lead + carbon dioxide**
>
> $PbO_2 + C \longrightarrow Pb + CO_2$

The Reactivity Series

- When metals react, their atoms lose electrons to form positive metal ions.
- Some metals lose electrons more easily than others.
- The more easily a metal atom loses electrons, the more reactive it is.
- The reaction of metals with acid and water can be used to place them in order of reactivity. This is called the **reactivity series**.
- Metals react with acids to produce metal salts and hydrogen.
- Lithium, sodium and potassium are very reactive metals – they react vigorously with water to produce a metal hydroxide solution and hydrogen.
- These metals are placed at the top of the reactivity series.
- Lithium, sodium and potassium would react so vigorously with dilute acids that it would not be safe to carry out the reactions.
- Calcium, magnesium, zinc and iron are fairly reactive metals – they react quickly with acids and slowly with water.
- Very unreactive metals, like copper and gold, do not react with acids or water and are placed at the bottom of the periodic table.
- Reactivity series often include carbon and hydrogen for comparison.

> ### Key Point
>
> Oxidation and reduction always occur together.

Potassium, K
Sodium, Na
Calcium, Ca
Magnesium, Mg
Aluminium, Al
Carbon, C
Zinc, Zn
Iron, Fe
Tin, Sn
Lead, Pb
Hydrogen, H
Copper, Cu
Silver, Ag
Gold, Au
Platinum, Pt

Most reactive

> ### Key Point
>
> Some metals lose electrons more easily than others. The more easily a metal atom loses electrons, the more reactive it is.

Displacement Reactions

- In a **displacement reaction** a more reactive metal will displace a less reactive metal from a solution of its salt.
- Magnesium is more reactive than copper, so magnesium will displace copper from a solution of copper sulfate:

> magnesium + copper sulfate ⟶ copper + magnesium sulfate
>
> $Mg(s) + CuSO_4(aq) \longrightarrow Cu(s) + MgSO_4(aq)$

Extraction of Metals

- The method of **extraction** of a metal depends on how reactive it is.
- Unreactive metals (e.g. gold) exist as elements at the Earth's surface.
- However, most metals are found as metal oxides, or as compounds that can be easily changed into metal oxides.
- Metals that are less reactive than carbon (e.g. iron and lead) can be extracted from their oxides by heating with carbon:

> iron oxide + carbon ⟶ iron + carbon dioxide

- The iron oxide loses oxygen, so it is reduced.
- The carbon gains oxygen, so it is oxidised.

> **Key Point**
>
> Metals that are more reactive than carbon (e.g. aluminium) are extracted from molten compounds by electrolysis.

HT Losing or Gaining Electrons

- Not all reduction and oxidation reactions involve oxygen.
- Because of this, scientists use the following rules:
  - Oxidation is the loss of electrons.
  - Reduction is the gain of electrons.
- The balanced symbol equation for the reaction between magnesium and oxygen can be split into two ionic equations:

> $2Mg \longrightarrow 2Mg^{2+} + 4e^-$
>
> $O_2 + 4e^- \longrightarrow 2O^{2-}$

The magnesium atoms lose electrons to become magnesium ions – the magnesium is oxidised. The oxygen atoms gain electrons to become oxide ions – the oxygen is reduced.

> **Quick Test**
>
> 1. calcium + oxygen → calcium oxide
>   Which substance is oxidised in this reaction?
> 2. Complete the word equation:
>   iron + copper sulfate → _____ + _____
> 3. By what method should lead be extracted from lead oxide?
> 4. HT Calcium reacts with hydrochloric acid to form calcium chloride and hydrogen. The equation for this reaction can be split into two ionic equations:
> $Ca \rightarrow Ca^{2+} + 2e^-$
> $2H^+ + 2e^- \rightarrow H_2$
>   a) Which species is oxidised? Explain your answer.
>   b) Which species is reduced? Explain your answer.

> **Key Words**
>
> oxidation
> reduction
> reactivity series
> displacement reaction
> extraction

The pH Scale and Salts

You must be able to:

- Use the pH scale to show the acidity or alkalinity of a solution
- Explain that in neutralisation reactions H^+ ions react with OH^- ions to produce water, H_2O
- Recall how soluble salts can be made from soluble and insoluble bases
- **HT** Describe the difference between a weak acid and a strong acid.

The pH Scale

- When substances dissolve in water, they **dissociate** into their individual ions:
  - Hydroxide ions, $OH^-(aq)$, make solutions alkaline.
  - Hydrogen ions, $H^+(aq)$, make solutions acidic.
- The pH scale is a measure of the acidity or alkalinity of an **aqueous** solution:
  - A solution with a pH of 7 is neutral.
  - Aqueous solutions with a pH less than 7 are acidic.
  - The closer to a pH of zero, the stronger the acid.
  - Aqueous solutions with a pH of more than 7 are alkaline.
  - The closer to a pH of 14, the stronger the alkali.
- The pH of a solution can be measured using a pH probe or universal indicator.
- **Indicators** are dyes that change colour depending on whether they are in acidic or alkaline solutions:
  - Litmus changes colour from red to blue or vice versa.
  - Universal indicator is a mixture of dyes that shows a range of colours to indicate how acidic or alkaline a substance is.

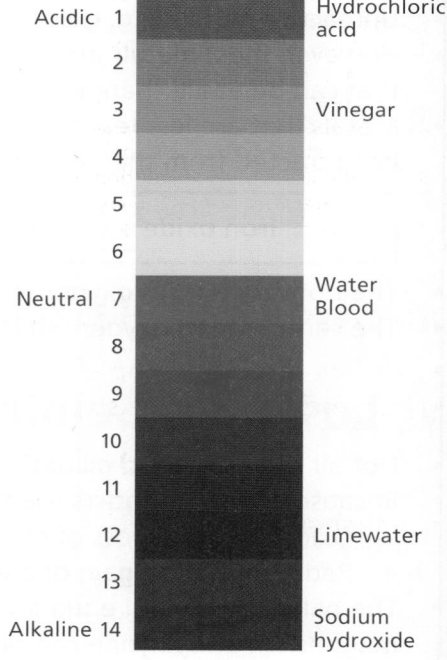

Acidic 1 — Hydrochloric acid
2
3 — Vinegar
4
5
6
Neutral 7 — Water / Blood
8
9
10
11
12 — Limewater
13
Alkaline 14 — Sodium hydroxide

Neutralisation of Acids

- Soluble bases are called alkalis.
- Acids are neutralised by bases.

> **LEARN**
>
> acid + metal hydroxide ⟶ salt + water

- Acids contain hydrogen ions, $H^+(aq)$.
- Alkalis contain hydroxide ions, $OH^-(aq)$.
- When an acid reacts with an alkali, the H^+ and OH^- ions react together to produce water, H_2O, which has a pH of 7.

$$H^+(aq) + OH^-(aq) \longrightarrow H_2O(l)$$

- This type of reaction is called **neutralisation** because:
  - acid is neutralised by an alkali
  - the solution that remains has a pH of 7, showing it is neutral.
- Acids can also be neutralised by metal oxides and metal carbonates:

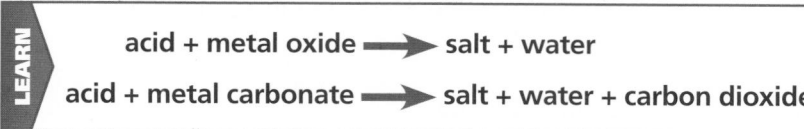

> **LEARN**
>
> acid + metal oxide ⟶ salt + water
>
> acid + metal carbonate ⟶ salt + water + carbon dioxide

- A salt is produced when the hydrogen in the acid is replaced by a metal ion.
- The name of the salt produced depends on the acid used:
  - Hydrochloric acid produces chloride salts.
  - Nitric acid produces nitrate salts.
  - Sulfuric acid produces sulfate salts.

Soluble Salts from Insoluble Bases

- Soluble salts can be made by reacting acids with insoluble bases, such as metal oxides, metal hydroxides and metal carbonates.

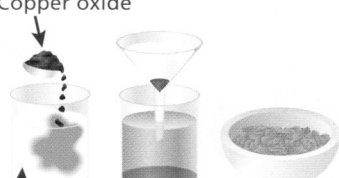

REQUIRED PRACTICAL	
Preparation of a pure, dry sample of a soluble salt from an insoluble oxide or carbonate.	
Sample Method 1. Add the metal oxide or carbonate to a warm solution of acid until no more will react. 2. Filter the excess metal oxide or carbonate to leave a solution of the salt. 3. Gently warm the salt solution so that the water evaporates and crystals of salt are formed.	**Hazards and Risks** • Corrosive acid can cause damage to eyes, so eye protection must be used. • Hot equipment can cause burns, so care must be taken when the salt solution is warmed.

Copper oxide

Sulfuric acid

Add copper oxide to sulfuric acid → Filter to remove any unreacted copper oxide → Evaporate using a water bath or electric heater to leave behind blue crystals of the 'salt' copper sulfate

HT Strong and Weak Acids

- **Strong acids** are completely **ionised** (split up into ions) in water.
- Hydrochloric acid is a strong acid:

$$HCl(g) + aq \longrightarrow H^+(aq) + Cl^-(aq)$$

- Ethanoic acid is a **weak acid**:

$$CH_3COOH(l) + aq \rightleftharpoons CH_3COO^-(aq) + H^+(aq)$$

- The pH of a solution is a measure of the concentration of H^+ ions.
- A pH decrease of one unit indicates that the concentration of hydrogen ions has increased by a factor of 10.
- For a given concentration of acid, a strong acid will have a higher concentration of hydrogen ions and, therefore, a lower pH.
- The terms 'dilute' and 'concentrated' are also applied to acids sometimes.
- An acid that has a concentration of $2mol/dm^3$ is more concentrated than an acid that has a concentration of $0.5mol/dm^3$.

HT Key Point

Strong acids such as hydrochloric acid, nitric acid and sulfuric acid are completely ionised in water.

Weak acids such as ethanoic acid, citric acid and carbonic acid are only partially ionised in water.

The 'aq' in the equation indicates water.

Notice how the sign for a reversible reaction is used in the equation.

Quick Test

1. What is the pH of a neutral solution?
2. Complete the general equation below:
   acid + alkali → _____ + _____
3. HT What is a strong acid?
4. HT The pH of a solution changes from 6 to 4. What happens to the concentration of hydrogen ions?

Key Words

dissociate
aqueous
indicator
neutralisation
HT strong acid
HT ionised
HT weak acid

Electrolysis

You must be able to:

- Explain why ionic compounds conduct electricity when molten or in aqueous solution
- Predict the products of the electrolysis of simple ionic compounds and explain how electrolysis can be used to extract reactive metals
- **HT** Write half equations for the reactions that take place at the electrodes during electrolysis.

Electrolysis

- **Electrolysis** is the use of an electrical current to break down compounds containing ions into their constituent elements.
- The substance being broken down is called the **electrolyte**.
- The **electrodes** are made from solids that conduct electricity.
- During electrolysis:
  - negatively charged ions move to the **anode** (positive electrode)
  - positively charged ions move to the **cathode** (negative electrode).
- Electrolysis can be used to separate ionic compounds into elements.
- For example, lead bromide can be split into lead and bromine:
  - The lead bromide is heated until it melts.
  - The positively charged lead ions move to the negative electrode (cathode).
  - Here they gain electrons to form lead atoms – pure lead is produced at this electrode.
  - The negatively charged bromide ions move to the positive electrode (anode).
  - Here they lose electrons to form bromine atoms, which join together to form bromine molecules – bromine is released at this electrode.

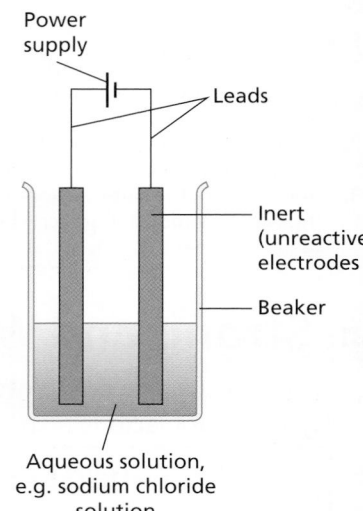

Power supply

Leads

Inert (unreactive) electrodes

Beaker

Aqueous solution, e.g. sodium chloride solution

> **HT** At the cathode: $Pb^{2+} + 2e^- \longrightarrow Pb$
> At the anode: $2Br^- \longrightarrow Br_2 + 2e^-$

- Ionic substances can only conduct electricity when they are molten or dissolved in water.

Key Point

For the electrolysis of molten ionic compounds, the electrodes used must be inert so that they do not react with the electrolyte or the products.

HT Oxidation and Reduction

- Reduction occurs when positively charged ions gain electrons at the negative electrode.
- Oxidation occurs when negatively charged ions lose electrons at the positive electrode.
- In a redox reaction both reduction and oxidation occur.
- Reactions that take place at the electrodes during electrolysis can be represented by half-equations.
- For example, in the electrolysis of molten copper chloride:
  - Copper is deposited at the negative electrode.

$$Cu^{2+} + 2e^- \longrightarrow Cu$$

HT Key Point

You can remember this by thinking of the word OILRIG:

- Oxidation Is Loss of electrons (OIL)
- Reduction Is Gain of electrons (RIG).

The copper ions gain electrons so they are reduced.

- Chlorine gas is given off at the positive electrode.

$$2Cl^- \longrightarrow Cl_2 + 2e^-$$

The chloride ions lose electrons so they are oxidised.

Remember that chlorine exists as molecules.

Extraction of Metals

- Metals that are more reactive than carbon can be extracted from their ores using electrolysis.
- Electrolysis requires lots of heat and electrical energy, making it an expensive process.
- Aluminium is obtained by the electrolysis of aluminium oxide that has been mixed with **cryolite** (a compound of aluminium).
- Cryolite lowers the melting point of the aluminium oxide, meaning less energy is needed (cheaper energy costs).
- Aluminium forms at the negative electrode.
- Oxygen gas forms at the positive carbon electrode and reacts with the carbon, forming carbon dioxide.
- This wears away the positive electrode, which is replaced regularly.

HT
At the cathode: $Al^{3+} + 3e^- \longrightarrow Al$
At the anode: $2O^{2-} \longrightarrow O_2 + 4e^-$

Key Point

In the exam you could be asked to suggest a hypothesis to explain given data.

Electrolysis of Aqueous Solutions

- When ionic compounds are dissolved in water to form aqueous solutions, it is slightly harder to predict the products of electrolysis.
- The water molecules break down to form hydroxide ions, OH^-, and hydrogen ions, H^+.
- At the negative electrode:
  - Hydrogen is produced if the metal is more reactive than hydrogen.
  - The metal is produced if the metal is less reactive than hydrogen.
- At the positive electrode:
  - Oxygen is produced unless the solution contains halide ions.
  - If halide ions are present, then the halogen is produced.
- In the electrolysis of sodium chloride solution:
  - Hydrogen is released at the negative electrode.
  - Chlorine gas is released at the positive electrode.

Key Point

When electrolysis is used to extract metal, the positive electrode is made of carbon.

REQUIRED PRACTICAL	
Investigate what happens when aqueous solutions are electrolysed using inert electrodes.	
Sample Method	**Hazards and Risks**
1. Set up the equipment as shown in the diagram on page 118. 2. Pass an electric current through the aqueous solution. 3. Observe the products formed at each inert electrode.	• A low voltage must be used to prevent an electric shock. • The room must be well ventilated, and the experiment must only be carried out for a short period of time, to prevent exposure to dangerous levels of chlorine gas.

Key Words

electrolysis
electrolyte
electrode
anode
cathode
cryolite

Quick Test

1. Predict the products of the electrolysis of aqueous sodium bromide.
2. **HT** Write down the half equations for the reactions that take place at each electrode in the electrolysis of aqueous sodium bromide.

Exothermic and Endothermic Reactions

You must be able to:

- Recall that exothermic reactions transfer energy to the surroundings and result in an increase in temperature
- Recall that endothermic reactions take in energy from the surroundings and result in a decrease in temperature
- Give some examples of exothermic and endothermic reactions.

Energy Transfers

- When chemical reactions occur, energy is transferred from the chemicals to or from the surroundings. Therefore, many reactions are accompanied by a temperature change.
- **Exothermic reactions** are accompanied by a temperature rise.
- They transfer heat energy from the chemicals to the surroundings, i.e. they give out heat energy.
- Exothermic reactions are used in products like self-heating cans (for coffee) and hand warmers.
- **Endothermic reactions** are accompanied by a fall in temperature.
- Heat energy is transferred from the surroundings to the chemicals, i.e. they take in heat energy.
- Some sports injury packs use endothermic reactions.
- If a **reversible reaction** is exothermic in one direction, then it is endothermic in the opposite direction.
- The same amount of energy is transferred in each case.

REQUIRED PRACTICAL	
Investigate the variables that affect temperature changes in reacting solutions.	
Sample Method 1. Set up the equipment as shown. 2. Take the temperature of the acid. 3. Add the metal powder and stir. 4. Record the highest temperature the reaction mixture reaches. 5. Calculate the temperature change for the reaction. 6. Repeat the experiment using a different metal.	**Considerations, Mistakes and Errors** • There should be a correlation between the reactivity of the metal and the temperature change, i.e. the more reactive the metal, the greater the temperature change. • When a measurement is made there is always some uncertainty about the results obtained. For example, if the experiment is repeated three times and temperature changes of 3°C, 4°C and 5°C are recorded: - the range of results is from 3°C to 5°C - the mean (average) = $\frac{(3 + 4 + 5)}{3}$ = 4°C
Variables • The independent variable is the metal used. • The dependent variable is the temperature change. • The control variables are the type, concentration and volume of acid.	**Hazards and Risks** • There is a low risk of a corrosive acid damaging the experimenter's eye, so eye protection must be used.

Energy Level Diagrams

- In chemical reactions, atoms are rearranged as old bonds are broken and new bonds are formed.

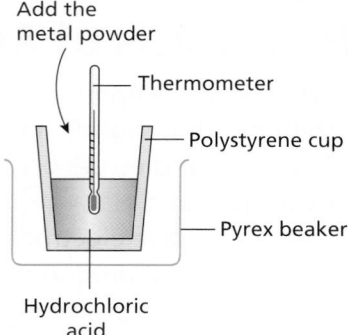

- For bonds to be broken, reacting particles must collide with sufficient energy.
- The minimum amount of energy that the particles must have for a reaction to take place is called the **activation energy**.
- The energy changes in a chemical reaction can be shown using an **energy level diagram** or **reaction profile**.
- In an exothermic reaction:
  - energy is given out to the surroundings
  - the products have less energy than the reactants.
- In an endothermic reaction:
  - energy is being taken in from the surroundings
  - the products have more energy than the reactants.

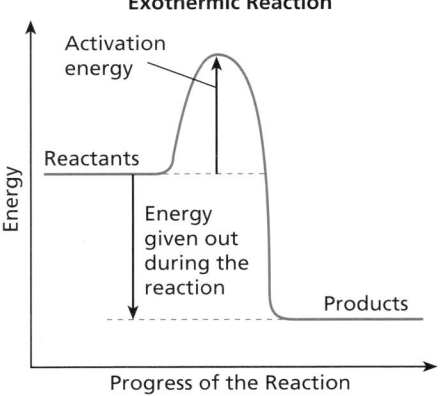

- **Catalysts** reduce the activation energy needed for a reaction. This makes the reaction go faster.

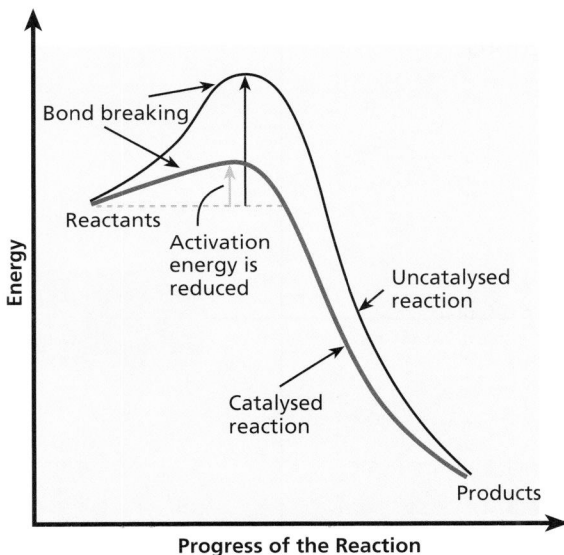

Quick Test

1. What is an endothermic reaction?
2. A white powder dissolves in a solution. The temperature falls by 2°C. State and explain the type of reaction that has taken place.
3. Sketch the energy level diagram for an:
   a) exothermic reaction
   b) endothermic reaction.

Measuring Energy Changes

You must be able to:

- Recall that energy is transferred in chemical reactions
- HT Recall that energy is released when new bonds are made
- HT Calculate the energy transferred in reactions and use it to deduce whether a reaction is exothermic or endothermic.

Measuring Energy Changes

- Energy is transferred in chemical reactions.
- In exothermic reactions heat energy is transferred to the surroundings, so the temperature of the surroundings increases.
- Exothermic reactions include combustion, many oxidation reactions and neutralisation.
- In endothermic reactions heat energy is taken in from the surroundings, so the temperature of the surroundings decreases.
- Endothermic reactions include the thermal decomposition of metal carbonates.
- The amount of energy produced in a chemical reaction in solution can be measured by mixing the reactants in an insulated container (see page 120).
- This enables the temperature change to be measured before heat is lost to the surroundings.
- This method would be suitable for neutralisation reactions and reactions involving solids, e.g. metal and acid reactions.

Self-Heating Can

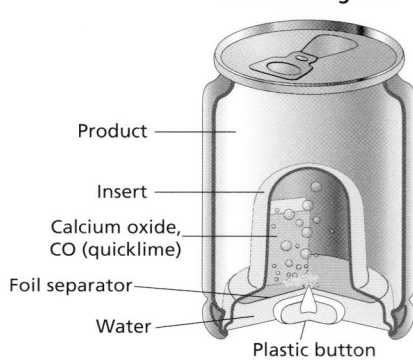

Product

Insert

Calcium oxide, CO (quicklime)

Foil separator

Water

Plastic button

Self-heating cans use exothermic reactions, e.g. $CaO + H_2O \rightarrow Ca(OH)_2$, to heat the product.

HT Energy Change of Reactions

- In a chemical reaction, new substances are produced:
  - The bonds in the reactants are broken.
  - New bonds are made to form the products.
- Breaking a chemical bond requires energy – it is an endothermic process.
- When a new chemical bond is formed, energy is given out – it is an exothermic process.
- If more energy is required to break bonds than is released when bonds are formed, the reaction must be endothermic.
- If more energy is released when bonds are formed than is needed to break bonds, the reaction must be exothermic.

HT Key Point

Endothermic: energy required to break old bonds > energy released when new bonds are formed

Exothermic: energy required to break old bonds < energy released when new bonds are formed

HT Energy Calculations

Calculate the energy transferred in the following reaction:

methane + oxygen → carbon dioxide + water
$CH_4(g) + 2O_2(g) \rightarrow CO_2(g) + 2H_2O(g)$

The bond energies needed for this are:

C–H is 412kJ/mol, O=O is 496kJ/mol
C=O is 805kJ/mol, H–O is 463kJ/mol

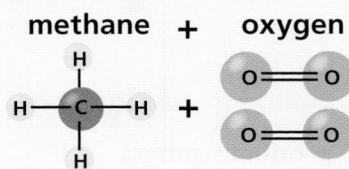

Break old bonds (energy in)

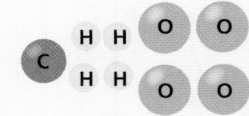

Make new bonds (energy out)

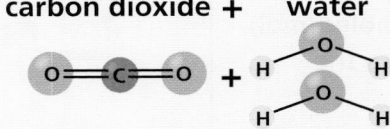

Energy used to break bonds is:
4 C–H = 4 × 412 = 1648kJ
2 O=O = 2 × 496 = 992kJ
Total = 1648kJ + 992kJ = 2640kJ

Energy given out by making bonds:
2 C=O = 2 × 805 = 1610kJ
4 H–O = 4 × 463 = 1852kJ
Total = 1610kJ + 1852kJ = 3462kJ

energy change =
energy used to break bonds − energy given out by making bonds
= 2640kJ − 3462kJ
= −822kJ

HT Key Point

You will always be given any bond energies that you need for questions like this. You are not expected to be able to recall them.

Quick Test

1. **HT** Describe how energy is transferred when **a)** bonds are broken and **b)** bonds are made.
2. Why must the temperature be recorded both before and after the reaction or change takes place?
3. **HT** A chemical reaction gives out more energy when bonds are made than it takes in to break bonds. What sort of reaction is it?

Key Words

endothermic reaction
exothermic reaction

Rate of Reaction

You must be able to:

- Describe how the rate of a chemical reaction can be found
- Use collision theory to explain how factors affect the rate of reactions
- HT Calculate the rate of a reaction from graphs.

Calculating the Rate of Reaction

LEARN

$$\text{mean rate of reaction} = \frac{\text{amount of reactant used OR product formed}}{\text{time taken}}$$

- The rate of reaction can be found in different ways.
- **Measuring the amount of reactants used:**
  - If one of the products is a gas, measure the mass in grams (g) of the reaction mixture before and after the reaction takes place and the time it takes for the reaction to happen.
  - The mass of the mixture will decrease.
  - The units for the rate of reaction may then be given as g/s.

  - HT The amount of a reactant can also be measured in moles (mol).
  - HT As the reaction takes place the reactant is used up, so the amount of reactant remaining decreases.
  - HT The concentration of the reactant is calculated as the amount (mol) divided by the volume of the reaction mixture (dm^3). It is measured in units of mol/dm^3.

- **Measuring the amount of products formed:**
  - If one of the products is a gas, measure the total volume of gas produced in cubic centimetres (cm^3) with a gas syringe and the time it takes for the reaction to happen.
  - The units for the rate of reaction may then be given as cm^3/s.
- **Measuring the time it takes for a reaction mixture to become opaque or change colour:**
  - Time how long it takes for the mixture to change colour.
  - Rate of reaction $\approx \dfrac{1}{\text{time taken for solution to change colour}}$.

Measuring the Amount of Reactants Used

Measuring the Amount of Products Formed (With a Catalyst)

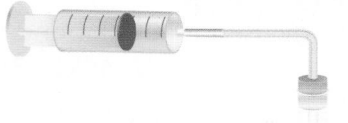

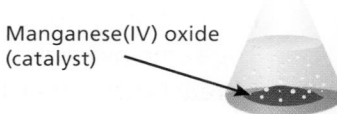

Manganese(IV) oxide (catalyst)

Collision Theory

- Chemical reactions only occur when reacting particles collide with each other with sufficient energy.
- The minimum amount of energy required to cause a reaction is called the activation energy.
- There are four important factors that affect the rate of reaction: temperature, concentration, surface area and catalysts (see page 126).
- **Temperature:**
  - In a hot reaction mixture the particles move more quickly – they collide more often and with greater energy, so more collisions are successful.

- **Concentration**:
  - At higher concentrations, the particles are crowded closer together – they collide more often, so there are more successful collisions.
  - Increasing the pressure of reacting gases also increases the frequency of collisions.

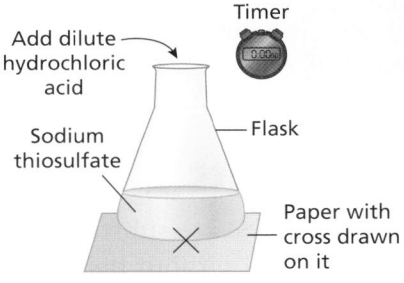

REQUIRED PRACTICAL	
Investigate how changes in concentration affect the rates of reactions by methods involving the production of gas or a colour change.	
This investigation uses the reaction between sodium thiosulfate and hydrochloric acid. **Sample Method** 1. Set up the equipment as shown. 2. Add the hydrochloric acid to the flask and swirl to mix the reactants. 3. Start the timer. 4. Watch the cross through the flask. 5. When the cross is no longer visible stop the timer. 6. Repeat the experiment using hydrochloric acid of a different concentration.	**Considerations, Mistakes and Errors** • There should be a correlation between the concentration of the acid and the time taken for the cross to 'disappear'. • The higher the concentration of the acid, the faster the rate of reaction, and the shorter the time for the cross to 'disappear'.
Variables • The independent variable is the concentration of the acid. • The dependent variable is the time it takes for the cross to 'disappear'. • The control variables are the volume of acid and the concentration and volume of sodium thiosulfate.	**Hazards and Risks** • Corrosive acid can damage eyes, so eye protection must be used. • Sulfur dioxide gas can trigger an asthma attack, so the temperature must always be kept below 50°C.

- **Surface area**:
  - Small pieces of a solid reactant have a large surface area in relation to their volume.
  - More particles are exposed and available for collisions, so there are more collisions and a faster reaction.

Plotting Reaction Rates

- Graphs can be plotted to show the progress of a chemical reaction.
- There are three key things to remember:
  - The steeper the line, the faster the reaction.
  - When one of the reactants is used up, the reaction stops (the line becomes horizontal).
  - The same amount of product is formed from the same amount of reactants, regardless of rate.

HT The rate of reaction at a particular time is given by graphs:
  - Draw a **tangent** to the curve at that time.
  - Find the **gradient** of the tangent.
  - The gradient is equal to the rate of reaction at that time.

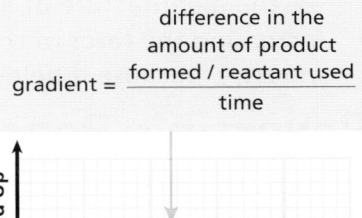

$$\text{gradient} = \frac{\text{difference in the amount of product formed / reactant used}}{\text{time}}$$

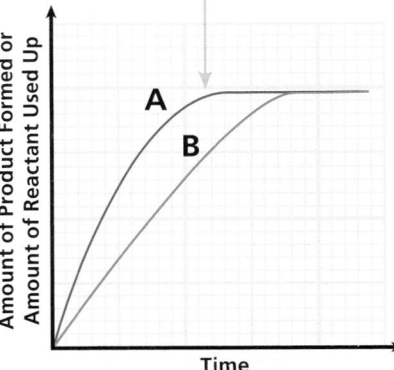

The graph shows that reaction A is faster than reaction B.

Quick Test

1. Why does increasing temperature increase the rate of reaction?
2. What is the name given to the minimum amount of energy that reacting particles must have to react?
3. How does surface area affect the rate of reaction?

Reversible Reactions

You must be able to:

- Explain how and why catalysts can affect the rate of reaction
- Explain what a reversible reaction is
- Define the term 'equilibrium'
- HT Predict the effect of changing the conditions on a system at equilibrium.

Catalysts

- A **catalyst** is a substance that increases the rate of a chemical reaction without being used up in the process.
- Catalysts are not included in the chemical equation for the reaction.
- A catalyst:
  - reduces the amount of energy needed for a successful collision
  - makes more collisions successful
  - speeds up the reaction
  - provides a surface for the molecules to attach to, which increases their chances of bumping into each other.
- Enzymes act as catalysts in biological systems.
- Different reactions need different catalysts, e.g.
  - the cracking of hydrocarbons uses broken pottery
  - the manufacture of ammonia uses iron.
- Increasing the rates of chemical reactions is important in industry, because it helps to reduce costs.

Catalysts Used in Industrial Reactions

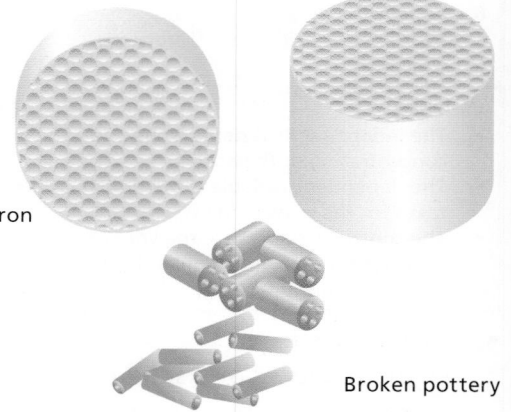

Iron

Broken pottery

Reversible Reactions

- Some chemical reactions are reversible, they can go forwards or backwards.
- In a **reversible reaction**, the products can react to produce the original reactants.
- These reactions are represented as:

$$A(g) + B(g) \rightleftharpoons C(g) + D(g)$$

- This means that:
  - A and B can react to produce C and D.
  - C and D can react to produce A and B.
- For example:
  - Solid ammonium chloride decomposes when heated to produce ammonia and hydrogen chloride gas (both colourless).
  - Ammonia reacts with hydrogen chloride gas to produce clouds of white ammonium chloride.

$$\text{ammonium chloride} \rightleftharpoons \text{ammonia} + \text{hydrogen chloride}$$
$$NH_4Cl(s) \rightleftharpoons NH_3(g) + HCl(g)$$

> ### Key Point
>
> Some chemical reactions are reversible, they can go forwards or backwards.

- The direction of reversible reactions can be changed by changing the conditions.

Closed Systems

- In a closed system, no reactants are added and no products are removed.
- When a reversible reaction occurs in a closed system, an equilibrium is achieved when the rate of the forward reaction is equal to the rate of the backward reaction.
- The relative amounts of all the reacting substances at equilibrium depend on the conditions of the reaction.

HT Changing Reaction Conditions

- Le Chatelier's Principle states that if a system in equilibrium is subjected to a change in conditions, then the system shifts to resist the change.
- In an exothermic reaction:
  - If the temperature is raised, the yield decreases.
  - If the temperature is lowered, the yield increases.
- In an endothermic reaction:
  - If the temperature is raised, the yield increases.
  - If the temperature is lowered, the yield decreases.
- In reactions involving gases:
  - An increase in pressure favours the reaction that produces the least number of gas molecules.
  - A decrease in pressure favours the reaction that produces the greater number of gas molecules.
- If the concentration of one of the reactants or products is changed:
  - the system is no longer in equilibrium
  - the system adjusts until it can reach equilibrium once more.
- If the concentration of one of the reactants is increased, the position of equilibrium shifts so that more products are formed until equilibrium is reached again.
- In contrast, if the concentration of one of the reactants is decreased, the position of equilibrium shifts so that more reactants are formed until equilibrium is reached again.
- These factors, together with reaction rates, determine the optimum conditions in industrial processes.

> **Key Point**
>
> When a reversible reaction occurs in a closed system, an equilibrium is achieved when the rate of the forward reaction is exactly the same rate as the backward reaction.

> **Key Words**
>
> catalyst
> reversible reaction
> closed system
> equilibrium
> HT Le Chatelier's Principle

Review Questions

Atoms, Elements, Compounds and Mixtures

1 When magnesium metal is heated, it reacts with oxygen to form magnesium oxide.

Write a word equation for this reaction. **[1]**

2 When magnesium is added to hydrochloric acid a chemical reaction takes place.
The products are magnesium chloride and hydrogen.

Why is the total mass of the reactants equal to the total mass of the products in this reaction? **[1]**

3 Complete the following symbol equations:

a) $H_2 + Br_2 \rightarrow \underline{\quad} HBr$

c) $CH_4 + \underline{\quad} O_2 \rightarrow CO_2 + \underline{\quad} H_2O$ **[3]**

b) $\underline{\quad} SO_2 + O_2 \rightarrow \underline{\quad} SO_3$

d) $N_2 + \underline{\quad} H_2 \rightarrow \underline{\quad} NH_3$ **[4]**

> **Total Marks** _____ / 9

Atoms and the Periodic Table

1 An atom of aluminium can be represented by: $^{27}_{13}Al$

How many **a)** protons, **b)** neutrons and **c)** electrons does this atom of aluminium have? **[3]**

2 What is the typical radius of an atom?
Tick **one** box.

0.1nm ☐   10nm ☐

1mm ☐   0.01mm ☐ **[1]**

3 An oxide ion can be represented by: $^{16}_{8}O^{2-}$

How many **a)** protons, **b)** neutrons and **c)** electrons does this oxide ion have? **[3]**

> **Total Marks** _____ / 7

The Periodic Table

1 Chlorine is a reactive non-metal with two common isotopes: chlorine-37 and chlorine-35.

a) Which group of the periodic table does chlorine belong to? **[1]**

b) What is the atomic number of chlorine? **[1]**

c) Define the word 'isotope' in terms of atomic number and mass number. **[1]**

d) In terms of the subatomic particles, explain the similarities and differences between an atom of chlorine-35 and an atom of chlorine-37. **[3]**

Total Marks _____ / 6

States of Matter

1 What do the state symbols (s) and (l) mean? **[2]**

2 Explain why water cannot be easily compressed. **[2]**

Total Marks _____ / 4

Ionic Compounds

1 **Table 1** shows the charge of some metal ions and non-metal ions.

Table 1	
Metal Ions	**Non-Metal Ions**
Lithium, Li^+	Oxide, O^{2-}
Strontium, Sr^{2+}	Chloride, Cl^-
Potassium, K^+	Bromide, Br^-
Magnesium, Mg^{2+}	Sulfide, S^{2-}

a) Define the term 'ion'. **[2]**

b) In terms of electron transfer, explain why lithium ions have a 1+ charge. **[2]**

c) In terms of electron transfer, explain why oxide ions have a 2– charge. **[2]**

d) Use the table above to suggest the formula of:

    i) Strontium chloride.   **ii)** Potassium bromide.   **iii)** Magnesium sulfide. **[3]**

Total Marks _____ / 9

Review Questions

Covalent Compounds

1 **Figure 1** shows the structure of graphite.

Figure 1

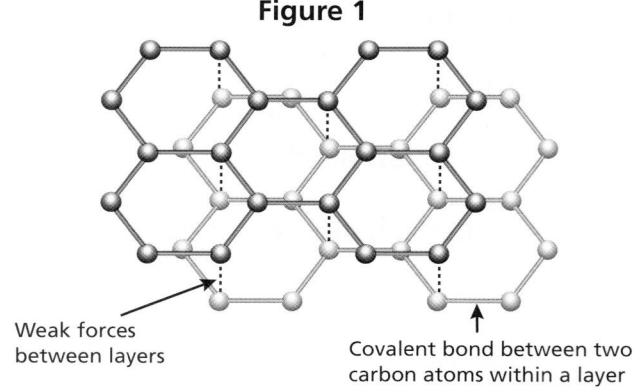

Weak forces between layers

Covalent bond between two carbon atoms within a layer

a) Graphite is a form of the element carbon.

Name **one** other form of carbon that is solid at room temperature. **[1]**

b) Within each layer of graphite, each carbon atom is bonded to other carbon atoms by strong bonds.

How many carbon atoms is each carbon atom joined to by strong bonds? **[1]**

c) Carbon in the form of graphite is the only non-metal that conducts electricity.

Explain why graphite can conduct electricity. **[2]**

d) Explain why graphite has a very high melting point. **[2]**

> **Total Marks** .... / 6

Metals and Special Materials

1 Lithium reacts with chlorine to produce lithium chloride: $2Li(s) + Cl_2(g) \rightarrow 2LiCl(s)$

a) What type of bonding is present in lithium? **[1]**

b) Why is lithium a good electrical conductor? **[2]**

c) What sort of bonding is present in chlorine molecules? **[1]**

d) Why is chlorine a gas at room temperature? **[1]**

e) The reaction produces lithium chloride.

What type of bonding is present in lithium chloride? **[1]**

f) Lithium chloride does not conduct electricity when it is solid, but it does conduct electricity when it is molten. Explain why. **[2]**

> **Total Marks** .... / 8

Conservation of Mass

1 When calcium carbonate is heated fiercely, it decomposes to form calcium oxide and carbon dioxide:

$$CaCO_3(s) \rightarrow CaO(s) + CO_2(g)$$

a) Why is the total mass of the reactants before the reaction equal to the total mass of reactants after the reaction? **[1]**

b) 10.0g of calcium carbonate was heated until it had all reacted.
5.6g of calcium oxide was produced.

Why has the mass of the solid gone down? **[2]**

c) Predict the mass of carbon dioxide produced in this reaction. **[1]**

2 Magnesium nitrate is an ionic compound.
Relative atomic masses (A_r): Mg = 24, N = 14, O = 16

a) Calculate the relative formula mass of magnesium nitrate, $Mg(NO_3)_2$. **[2]**

b) HT Calculate the mass of 1.00 mole of magnesium nitrate, $Mg(NO_3)_2$. **[1]**

Total Marks _____ / 7

Amount of Substance

1 HT Many fuels contain small amounts of sulfur.
When sulfur is burned, sulfur dioxide, SO_2, is produced

$$S + O_2 \rightarrow SO_2$$

1.6g of sulfur was completely burned in oxygen to produce sulfur dioxide.
Relative atomic masses (A_r): S = 32, O = 16

a) How many moles of sulfur were burned? **[2]**

b) Calculate the mass of sulfur dioxide, SO_2, produced in this reaction. **[2]**

Total Marks _____ / 4

Reactivity of Metals

1 Calcium can be burned to produce calcium oxide:

$$2Ca + O_2 \rightarrow 2CaO$$

a) Write a word equation for this reaction. [1]

b) HT The balanced symbol equation can be broken into two ionic equations:

$$2Ca \rightarrow 2Ca^{2+} + 4e^-$$
$$O_2 + 4e^- \rightarrow 2O^{2-}$$

In terms of oxidation and reduction, explain what happens to the calcium and the oxygen in the reaction. [4]

> Total Marks .... / 5

The pH Scale and Salts

1 Which of these values shows the pH of a strong alkali?
Tick **one** box.

7 ☐   14 ☐   8 ☐   1 ☐ [1]

2 Which of these ions is found in excess in acidic solutions?
Tick **one** box.

H^+ ☐ H^- ☐ OH^+ ☐ OH^- [1]

3 Sulfuric acid is a strong acid.

a) A solution of sulfuric acid has a pH of 1.

What does the pH scale measure? [2]

b) Complete the equation to show the neutralisation reaction between sulfuric acid and potassium hydroxide.

sulfuric acid + potassium hydroxide → _____ + _____ [2]

> Total Marks .... / 6

Electrolysis

1. Copper chloride is an ionic compound that can be separated by electrolysis.

   a) Name the element formed during the electrolysis of molten copper chloride:

      i) At the positive electrode. [1]

      ii) At the negative electrode. [1]

   b) HT Complete the half equations to show the reactions that take place at each electrode.

      i) At the anode: $2Cl^- \rightarrow$ _____ $+ 2e^-$ [1]

      ii) At the cathode: $Cu^{2+} +$ _____ $\rightarrow Cu$ [1]

2. A student carries out an experiment to find out what happens when an aqueous solution of sodium chloride is electrolysed.

   a) Identify the two positive ions present in an aqueous solution of sodium chloride. [2]

   b) Name the substance produced at the negative electrode. [2]
      You must give a reason for your answer.

   c) Name the substance produced at the positive electrode. [2]
      You must give a reason for your answer.

3. During the electrolysis of molten lead chloride, lead and chlorine are produced.

   At the anode: $2Cl^- \rightarrow Cl_2 + 2e^-$
   At the cathode: $Pb^{2+} + 2e^- \rightarrow Pb$

   a) HT During the electrolysis of lead chloride, oxidation and reduction reactions take place.

      i) In terms of oxidation and reduction, describe what happens to the lead ions. [2]

      ii) In terms of oxidation and reduction, describe what happens to the chloride ions. [2]

   b) Why does the lead chloride have to be molten? [1]

   Total Marks _____ / 15

Practice Questions

Exothermic and Endothermic Reactions

1 Which of the following types of reaction is an endothermic reaction?
Tick **one** box.

Combustion ☐   Oxidation ☐

Neutralisation ☐   Thermal decomposition ☐   **[1]**

2 Sketch an energy profile diagram to show an exothermic reaction.
Include the:
- Reactants
- Products
- Activation energy
- Energy change of the reaction.   **[4]**

3 A student carries out an experiment to find out whether changing the metal powder added to dilute hydrochloric acid affects the temperature change for the reaction.

   a) Why must the student take the temperature at the start and the end of each reaction?   **[1]**

   b) Why must the student stir the mixture of metal powder and acid?   **[1]**

   c) Name a control variable in this experiment.   **[1]**

   d) Name a hazard in this experiment and describe a control measure the student must take to reduce the risk of an accident happening.   **[2]**

Total Marks _____ / 10

Measuring Energy Changes

1 Which of these chemical reactions is an example of an **endothermic** reaction?
Tick **one** box.

The combustion of methane ☐

The neutralisation of hydrochloric acid by sodium hydroxide ☐

The oxidation of iron sulfide ☐

The thermal decomposition of strontium carbonate ☐   **[1]**

Total Marks _____ / 1

Rate of Reaction

1 Which of the following changes would increase the rate of a chemical change?
Tick **one** box.

Adding a catalyst ☐   Increasing the size of particles ☐

Reducing the temperature by 10°C ☐   Decreasing the concentration ☐ **[1]**

2 Collision theory can be used to explain the rate of a chemical reaction.

    a) What **two** things must happen for a chemical reaction to take place? **[2]**

    b) Explain, in terms of collision theory, why increasing the temperature increases the
rate of a chemical change. **[2]**

3 In the 'disappearing cross' experiment, hydrochloric acid reacts with sodium thiosulfate.
One of the products of the reaction is sulfur, which is insoluble.
A student carries out the experiment using 1.00mol/dm^3 acid.
She adds the acid to the sodium thiosulfate and times how long it
takes for the cross to 'disappear'.
She then repeats the experiment using 0.500mol/dm^3 and then
0.250mol/dm^3 hydrochloric acid.

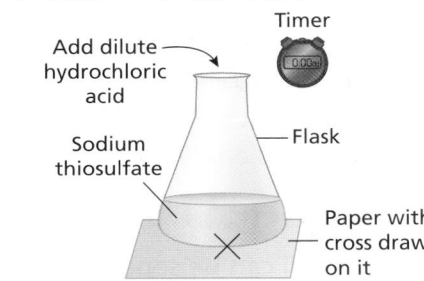

    a) Name the independent variable in this investigation. **[1]**

    b) Temperature could affect the rate of reaction, so the same temperature must be
used for each part of the investigation.

       Predict and explain how increasing the temperature would affect the time taken
for the cross to disappear. **[2]**

> **Total Marks** .... / 8

Reversible Reactions

1 If a reversible reaction is carried out in a closed system, equilibrium can be reached.

    a) What is a closed system? **[1]**

    b) Explain how you can tell that a system is in equilibrium. **[1]**

> **Total Marks** .... / 2

Organic Chemistry

Alkanes

You must be able to:

- Describe how crude oil is formed
- Recall the general formula for alkanes
- Understand how the fractional distillation of crude oil works
- Recall how the properties of hydrocarbons are linked to their size.

Crude Oil and Hydrocarbons

- Crude oil is:
  - formed over millions of years from the fossilised remains of plankton
  - found in porous rocks in the Earth's crust
  - a finite (non-renewable) resource that is used to produce fuels and other chemicals.
- Most of the compounds in crude oil are hydrocarbons (molecules made of only carbon and hydrogen atoms).
- Hydrocarbon molecules vary in size, which affects their properties and how they can be used as fuels.
- The larger the hydrocarbon:
  - the more viscous it is (i.e. the less easily it flows)
  - the higher its boiling point
  - the less volatile it is
  - the less easily it ignites.

Fractional Distillation

- Crude oil can be separated into different fractions (parts) by fractional distillation.
- Each fraction contains hydrocarbon molecules with a similar number of carbon atoms.
- Most of the hydrocarbons obtained are alkanes (see below).
- First, the crude oil is heated until it evaporates.
- The vapour moves up the fractionating column.
- The top of the column is much colder than the bottom.
- Shorter hydrocarbon molecules can reach the top of the fractionating column before they condense and are collected.
- Longer hydrocarbon molecules condense at higher temperatures and are collected lower down the column.

Alkanes

- Carbon atoms are linked to four other atoms by single bonds.
- Alkanes only contain single bonds and are described as saturated hydrocarbons (because they contain the maximum number of bonds possible).
- Alkanes are fairly unreactive, but they burn well.
- The general formula for alkanes is:

LEARN

$$C_nH_{2n+2}$$

> **Key Point**
>
> Fuels are substances that can be burned to release energy.

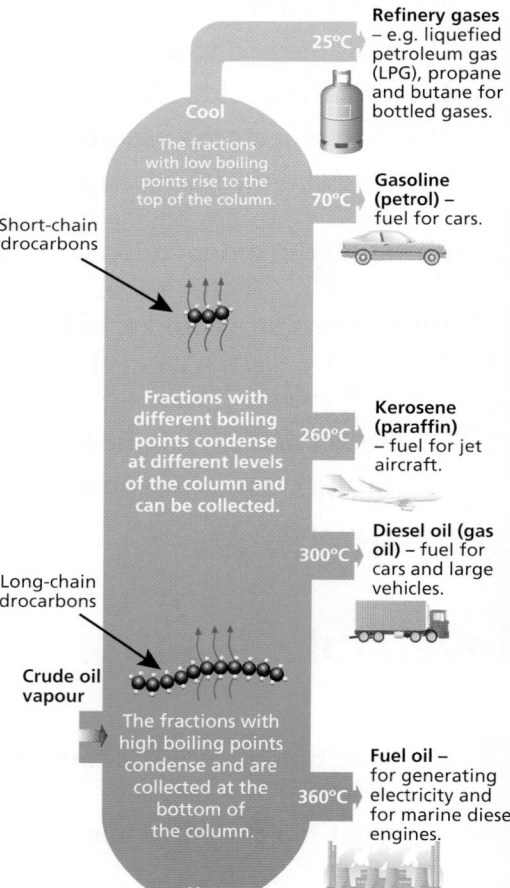

Fractional Distillation

25°C — **Refinery gases** – e.g. liquefied petroleum gas (LPG), propane and butane for bottled gases.

Cool — The fractions with low boiling points rise to the top of the column.

Short-chain hydrocarbons

70°C — **Gasoline (petrol)** – fuel for cars.

Fractions with different boiling points condense at different levels of the column and can be collected.

260°C — **Kerosene (paraffin)** – fuel for jet aircraft.

300°C — **Diesel oil (gas oil)** – fuel for cars and large vehicles.

Long-chain hydrocarbons

Crude oil vapour

The fractions with high boiling points condense and are collected at the bottom of the column.

360°C — **Fuel oil** – for generating electricity and for marine diesel engines.

Hot

over 400°C — **Bitumen** – to make roads.

- Alkanes can be drawn with a single line between atoms, which represents a single covalent bond:

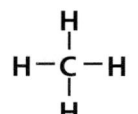
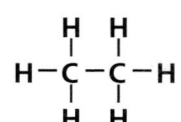
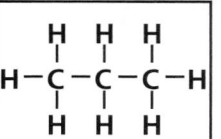

Hydrogen atoms can make 1 bond each.	Carbon atoms can make 4 bonds each.	The simplest alkane, **methane**, CH_4, is made up of 4 hydrogen atoms and 1 carbon atom.	**Ethane, C_2H_6** A molecule made up of 2 carbon atoms and 6 hydrogen atoms.	**Propane, C_3H_8** A molecule made up of 3 carbon atoms and 8 hydrogen atoms.

- The shorter-chain alkanes release energy more quickly by burning, so there is greater demand for them as fuels.

Burning Fuels

- Most fuels are compounds of carbon and hydrogen. Many also contain sulfur.
- During the combustion (burning) of hydrocarbon fuels:
  - both carbon and hydrogen are oxidised
  - energy is released
  - waste products are produced, which are released into the atmosphere.
- If combustion is not complete, then carbon monoxide, unburnt fuels and solid particles containing soot (carbon) may be released.
- Carbon monoxide is a colourless, odourless and toxic gas.
- Solid particles in the air, called particulates, can cause global dimming by reducing the amount of sunlight reaching the Earth's surface and cause damage to people's lungs.
- Due to the high temperature reached when fuels burn, nitrogen in the air can react with oxygen to form nitrogen oxides.
- These gases can cause respiratory problems in people and react with rain water (in the same way as sulfur dioxide) to form acid rain, which can damage plants and buildings.
- Sulfur can be removed from fuels before burning (in motor vehicles) and removed from the waste gases after combustion (in power stations).

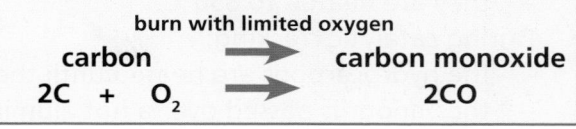

burn with plenty of oxygen
carbon → carbon dioxide
$C + O_2$ → CO_2

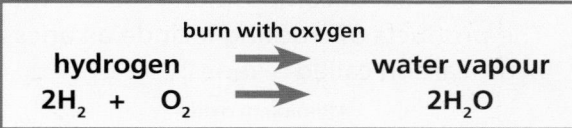

burn with limited oxygen
carbon → carbon monoxide
$2C + O_2$ → $2CO$

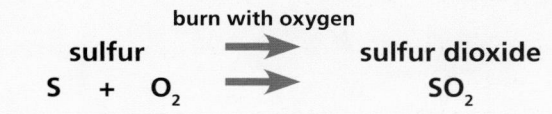

burn with oxygen
hydrogen → water vapour
$2H_2 + O_2$ → $2H_2O$

burn with oxygen
sulfur → sulfur dioxide
$S + O_2$ → SO_2

Quick Test

1. What is a hydrocarbon?
2. How many bonds does a carbon atom form?
3. What is the chemical formula for propane?
4. What is the general formula of an alkane?
5. What is the balanced symbol equation for the complete combustion of methane?

Key Words

crude oil
hydrocarbon
viscous
fractional distillation
alkanes
saturated
combustion

Cracking Hydrocarbons

You must be able to:

- Describe the cracking of alkanes
- Describe how bromine water can be used to differentiate between an alkane and an alkene.

Cracking Hydrocarbons

- Longer-chain hydrocarbons can be broken down into shorter, more useful hydrocarbons. This process is called **cracking**.
- Cracking is a useful industrial process.
- The two main methods of cracking are steam cracking and catalytic cracking.
- During steam cracking:
  - the hydrocarbons are mixed with steam
  - they are heated to 850°C.
- During catalytic cracking:
  - the hydrocarbons are heated until they vaporise
  - the vapour is passed over a hot aluminium oxide catalyst
  - a thermal decomposition reaction then takes place.
- The products of cracking include alkanes and another type of hydrocarbon called **alkenes**.

> ### Key Point
>
> Cracking breaks down long-chain hydrocarbons into useful short-chain alkanes and reactive alkene molecules.

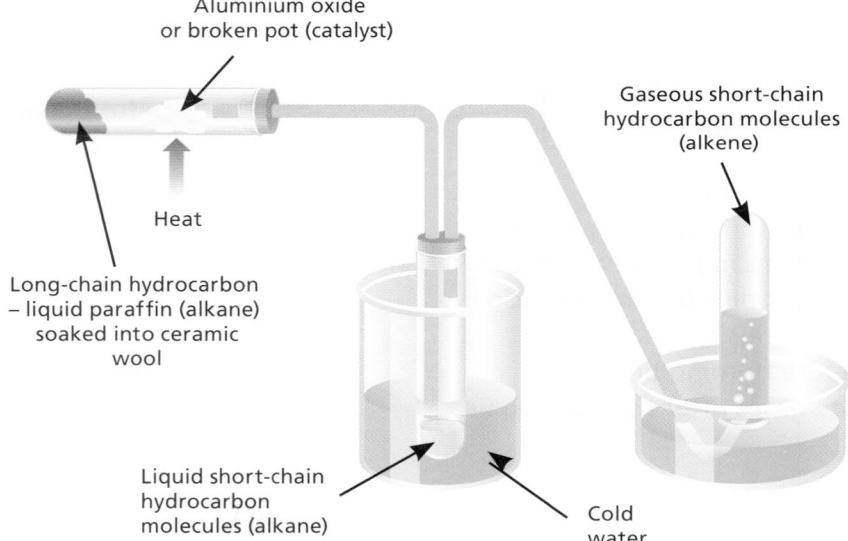

Aluminium oxide or broken pot (catalyst)

Gaseous short-chain hydrocarbon molecules (alkene)

Heat

Long-chain hydrocarbon – liquid paraffin (alkane) soaked into ceramic wool

Liquid short-chain hydrocarbon molecules (alkane)

Cold water

- No atoms are made or destroyed during the cracking reaction, so the chemical equation must always balance.

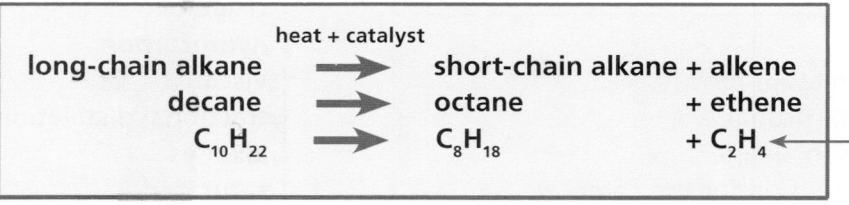

	heat + catalyst	
long-chain alkane	⟶	short-chain alkane + alkene
decane	⟶	octane + ethene
$C_{10}H_{22}$	⟶	C_8H_{18} + C_2H_4

The total number of carbon atoms and the total number of hydrogen atoms must be the same on both sides of the equation.

- The alkanes produced are valuable as fuels.
- There is a high demand for fuels with small chains of carbon atoms because they are easy to ignite and have low boiling points.
- The alkenes produced can be used to make a range of new compounds, including polymers and industrial alcohol.

Bromine Water

- Alkenes are more reactive than alkanes.
- They react when shaken with bromine water, turning it from orange to colourless.
- This can be used to differentiate between alkanes and alkenes.

> ethene (colourless) + bromine water (orange brown)
> $\longrightarrow$ colourless solution
> ethane (colourless) + bromine water (orange brown)
> $\longrightarrow$ orange brown solution

- Ethene reacts with bromine to form dibromoethane in an addition reaction:

$$C_2H_4 + Br_2 \longrightarrow CH_2BrCH_2Br$$

Unsaturated Alkene (C=C)   **Saturated Alkane (C–C)**

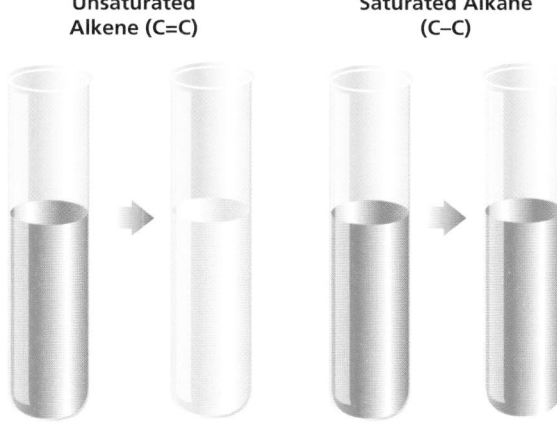

Key Point

Other halogens react with alkenes in a similar way to bromine.

Quick Test

1. Complete the equation for the cracking of octane:
   $C_8H_{18} \rightarrow C_6H_{14} +$ _____
2. Explain how a sample of butene can be told apart from a sample of butane.

Key Words

cracking
alkenes

Chemical Analysis

You must be able to:

- Explain what the term 'pure' means in chemistry
- Explain what a formulation is
- Explain how chromatography can be used to separate mixtures
- Identify a range of gases.

Pure and Impure Substances

- In chemistry, the word **pure** has a special meaning – a pure substance contains only one type of element or one type of compound.
- This means that pure substances:
  - melt and solidify at one temperature called the melting point
  - boil and condense at one temperature called the boiling point.
- Impure substances are mixtures. They do not melt and boil at one temperature – they change state over a range of temperatures.

Formulations

- **Formulations** are mixtures that have been carefully designed to have specific properties.
- The components in a formulation are carefully controlled.
- Examples of formulations include fuels, cleaning agents, paints, medicines, alloys, fertilisers and foods.

Chromatography

- **Chromatography** involves:
  - a **stationary phase**, which does not move
  - a **mobile phase**, which does move.
- In paper chromatography:
  - the stationary phase is the absorbent paper
  - the mobile phase is the solvent, which is often water.
- During chromatography, mixtures are separated into their constituent components.
- The solvent dissolves the samples and carries them up the paper.
- Each component moves a different distance up the paper depending on its attraction for the paper and for the solvent.
- Chromatography can be used to identify artificial colours (e.g. in food) by comparing them to the results obtained from known substances.

> ### Key Point
>
> A pure substance contains only one type of element or one type of compound.

REQUIRED PRACTICAL
Investigate how paper chromatography can be used to separate and tell the difference between coloured substances.

Sample Method	Considerations, Mistakes and Errors
1. Draw a 'start line', in pencil, on a piece of absorbent paper. 2. Put samples of five known food colourings (A, B, C, D and E), and the unknown substance (X), on the 'start line'. 3. Dip the paper into a solvent. 4. Wait for the solvent to travel to the top of the paper. 5. Identify substance X by comparing the horizontal spots with the results of A, B, C, D and E.	• Pure substances produce a single spot in all solvents. • Only ever use pencil to draw the start line, as ink will dissolve and affect your results.

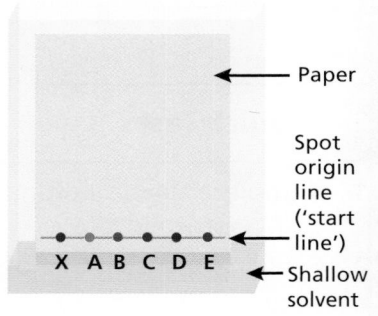

Paper

Spot origin line ('start line')

X A B C D E

Shallow solvent

- R_f values can be used to identify the components in a mixture.

LEARN

$$R_f = \frac{\text{distance moved by substance}}{\text{distance moved by solvent}}$$

- Different components have different R_f values.
- Providing the same temperature and solvent are used, the R_f value for a particular component is constant and can be used to identify the component.

A student produces a chromatogram of the ink in a red pen.
The solvent front moves 12.0cm.
One component moves 6.0cm.

Calculate the R_f value of the component.

$$R_f = \frac{6.0}{12.0}$$
$$= 0.5$$

Gas Tests

Gas	Properties	Test for Gas
Hydrogen, H_2	A colourless gas. It combines violently with oxygen when ignited.	When mixed with air, burns with a squeaky pop.
Chlorine, Cl_2	A green poisonous gas that bleaches dyes.	Turns damp indicator paper white.
Oxygen, O_2	A colourless gas that helps fuels burn more readily than in air.	Relights a glowing splint.
Carbon dioxide, CO_2	A colourless gas.	When bubbled through limewater (a solution of calcium hydroxide), turns the limewater cloudy.

Quick Test

1. In chemistry what does the term 'pure' mean?
2. In paper chromatography what is the:
   a) Stationary phase?
   b) Mobile phase?
3. Describe the gas test for hydrogen and the results of a positive test.

Key Words

pure
formulation
chromatography
stationary phase
mobile phase

The Earth's Atmosphere

You must be able to:

- Recall the present day composition of the Earth's atmosphere
- Describe how and why the Earth's atmosphere has changed over time
- Explain why the levels of oxygen have increased over time
- Explain why the levels of carbon dioxide have decreased over time.

The Earth's Atmosphere

- The atmosphere has changed a lot since the formation of the Earth 4.6 **billion** years ago.

Timescale	Condition of the Atmosphere	Key Factors and Events that Shaped the Atmosphere
Formation of the Earth / 4 billion years ago	Other gases / CO_2 / Decrease in carbon dioxide and other gases / Increase in oxygen and nitrogen	• Intense volcanic activity releases: – mainly carbon dioxide (like the atmospheres of Mars and Venus today) – water vapour (which condenses to form the oceans) – small proportions of methane and ammonia.
3 billion years ago / 2 billion years ago / 1 billion years ago	Other gases / CO_2 / N_2 / O_2 / Decrease in carbon dioxide / Increase in nitrogen and oxygen	• Green plants and algae evolve and: – carbon dioxide is reduced as the plants take it in and give out oxygen – microorganisms that can't tolerate oxygen are killed off – carbon from carbon dioxide becomes locked up in sedimentary rocks formed from the shells and skeletons of marine organisms – other gases react with oxygen to release nitrogen – nitrogen is also produced by bacteria removing nitrates from decaying plant material.
Now	Other gases / CO_2 / N_2 / O_2	• There is now about 20% oxygen and about 80% nitrogen in the atmosphere. • The amount of carbon dioxide has decreased significantly.

The Atmosphere Today

- The proportions of gases in the atmosphere have been more or less the same for about 200 million years.
- Water vapour may also be present in varying quantities (0–3%).

Increase of Oxygen Levels

- **Algae** and plants photosynthesise.
- During **photosynthesis**, carbon dioxide and water react to produce glucose and oxygen:

$$\text{carbon dioxide} + \text{water} \longrightarrow \text{glucose} + \text{oxygen}$$
$$6CO_2 + 6H_2O \longrightarrow C_6H_{12}O_6 + 6O_2$$

- Algae first started producing oxygen about 2.7 billion years ago.
- Over the next billion years, plants evolved and the amount of oxygen in the atmosphere increased.
- Eventually the level of oxygen in the atmosphere increased enough to allow animals to evolve.

Decrease of Carbon Dioxide Levels

- As plants and algae have evolved, the level of carbon dioxide in the atmosphere has decreased. This is because plants use carbon dioxide during photosynthesis.
- Carbon dioxide has also decreased as carbon becomes locked up in sedimentary rocks (e.g. limestone) and fossil fuels (e.g. coal, crude oil and natural gas).
- Limestone contains calcium carbonate and can be formed from the shells and skeletons of sea creatures.
- Coal is a sedimentary rock formed from plant deposits that were buried and compressed over millions of years.
- The level of carbon dioxide in the atmosphere has also been reduced by the reaction between carbon dioxide and sea water. This reaction produces:
  - insoluble carbonates that are deposited as sediment
  - soluble hydrogen carbonates.
- However, too much carbon dioxide dissolving in the oceans can harm marine life, such as coral reefs.

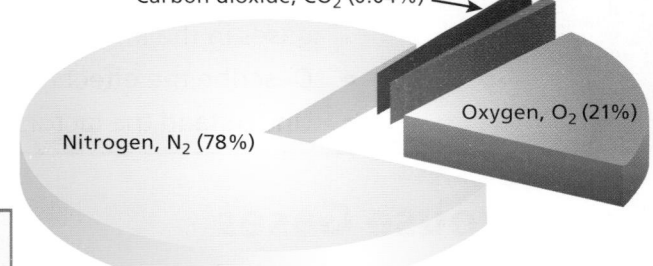

Mainly argon, plus other noble gases (1%)

Carbon dioxide, CO_2 (0.04%)

Oxygen, O_2 (21%)

Nitrogen, N_2 (78%)

> ### Key Point
>
> The Earth's early atmosphere was mainly carbon dioxide. Over time the levels of oxygen have increased and the levels of carbon dioxide have decreased.

Quick Test

1. What is the main gas in the Earth's atmosphere today?
2. Name the main gas in the Earth's early atmosphere.
3. The atmospheres of which planets are thought to be like the Earth's early atmosphere?
4. What is the main compound in limestone rocks?
5. Write the balanced symbol equation for photosynthesis.

Key Words

billion
algae
photosynthesis

Greenhouse Gases

You must be able to:

- Recall the names of some greenhouse gases
- Explain how greenhouse gases increase the Earth's temperature
- Describe how human activities can increase the levels of greenhouse gases in the atmosphere
- Describe the effects of global climate change
- Understand the factors that contribute to and affect a carbon footprint.

Greenhouse Gases

- High energy, short wavelength **infrared** radiation from the Sun passes through the atmosphere and reaches the Earth's surface.
- Some of this radiation is absorbed by the Earth.
- However, lower energy, longer wavelength infrared radiation is reflected by the Earth's surface.
- **Greenhouse gases** in the Earth's atmosphere absorb this outgoing infrared radiation, which increases the Earth's temperature.
- Without some greenhouse gases, the Earth would be too cold for water to be a liquid and would not be able to support life.
- Greenhouse gases include carbon dioxide, water vapour and methane.

The Impact of Human Activities

- Human activities can increase the levels of greenhouse gases in the atmosphere.
- The amount of carbon dioxide in the atmosphere has increased over the last 100 years. This increase correlates with an increase in the amount of fossil fuels being burned.
- Fossil fuels contain carbon that has been locked up for millions of years.
- Burning them releases carbon dioxide into the atmosphere.
- Deforestation also leads to an increase in carbon dioxide in the atmosphere as there are fewer trees taking up the gas for photosynthesis.
- Activities that increase the levels of methane in the atmosphere include:
  - the decomposition of rubbish in landfill sites
  - the increase in animal farming – it is produced by animals during digestion and by the decomposition of their waste materials.
- Many scientists believe that the increase in the levels of greenhouse gases in the atmosphere will increase the temperature of the Earth's surface and could result in global climate change.
- However, with so many different factors involved, it is difficult to produce an accurate model for such a complicated system.
- As a result, people may use simplified models.
- This can lead to speculation and opinions being expressed to the media that may be based on only part of the evidence.
- In addition, some people may have views that are **biased**, e.g. people being paid by companies that produce greenhouse gases and who have a vested interest in these issues.

> ### Key Point
>
> Climate change is a good example of the popular media reporting on scientific ideas in a way that may be oversimplified, inaccurate or biased.
>
> It is also a good example of an area where scientists can work to tackle problems caused by human impacts on the environment.

Global Climate Change

- If the average global temperature increases, this could cause global climate change. The impact of this could include:
  - a rise in sea level, which could cause devastating floods and more coastal erosion
  - more frequent and severe storm events
  - changes in the amount and timing of rainfall, with some areas receiving more rain and other areas receiving much less
  - an increased number of heatwave events, which can be harmful to people and wildlife
  - more droughts
  - changes to the distribution of plants and animals, as some areas become too hot for species to survive and other areas warm up enough to become habitable
  - food shortages in some areas, due to changes in the amount of food that countries can produce.

Carbon Footprints

- The carbon footprint of a product, service or event is the total amount of carbon dioxide and other greenhouse gases, such as methane, that are emitted over its full life cycle.
- For a product, this includes the production, use and disposal of the item.
- The carbon footprint can be reduced through:
  - using more alternative energy supplies, e.g. solar power
  - wasting less energy
  - carbon capture and storage (CCS), to prevent carbon dioxide being released into the atmosphere
  - carbon taxes and licences, to deter companies and individuals from choosing options that release lots of greenhouse gas
  - carbon off-setting, through activities such as tree planting
  - encouraging people to choose carbon-neutral products.
- However, reducing the carbon footprint is not straightforward.
- Problems include:
  - disagreement between scientists over the causes and consequences of global climate change
  - lack of information and knowledge in the general population
  - the reluctance of people to make lifestyle changes
  - economic considerations, such as the high cost of producing electricity from alternative energy resources rather than using cheaper fossil fuels
  - disagreement between countries as to what should be done.

Key Point

Climate change is an area where governments and individuals need to make decisions about the best course of action by evaluating evidence and considering all arguments.

Key Point

Trees use carbon dioxide for photosynthesis and, therefore, reduce the net amount of carbon dioxide reaching the atmosphere. The idea of carbon off-setting is to plant enough trees to balance out the carbon dioxide being produced by manufacturing processes / product use.

Key Point

Carbon-neutral products lead to no overall increase in the amount of carbon dioxide in the atmosphere.

Quick Test

1. Name **three** greenhouse gases.
2. Explain why, without greenhouse gases, the Earth could not support life.
3. What human activities are increasing the levels of carbon dioxide in the atmosphere?
4. What human activities are increasing the levels of methane in the atmosphere?
5. What does CCS stand for?

Key Words

infrared
greenhouse gases
biased
carbon footprint

Earth's Resources

You must be able to:

- Recall the resources that humans need to survive
- Understand how chemists can contribute towards sustainable development
- Describe how drinking water is produced
- Describe how waste water is treated
- **HT** Understand how and why copper is extracted from low-grade ores.

Sustainable Development

- Humans rely on the Earth's resources to provide them with warmth, shelter, food and transport.
- All our resources come from the Earth's crust, oceans or atmosphere.
- These resources can be renewable, such as timber, or finite (non-renewable), such as metal ores.
- Finite resources must be used with great care.
- Care must also be taken to ensure that the planet does not become too polluted.
- In the past, natural resources were sufficient to provide the human population with food, timber, clothing and fuels.
- However, as the population has increased, humans have come to rely on agriculture to supplement or even replace such resources.
- Chemistry plays an important role in improving agricultural and industrial processes – allowing new products to be developed and contributing towards sustainable development.

Drinking Water

- Water of the correct quality is essential for life.
- Water naturally contains microorganisms and dissolved salts.
- These need to be at low levels for the water to be safe for humans to drink.
- Fresh water contains low levels of dissolved salts.
- Water that is good quality and safe to drink is called potable.
- In the UK potable water is produced in the following way:
  1. Fresh water from a suitable source, e.g. a lake or river away from polluted areas, is collected.
  2. It is passed through a filter bed to remove solid particles.
  3. Chlorine gas is added to kill any harmful microorganisms.
  4. Fluoride is added to drinking water to reduce tooth decay (although too much fluoride can cause discolouration of teeth).
- Ozone and ultraviolet can also be used sterilise water.
- To improve the taste and quality of tap water, more dissolved substances can be removed by passing the water through a filter containing carbon, silver and ion exchange resins.
- If fresh water supplies are limited, seawater can be desalinated to produce pure water. This can be done by distillation or reverse osmosis.
- Both of these processes use a lot of energy, making them very expensive.
- During distillation:
  - the water is boiled to produce steam
  - the steam is condensed to produce pure liquid water.

> **Key Point**
>
> Sustainable development meets the needs of the current generation without compromising the ability of future generations to meet their own needs.

Waste Water Treatment

- Large amounts of waste water are produced by homes, agricultural processes and industrial processes.
- This waste water must be treated before it can be safely released back into the environment.
- Organic matter, harmful microorganisms and toxic chemicals have to be removed from sewage and agricultural and industrial waste water.
- Sewage treatment includes:
  - screening and grit removal
  - sedimentation to produce sewage sludge and effluent
  - anaerobic digestion of sewage sludge
  - aerobic biological treatment of effluent.

> **Key Point**
>
> Pure water contains no dissolved substances.

HT Alternative Methods of Extracting Metals

- Copper is a useful metal because:
  - it is a good conductor of electricity and heat
  - it is easily bent, yet hard enough to make water pipes and tanks
  - it does not react with water, so lasts for a long time.
- Copper can be extracted from copper-rich ores by heating the ores with carbon in a furnace. This process is known as smelting.
- The copper can then be purified by electrolysis.
- Copper can also be obtained:
  - from solutions of copper salts by electrolysis
  - by displacement using scrap iron.
- During electrolysis the positive copper ions move towards the negative electrode and form pure copper.
- The extensive mining of copper in the past means that we are running out of copper-rich ores.
- As a result, new methods have been developed to extract it from ores that contain less copper.
- Copper can be extracted from:
  - low-grade ores (ores that contain small amounts of copper)
  - contaminated land by biological methods.
- **Phytomining** is a method that uses plants to absorb copper:
  - As the plants grow they absorb (and store) copper.
  - The plants are then burned and the ash produced contains copper in relatively high quantities.
- **Bioleaching** uses bacteria to extract metals from low-grade ores:
  - A solution containing bacteria is mixed with a low-grade ore.
  - The bacteria convert the copper into a solution (known as a leachate solution), from which copper can be easily extracted.

> **Key Point**
>
> Phytomining and bioleaching are more environmentally friendly than traditional mining methods, which involve digging up and moving large quantities of rock and having to dispose of large amounts of waste materials.

> **Key Words**
>
> agriculture
> sustainable development
> fresh water
> potable
> desalinated
> HT phytomining
> HT bioleaching

> **Quick Test**
>
> 1. Why is chlorine added to drinking water?
> 2. HT What is phytomining?
> 3. HT What is bioleaching?

Using Resources

You must be able to:

- Explain what the life cycle assessment, LCA, of a product is
- Explain how an LCA can be used to help people make good decisions about which product to buy.

Life Cycle Assessment (LCA)

- A **life cycle assessment (LCA)** is used to assess the environmental impact a product has over its whole lifetime.
- They provide a way of comparing several alternative products to see which one causes the least damage to the environment.
- For example, comparing the LCA of aluminium cans and plastic bottles will allow a drinks manufacturer to make an informed choice about what packaging to use.

- To carry out an LCA, scientists measure the impact of:
  - extracting the raw materials
  - processing the raw materials
  - manufacturing the product
  - how the product is used
  - how the product is transported
  - how the product is disposed of at the end of its life.
- Some aspects of the LCA are quite easy to quantify, e.g. the amounts of energy, water and raw materials used.
- However, some aspects of the LCA are difficult to quantify and involve value judgements, e.g. the impact of a pollutant on the environment, meaning a LCA is not completely objective.

The Nuclear Model

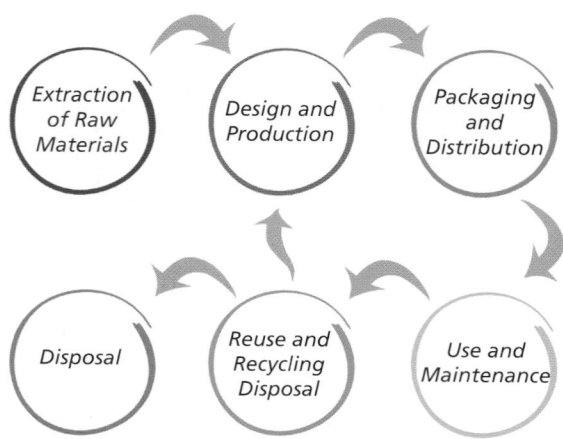

Reducing the Use of Resources

- Materials such as glass, metals and plastics are important to our standard of living. However, they must be used wisely and reused and recycled wherever possible to:
  - save money and energy
  - make sure natural resources are not used up unnecessarily
  - reduce the amount of waste produced
  - reduce damage to the environment caused by extraction.
- Metal, glass, building materials and plastics made from crude oil are produced from limited resources.
- Our supplies of these raw materials, and the fossil fuels often used to obtain them, are finite.
- The mining and quarrying processes used to extract these raw materials can have devastating environmental impacts.
- Some objects such as plastic bags and glass bottles can be reused:
  - Waste glass can be crushed, melted and reused.
  - Some waste plastic can be recycled to make fleece material.
  - Metals can be recycled by melting them down and then making them into new objects.
- Recycling generally uses far less energy than the initial extraction and production processes.
- As a result, less fossil fuel is burned and less greenhouse gases are released into the atmosphere.
- It also preserves our reserves of raw materials for the future.

Quick Test

1. How can LCAs help people make good decisions about which products to buy?
2. Suggest how a LCA might be misused.
3. How does recycling benefit:
   a) the current population?
   b) future generations?
   c) the environment?

Key Words

life cycle assessment (LCA)

Review Questions

Reactivity of Metals

1. Metals are usually extracted from their ores.

   a) What is an ore? [1]

   b) Explain how lead is extracted from lead oxide. [2]

   Total Marks _____ / 3

The pH Scale and Salts

1. Which of these ions is found in excess in alkaline solutions?
   Tick **one** box.

   H^+ ☐ H^- ☐ OH^+ ☐ OH^- ☐   [1]

2. HT Strong acids are completely ionised in water.

   a) Name a strong acid. [1]

   b) What does 'ionised' mean? [1]

3. Which of these pH values shows the pH of a strong acid?
   Tick **one** box.

   7 ☐   14 ☐   8 ☐   1 ☐   [1]

4. The pH scale is used to measure the acidity or alkalinity of aqueous solutions.
   The pH of a solution can be found by using an indicator.

   a) How does an indicator work? [1]

   b) Suggest **one** other way that the pH of a solution could be found. [1]

   c) Hydrochloric acid is a strong acid.

   Complete the word equation to show the neutralisation reaction between hydrochloric acid and potassium hydroxide.

   hydrochloric acid + potassium hydroxide → _____ + _____ [2]

   d) What is the pH of a neutral solution? [1]

   Total Marks _____ / 9

Electrolysis

1 Lead bromide is an ionic compound that can be separated into lead and bromine.

a) Name the process used to separate the lead bromide. [1]

b) Lead bromide contains Pb^{2+} and Br^- ions.

Name the elements formed:

i) At the positive electrode. [1]

ii) At the negative electrode. [1]

c) Explain why the lead bromide must be molten for this process. [1]

d) Explain why the electrodes used in the experiment must be inert. [1]

e) HT Complete the half equations to show the reactions that take place at each electrode:

i) At the anode: $2Br^- \rightarrow Br_2 +$ .... [1]

ii) At the cathode: $Pb^{2+} + 2e^- \rightarrow$ .... [1]

2 Aluminium is extracted from its ore, aluminium oxide, by electrolysis.
Electrolysis is a very expensive process.

a) Why is electrolysis expensive? [1]

b) During the extraction of aluminium, the main ore, bauxite (which contains aluminium oxide), is mixed with cryolite (another ore of aluminium).

Why is the bauxite mixed with cryolite? [1]

c) Name the substance formed at the negative electrode during the extraction of aluminium. [1]

d) HT Complete the half equation for the reaction that takes place at the anode during the electrolysis of aluminium oxide.

$$2O^{2-} \rightarrow O_2 +$$ .... [1]

Total Marks .... / 11

Review Questions

Exothermic and Endothermic Reactions

1 Complete the energy profile diagram in **Figure 1** to show an endothermic reaction.

Include the:
- Reactants
- Products
- Activation energy
- Energy change of the reaction.

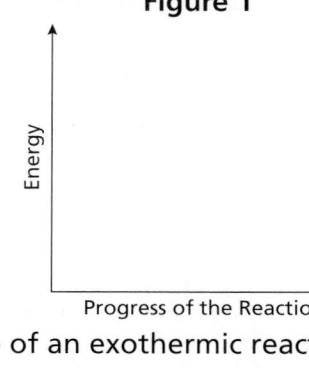

Figure 1

Energy

Progress of the Reaction

[4]

2 Which of the following types of reaction is an example of an exothermic reaction? Tick **one** box.

The reaction between citric acid and sodium hydrogen carbonate ☐

The neutralisation of hydrochloric acid by sodium hydroxide solution ☐

Dissolving ammonium nitrate crystals in water ☐

The thermal decomposition of calcium carbonate ☐ [1]

Total Marks _____ / 5

Measuring Energy Changes

1 A student carries out an experiment to find out whether changing the metal powder added to a solution of nitric acid affects the temperature change for the reaction.

a) Name the independent variable in this experiment. [1]

b) Why does the student use powdered metals in this experiment? [1]

c) Complete the results in **Table 1** below.

Table 1

Metal	Temperature at the Start (°C)	Temperature at the End (°C)	Temperature Change (°C)
magnesium	21.0	35.5	
zinc	22.0		11.0
iron		29.5	8.5

[3]

Total Marks _____ / 5

Rate of Reaction

1 Which of the following changes would decrease the rate of a chemical change?
Tick **one** box.

Increasing the concentration ☐   Increasing the temperature by 20°C ☐

Adding a catalyst ☐   Increasing the size of particles ☐ **[1]**

2 Explain, in terms of collision theory, why increasing the concentration of the
reactants increases the rate of a chemical change. **[2]**

3 Catalysts are very important in industry.

a) Name the catalyst used in the manufacture of ammonia. **[1]**

b) Catalysts increase the rate of chemical reactions.

Why do manufacturers want to increase the rate of chemical reactions? **[1]**

c) Explain how catalysts work. **[2]**

4 The rate of a chemical reaction can be calculated directly from graphs.

a) Write the equation for the mean rate of reaction. **Figure 1** **[1]**

b) The graph in **Figure 1** shows the rate of reaction is fastest
at the start of the experiment.

Explain why the rate of reaction is faster at the start of
the reaction. **[2]**

c) **HT** Explain how the rate of reaction can be calculated from
the graph in **Figure 1**. **[2]**

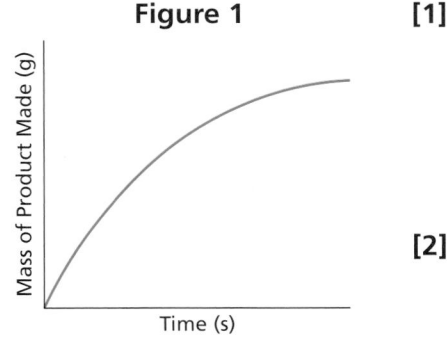

Total Marks .... / 12

Reversible Reactions

1 a) What does the symbol ⇌ mean? **[1]**

b) What is happening when equilibrium is achieved in a reaction? **[1]**

Total Marks .... / 2

Practice Questions

Alkanes

1 **a)** Complete the equation for the complete combustion of the fuel, ethane.

$$2C_2H_6 + \underline{\quad} O_2 \rightarrow \underline{\quad} CO_2 + \underline{\quad} H_2O$$ [3]

b) Carbon monoxide and water vapour are produced in the **incomplete** combustion of ethane.

Name **one** other product of the incomplete combustion of ethane. [1]

2 Methane, CH_4, is an alkane (a saturated hydrocarbon).

a) What is a saturated hydrocarbon? [2]

b) Give the general formula for alkanes. [1]

c) Give the formula for the fourth member of the alkane homologous series. [1]

3 Name the type of bonds found in alkane molecules.
Tick **one** box.

Metallic ☐   Ionic ☐   Single covalent ☐   Double covalent ☐   [1]

Total Marks .... / 9

Cracking Hydrocarbons

1 Bromine water can be used to identify alkenes, such as propene.

a) What is the formula of propene? [1]

b) The bromine water is added to a sample and the mixture is shaken.

Which of these observations would show that propene was present?
Tick **one** box.

White precipitate ☐   No colour change ☐

Cream precipitate ☐   Orange / brown to colourless ☐   [1]

Total Marks .... / 2

Chemical Analysis

1 **a)** Define the term 'pure'. [1]

  b) A substance melts at 34°C.

  Is this substance a mixture?
  You must explain your answer. [2]

2 A student analysed the colourings added to three fizzy drinks.
Figure 1 shows the results.

Figure 1

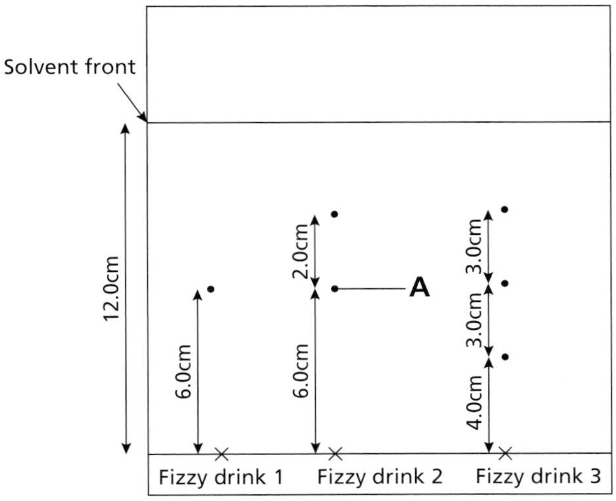

  a) Name the technique that the
    student used to analyse the
    colourings. [1]

  b) Which fizzy drink contains
    a pure colouring?
    Give a reason for your answer. [2]

  c) The R$_f$ value can be used to identify
    the components in the colouring
    added to fizzy drinks.

    What is the R$_f$ value of component **A**? [2]

  d) Which other fizzy drink also contains component **A**?
    Explain your answer. [2]

Solvent front

12.0cm

6.0cm

6.0cm   2.0cm   A

4.0cm 3.0cm 3.0cm

Fizzy drink 1   Fizzy drink 2   Fizzy drink 3

Total Marks .... / 10

The Earth's Atmosphere

1 How many years ago was the Earth formed?
  Tick **one** box.

  4.6 billion ☐   4.6 million ☐   2.7 billion ☐   200 million ☐ [1]

Practice Questions

2 The atmosphere of the Earth has changed over time.

   a) What was the main gas in the Earth's early atmosphere? **[1]**

   b) Which planets have an atmosphere similar to Earth's early atmosphere? **[2]**

   c) How has the evolution of algae and plants affected the level of oxygen in the Earth's atmosphere? **[2]**

3 During photosynthesis, carbon dioxide and water react to produce glucose and oxygen.

   a) Write a word equation for this reaction. **[1]**

   b) Explain how photosynthesis has changed the level of carbon dioxide in the Earth's atmosphere. **[2]**

   Total Marks .... / 9

Greenhouse Gases

1 Methane is a greenhouse gas. Name **one** other greenhouse gas. **[1]**

2 Global climate change could cause sea levels to rise.

   a) Suggest how a rise in sea levels could cause damage. **[1]**

   b) How could global climate change affect the distribution of wildlife? **[2]**

   c) What does 'CCS' stand for? **[1]**

   d) How could CCS help prevent global climate change? **[1]**

3 A cotton bag has a label, which says the bag is carbon-neutral.

   a) What is the 'carbon footprint' of a product? **[2]**

   b) What does the term 'carbon-neutral' mean? **[1]**

   c) Suggest why it is a good idea that the bag has the carbon-neutral label. **[1]**

   d) How might carbon taxes reduce the likelihood of global climate change? **[2]**

   Total Marks .... / 12

Earth's Resources

1 Water is essential for life.

a) What is 'fresh water'? [1]

b) What is good quality water that is safe to drink called? [1]

c) Water is collected from sources away from polluted areas and treated to make it safe to drink. In one of the steps, it is passed through filter beds.

Why is it passed through filter beds? [1]

d) Why is chlorine added to water? [1]

2 Copper is a very useful metal. It is extracted from its ores.

a) What is an 'ore'? [1]

b) Copper can be extracted by smelting.

Describe what happens in the smelting process. [2]

c) HT Copper can also be extracted by bioleaching.

Describe how copper is extracted by bioleaching. [2]

Total Marks .... / 9

Using Resources

1 a) Why are LCAs useful? [1]

b) Why are LCAs not completely objective? [1]

2 What are the benefits of recycling aluminium cans? [3]

Total Marks .... / 5

Forces – An Introduction

You must be able to:

- Describe the difference between a vector and a scalar quantity
- Use vectors to describe the forces involved when objects interact
- Calculate the resultant of two forces that act in a straight line
- Explain the difference between mass and weight
- Recognise and use the symbol for proportionality
- HT Use vector diagrams to show resolution and addition of forces.

Scalar and Vector Quantities

- A **scalar** quantity has magnitude (size) only, e.g. number of apples.
- A **vector** quantity has magnitude and direction, e.g. velocity, which shows the speed *and* the direction of travel.
- Arrows can be used to represent vector quantities :
  - the length of the arrow shows the magnitude
  - the arrow points in the direction that the vector quantity is acting.
- **Forces** are vector quantities.
- The diagram shows the forces acting on a boat. The arrows indicate the direction they are acting.

Contact and Non-Contact Forces

- A force occurs when two or more objects interact.
- Forces are either:
  - **contact forces** – the objects are actually touching, e.g. the tension as two people pull against one another
  - **non-contact forces** – the objects are not touching, e.g. the force of gravity acts even when the objects are not touching.

Contact Forces	Non-Contact Forces
Friction Air-resistance / drag Tension Normal contact force Upthrust	Gravitational force Electrostatic force Magnetic force

> ### Key Point
>
> A force is a vector quantity. It occurs when objects interact.

Gravity

- **Gravity** is a force of attraction between all masses.
- The force of gravity close to Earth is due to the gravitational field around the planet.
- The **mass** of an object is related to the amount of matter it contains and is constant.
- **Weight** is the force acting on an object due to gravity.
- The weight of an object depends on the gravitational field strength where the object is and is directly proportional to its mass.

- This symbol is used to indicate two things are proportional: ∝.

> **weight = mass × gravitational field strength**
>
> $W = mg$

Weight (W) is measured in newtons (N).
Mass (m) is measured in kilograms (kg).
Gravitational field strength (g) is measured in newtons per kilogram (N/kg).

Resultant Forces

- When more than one force acts on an object, these forces can be seen as a single force that has the same effect as all the forces acting together.
- This is called the resultant force.

HT Vector Diagrams

- A free body diagram can be used to show different forces acting on an object (see the diagram of the boat on page 158).
- Scale vector diagrams are used to illustrate the overall effect when more than one force acts on an object:
  - The forces are added together to find a single resultant force, including both magnitude and direction.
  - The vectors are added head to tail and a resultant force arrow is drawn.

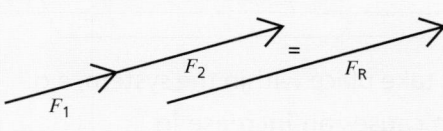

$$F_R = F_1 + F_2$$

- Scale vector diagrams can also be used when a force is acting in a diagonal direction:
  - Expressing the diagonal force as two forces at right-angles to each other can help to work out what effect the force will have.
  - The force F_R can be broken down into F_1 and F_2.
  - F_1 is the same length as the length of F_R in the horizontal direction.
  - F_2 is the same length as the length of F_R in the vertical direction.
  - F_R is also the vector found by adding F_1 and F_2 head to tail.

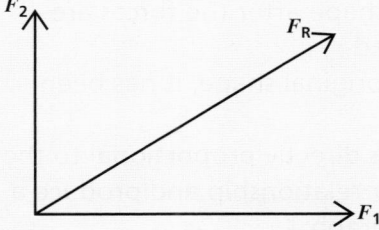

> **Key Point**
>
> The gravitational field strength (g) on Earth is 10N/kg, so a student with a mass of 50kg has a weight of (50 × 10 =) 500N.

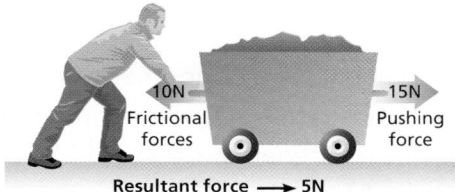

10N
Frictional forces

15N
Pushing force

Resultant force ⟶ 5N

> **Key Point**
>
> Weight is a force that can be measured using a newtonmeter (a calibrated spring-balance). The unit of measurement is newtons (N).
> weight ∝ mass

Quick Test

1. How is a scalar quantity different from a vector quantity?
2. Use an example to explain what is meant by 'non-contact force'.
3. An astronaut with a mass of 80kg stands on the Moon. What is his weight? (g = 1.6N/kg)
4. A car travels in a straight line to the east with a driving force of 500N. If total frictional forces are 400N, what is the resultant force and in which direction does it act?

> **Key Words**
>
> scalar
> vector
> force
> contact force
> non-contact force
> gravity
> mass
> weight
> resultant
> HT free body diagram

Forces in Action

You must be able to:

- Describe the energy transfers involved when work is done
- Explain why changing the shape of an object can only happen when more than one force is applied to the object
- Interpret data showing the relationship between force and extension
- Perform force calculations for balanced objects.

Work Done and Energy Transfer

- When a force causes an object to move, **work** is done on the object.
- This is because it requires energy to move the object.
- One joule of work is done when a force of one newton causes a displacement of one metre: 1 joule = 1 newton metre.

> **LEARN**
>
> **work done = force × distance (moved along the line of action of the force)**
>
> $W = Fs$

Work done (W) is measured in joules (J).
Force (F) is measured in newtons (N).
Distance (s) is measured in metres (m).

- When work is done, energy transfers take place within the system, e.g.
  - Work done to overcome friction causes an increase in heat energy.
  - An electric lift uses electrical energy to do work against gravity, which leads to an increase in gravitational potential energy.

> **Key Point**
>
> Overcoming forces requires energy. When a force is used to move an object, work is done on the object. The movement of the object is called displacement.

Forces and Elasticity

- To change the shape of an object, more than one force must be applied, e.g. a spring must be pulled from both ends to stretch it.
- If the object returns to its original shape after the forces are removed, it was **elastically deformed**.
- If the object does *not* return to its original shape, it has been **inelastically deformed**.
- The **extension** of an elastic object is directly proportional to the applied force, i.e. they have a linear relationship and produce a straight line on a force–extension graph.
- However, once the **limit of proportionality** has been exceeded:
  - doubling the force will no longer exactly double the extension
  - the relationship becomes non-linear
  - a force–extension graph will stop being a straight line.

The gradient of the linear section of the graph = the spring constant (*k*)

The area below the graph = work done to stretch or compress the spring

- This equation applies to the linear section of a force–extension graph:

LEARN

$$\text{force} = \text{spring constant} \times \text{extension}$$

$$F = ke$$

Force (F) is measured in newtons (N). Spring constant (k) is measured in newtons per metre (N/m). Extension (e) is measured in metres (m).

- This also applies to the **compression** of an elastic object.
- The **spring constant** indicates how easy it is to stretch or compress a spring – the higher the spring constant, the stiffer the spring.
- A force that stretches or compresses a spring stores elastic potential energy in the spring.
- The amount of work done and the energy stored are equal, provided the spring does not go past the limit of proportionality.

In a school experiment, a force of 2N causes a spring to stretch from 20mm to 36mm.

Calculate the spring constant.

Sometimes mass will be given instead of weight. In this type of question, the mass needs converting to the force from gravity using the equation $W = mg$ ($g = 10\text{N/kg}$).

$$\text{extension} = 36 - 20$$

Calculate the extension.

$$= 16\text{mm}$$

$$\text{extension} = \frac{16}{1000}$$

Convert the extension to metres.

$$= 0.016\text{m}$$

$$2 = k \times 0.016$$

Substitute your values into the equation $F = ke$

$$k = \frac{2}{0.016}$$

$$= 125\text{N/m}$$

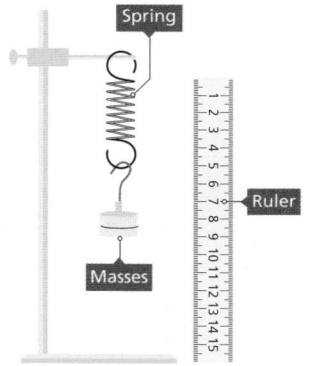

Spring

Ruler

Masses

REQUIRED PRACTICAL

Investigate the relationship between force and extension for a spring.

Sample Method	Considerations, Mistakes and Errors
1. Set up the equipment as shown. 2. Add 100g (1N) to the mass holder. 3. Measure the extension of the spring and record the result. 4. Repeat steps 2 to 3 for a range of masses from 1N to 10N.	• The extension is the total increase in length from the original unloaded length. It is *not* the total length or the increase each time. • Adding too many masses can stretch the spring too far, which means repeat measurements cannot be made.

Variables	Hazards and Risks
• The independent variable is the one deliberately changed – in this case, the force on the spring. • The dependent variable is the one that is measured – the extension.	• The biggest hazard in this experiment is masses falling onto the experimenter's feet. To minimise this risk, keep masses to the minimum needed for a good range of results.

 Key Point

The range in an experiment needs to be large enough to show a pattern.

 Quick Test

1. How much heat energy is produced when an object is pushed 2m against a 150N frictional force?
2. A 50kg person climbing a flight of stairs ascends a vertical height of 4m.
   a) Calculate the work done ($g = 10\text{N/kg}$).
   b) Suggest what energy transfers have taken place.
3. The springs on car shock absorbers have a high spring constant so that they behave elastically when the car is driven over bumps. Explain why this is important.

Key Words

work
elastically deformed
inelastically deformed
extension
limit of proportionality
compression
spring constant

Forces and Motion

You must be able to:

- Describe displacement in terms of both magnitude and direction
- Recall typical values for common speeds
- Calculate speed from measurements of distance and time
- HT Give examples of objects with constant speed but changing velocity
- Explain the motion of objects using Newton's first law
- Draw and interpret distance–time graphs.

Distance and Displacement

- **Distance** is a scalar quantity:
  - It is how far an object moves.
  - It does not take into account the direction an object is travelling in or even if it ends up back where it started.
- **Displacement** is a vector quantity:
  - It has a magnitude, which describes how far the object has travelled from the origin, measured in a straight line.
  - It has a direction, which is the direction of the straight line.

Path of travel

Displacement

> **Key Point**
>
> Distance and speed are scalar quantities, so they only have magnitude.
>
> Displacement is a vector quantity, so it has direction as well as magnitude.

Speed

- The **speed** of an object is a measure of how fast it is travelling.
- It is a scalar quantity measured in metres per second (m/s).
- The speed that a person can walk, run or cycle depends on factors like age, fitness, terrain and distance.
- Some typical speeds can be seen below:

Method of Travel	Speed (m/s)	Method of Travel	Speed (m/s)
Walking	1.5	Motorway driving	30
Running	3	High speed trains	75
Cycling	6	Commercial aircraft	250
City driving	12	Speed of sound in air	330

- Most things (including sound) do not travel at a constant speed, so it is often the average speed over a period of time that is used.

 **LEARN**

distance travelled = speed × time

$$s = vt$$

Distance (s) is measured in metres (m).
Speed (v) is measured in metres per second (m/s).
Time (t) is measured in seconds (s).

Velocity

- **Velocity** is a vector quantity.
- It is the speed of an object in a given direction.

HT When travelling in a straight line, an object with constant speed also has constant velocity.

> **Key Point**
>
> Velocity is a vector quantity.

HT If the object is *not* travelling in a straight line, e.g. it is turning a corner:
  – the speed can still be constant
  – the velocity will change, because the direction has changed.
HT An object moving in a circle:
  – is constantly changing direction, so it is constantly changing velocity
  – is **accelerating** even if it is travelling at constant speed.
HT Orbiting planets are an example of this – it is the force of gravity that causes the acceleration.

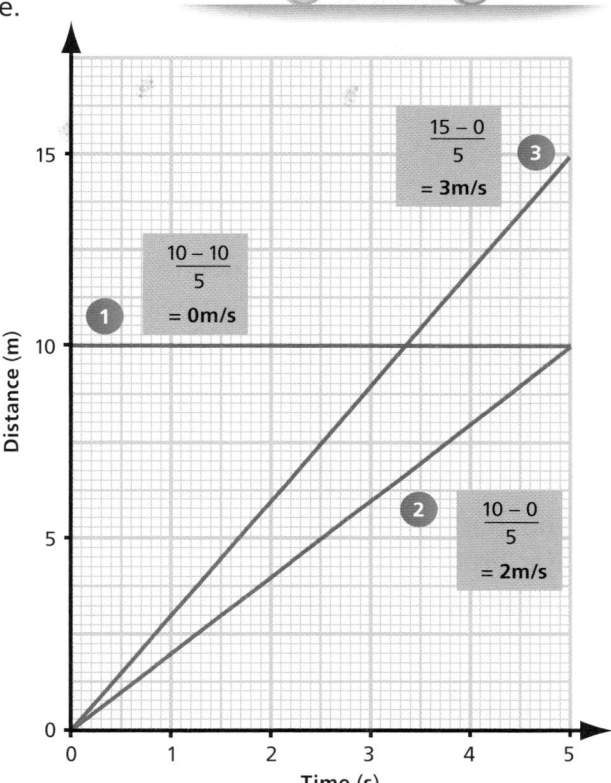

> **HT Key Point**
>
> An object travelling in a circle can have constant speed, but its velocity is still changing.

Newton's First Law

- Newton's first law is often stated as: an object will remain in the same state of motion unless acted on by an external force.
- When the resultant force acting on an object is zero:
  – if the object is stationary, it remains stationary
  – if the object is moving, it continues to move at the same speed and in the same direction, i.e. at constant velocity.

  > **HT** This tendency for objects to continue in the same state of motion is called **inertia**.

- The velocity (speed or direction) of an object will only change if there is a resultant force acting on it.
- For a car travelling at a steady speed, the driving force is balanced by the resistive forces.

Distance–Time Graphs

- A distance–time graph can be used to represent the motion of an object travelling in a straight line.
- The speed of the object is found from the gradient (slope) of the line.
- The graph on the right shows:
  ① A stationary object.
  ② An object moving at a constant speed of 2m/s.
  ③ An object moving at a greater constant speed of 3m/s.

HT If an object is accelerating, the distance–time graph will be a curve.
HT For an accelerating object, its speed at a particular time is found by:
  – drawing a tangent to the curve at the point in time
  – working out the gradient of the tangent.

> **Key Point**
>
> A distance–time graph can be used to calculate speed.

> **Key Words**
>
> distance
> displacement
> speed
> velocity
> **HT** accelerating
> **HT** inertia

> **Quick Test**
>
> 1. Describe the difference between distance and displacement.
> 2. Work out the speed of a car that travels 1km in 50 seconds.
> 3. What does the gradient of a distance–time graph show?

Forces and Acceleration

You must be able to:

- Apply Newton's second law to situations where objects are accelerating
- Estimate the magnitude of everyday accelerations
- Draw and interpret velocity–time graphs.

Acceleration

- The **acceleration** of an object is a measure of how quickly it speeds up, slows down or changes direction.

LEARN

$$\text{acceleration} = \frac{\text{change in velocity}}{\text{time taken}}$$
$$a = \frac{\Delta v}{t}$$

- When an object slows down, the change in velocity is negative, so it has a negative acceleration.
- Acceleration can also be can be calculated using the equation:

$$(\text{final velocity})^2 - (\text{initial velocity})^2 = 2 \times \text{acceleration} \times \text{distance}$$
$$v^2 - u^2 = 2as$$

Velocity–Time Graphs

- The **gradient** of a velocity–time graph can be used to find the acceleration of an object.

- **HT** The total distance travelled is equal to the area under the graph.

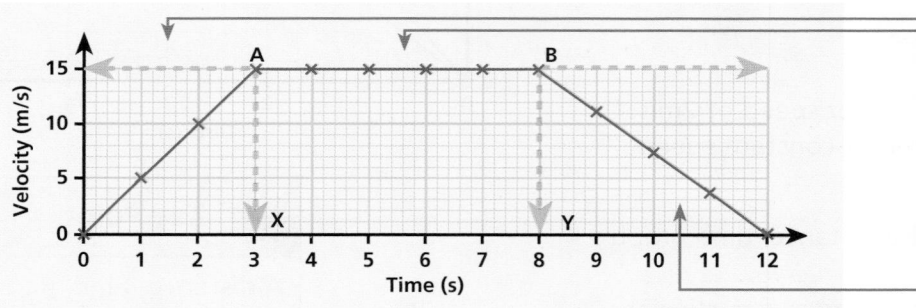

Distance travelled:

First section: $\frac{1}{2} \times 3 \times 15 = 22.5\text{m}$

Middle section: $15 \times 5 = 75\text{m}$

Final section: $\frac{1}{2} \times 4 \times 15 = 30\text{m}$

Total distance = 22.5 + 75 + 30

   = 127.5m

Key Point

Acceleration is a measure of rate of change of velocity.

Acceleration (a) is measured in metres per second squared (m/s²). Change in velocity (Δv) is found by subtracting initial velocity from final velocity ($v - u$) and is measured in metres per second (m/s). Time (t) is measured in seconds (s).

Final velocity (v) is measured in metres per second (m/s). Initial velocity (u) is measured in metres per second (m/s). Acceleration (a) is measured in metres per second squared (m/s²). Distance (s) is measured in metres (m).

The change in velocity is 15m/s over a 3 second period, so the acceleration is $\frac{15}{3} = 5\text{m/s}^2$

The velocity is constant, so the acceleration is zero.

The change in velocity is –15m/s over a 4 second period, so the acceleration is $\frac{-15}{4} = -3.75\text{m/s}^2$. The negative value shows that the object is slowing down.

Break down the area under the graph into smaller shapes.

Add together all of the areas to find the total distance.

Newton's Second Law

- Newton's second law is often stated as: the acceleration of an object is **proportional** to the resultant force acting on the object and **inversely proportional** to the mass of the object, i.e.
  - if the resultant force is doubled, the acceleration will be doubled
  - if the mass is doubled, the acceleration will be halved.
- This law can be summarised with the equation:

> **force = mass × acceleration**
>
> $$F = ma$$

Force (F) is measured in newtons (N).
Mass (m) is measured in kilograms (kg).
Acceleration (a) is measured in metres per second squared (m/s²).

HT Mass is a measure of **inertia**.

HT It describes how difficult it is to change the velocity of an object.

HT This inertial mass is given by the ratio of force over acceleration, i.e. $m = \dfrac{F}{a}$.

HT The larger the mass, the bigger the force needed to change the velocity.

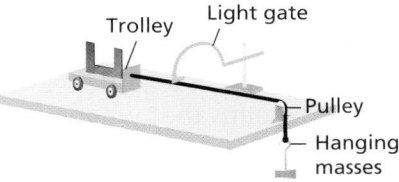

Trolley — Light gate — Pulley — Hanging masses

REQUIRED PRACTICAL

Investigate the effect of varying the force and / or the mass on the acceleration of an object.

Sample Method
1. Set up the equipment as shown.
2. Release the trolley and use light gates or a stopwatch to take the measurements needed to calculate acceleration.
3. Move 100g (1N) from the trolley onto the mass holder.
4. Repeat steps 2 and 3 until all the masses have been moved from the trolley onto the mass holder.

If investigating the mass, keep the force constant by removing a mass from the trolley but not adding it to the holder.

Considerations, Mistakes and Errors
- When changing the force it is important to keep the mass of the system constant. Masses are taken from the trolley to the holder. No extra masses are added.
- Fast events often result in timing errors. Repeating results and finding a mean can help reduce the effect of these errors.
- If the accelerating force is too low or the mass too high, then frictional effects will cause the results to be inaccurate.

Variables
- The independent variable is the force or the mass.
- The control variable is kept the same. In this case, the force if the mass is changed or the mass if the force is changed.

Hazards and Risks
- The biggest hazard in this experiment is masses falling onto the experimenter's feet. To minimise this risk, masses should be kept to the minimum needed for a good range of results.

Key Point

Calculating a mean helps to reduce the effect of random errors.

A result that is accurate is close to the true value.

Quick Test

1. An object accelerates from 2m/s to 6m/s over a distance of 8m. Use the equation $v^2 - u^2 = 2as$ to find the acceleration of the object.
2. Comparing two velocity–time graphs, it can be seen that graph A is twice as steep as graph B. What does this indicate?
3. Why is it important to carry out repeat readings during an experiment?

Key Words

acceleration
gradient
proportional
inversely proportional
HT inertia

Terminal Velocity and Momentum

You must be able to:

- Explain terminal velocity
- Give examples of Newton's third law in action
- HT Describe examples of conservation of momentum in collisions.

Terminal Velocity

- When an object falls through a fluid:
  - At first, the object accelerates due to the force of gravity.
  - As it speeds up, the resistive forces increase.
  - The resultant force reaches zero when the resistive forces balance the force of gravity. At this point the object will fall at a steady speed, called its terminal velocity.
- Near the Earth's surface, the acceleration due to gravity is $10\,m/s^2$.

Newton's Third Law

- Newton's third law is often stated as: for every action there is an equal and opposite reaction.
- This means that whenever one object exerts a force on another, the other object exerts an force back.
- This reaction force is of the same type and is equal in size but opposite in direction.

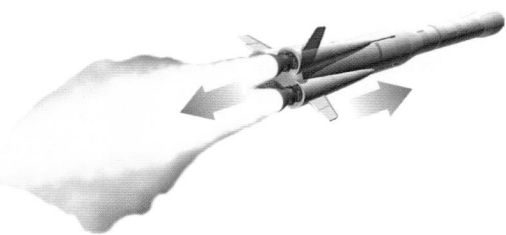

A rocket pushes fuel backwards, which in turn pushes the rocket forwards.

HT Momentum

- All moving objects have momentum.
- The momentum of a moving object depends on its mass and its velocity.
- The greater the momentum of an object, the greater the force needed to stop the object.
- At the same speed, a large mass will have more momentum than a small mass.
- With the same mass, a faster object will have more momentum than a slower object.
- Momentum can be calculated with the equation:

LEARN HT

$$\text{momentum} = \text{mass} \times \text{velocity}$$

$$p = mv$$

HT **Key Point**

Momentum is the product of mass and velocity.

Momentum (p) is measured in kilograms metres per second (kg m/s). Mass (m) is measured in kilograms (kg). Velocity (v) is measured in metres per second (m/s).

HT Conservation of Momentum

- In a closed system, the total momentum before an event is equal to the total momentum after the event.
- This conservation of momentum is most often referred to during **collisions**, but also applies to rockets and projectiles.
- With rockets, the momentum gained by the exhaust being expelled backwards is equal to the momentum gained by the rocket moving forwards.
- With projectiles, the recoil momentum of the gun is equal to the momentum of the bullet.

Two cars are travelling in the same direction along a road. Car A collides with the back of car B and they stick together.

Before

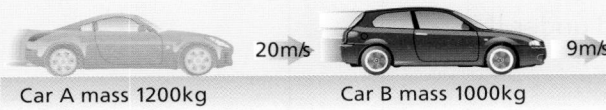

20m/s   9m/s

Car A mass 1200kg   Car B mass 1000kg

After

v m/s

Car A + Car B mass 2200kg

Calculate the velocity of the cars after the collision.

Momentum before collision = momentum A + momentum B
= (mass A × velocity of A) + (mass B × velocity of B)
= (1200 × 20) + (1000 × 9)
= 24000 + 9000 = 33000kg m/s

Momentum after collision = 33000kg m/s
Momentum after collision = (mass A + mass B) × (new combined velocity)
$33000 = (1200 + 1000)v$
$33000 = 2200v$
$v = \dfrac{33\,000}{2200} = 15$m/s

Revise

Conservation of Momentum

Recoil

Explosion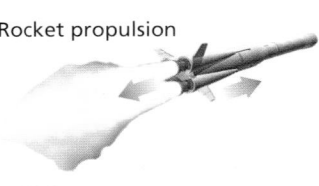

Rocket propulsion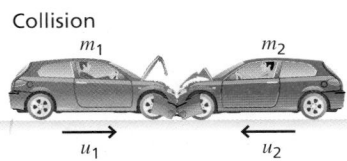

Collision

m_1 m_2

u_1 u_2

Start by calculating the momentum before the collision.

Remember, momentum is conserved so: momentum before collision = momentum after collision.

Substitute in the values for momentum and mass.

Rearrange the equation to find the velocity.

HT Key Point

In a closed system, momentum is conserved.

Quick Test

1. Calculate the momentum of a 2000kg car travelling at 20m/s.
2. Use Newton's third law and the idea of equal and opposite forces to explain how a fish propels itself through water.
3. HT a) Calculate the momentum of a horse and rider with a total mass of 600kg travelling at 8m/s.
   b) A motorcycle and rider travelling at 12m/s has the same momentum as the horse and rider in part a). Work out the combined mass of the motorcycle and rider.

Key Words

terminal velocity
HT momentum
HT collision

Stopping and Braking

You must be able to:

- Interpret graphs that relate speed to stopping distance
- Describe factors affecting reaction time and braking distance
- Explain the dangers of large decelerations.

Stopping Distance

- The stopping distance of a vehicle depends on:
  - the thinking distance (the distance travelled during the driver's reaction time)
  - the **braking distance** (the distance travelled under the braking force).

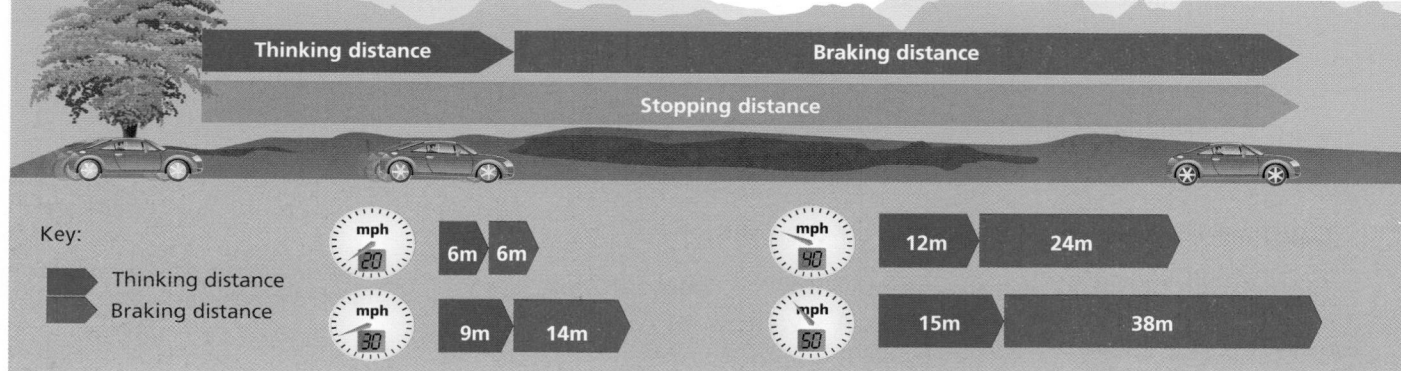

Thinking distance   Braking distance

Stopping distance

Key:

Thinking distance
Braking distance

mph 20	6m	6m
mph 30	9m	14m
mph 40	12m	24m
mph 50	15m	38m

- For a given braking force: the greater the speed of the vehicle, the longer the stopping distance.

Reaction Time

- Reaction times vary from person to person, but the typical human reaction time is in the range of 0.4–0.9 seconds.
- This means that a car travelling at 30m/s (≈70mph) will travel between 12 and 27 metres before the person even begins to brake.
- This reaction time can be affected by tiredness, drugs and alcohol.
- Distractions, e.g. mobile phone use, also affect a person's ability to react.
- Factors that cause a driver to react more slowly will increase the reaction time and, therefore, increase the overall stopping distance.
- To measure reaction time, use lights or sounds as a 'start' signal and an electronic timer to measure how long someone takes to react.
- In the classroom, reaction times can be measured by dropping a ruler vertically and catching it as it falls.
- The distance the ruler falls through a person's fingers can be used to calculate the time it took them to react.

> **Key Point**
>
> Stopping distance is the sum of the thinking distance and the braking distance.

> **Key Point**
>
> Reaction time can be affected by alcohol, drugs, fatigue and distractions.

Factors Affecting Braking Distance

- The braking distance of a vehicle can be affected by the condition of the road, the vehicle and the weather.
- Adverse weather conditions include wet or icy / snowy roads.
- Vehicle condition includes factors such as worn brakes or tyres and over-inflated or under-inflated tyres.
- To stop a vehicle, the brakes need to apply a force to the wheels.
- The greater the braking force, the greater the deceleration of the vehicle.
- Work done by this frictional force transfers the kinetic energy of the vehicle into heat energy, increasing the temperature of the brakes.
- If the braking force is too large, the brakes may overheat or the tyres may lose traction on the road resulting in the vehicle skidding.
- As well as the risk of skidding, large decelerations cause large forces to act on the people in the vehicle, which can result in injury.
- Overheating and loss of traction are more likely to occur if the brakes or tyres are in poor condition.
- When a vehicle is travelling faster, it needs a larger braking force to be able to stop it in a certain distance.
- For a given braking distance:
  - doubling the mass doubles the force required
  - doubling the speed quadruples the force required.

HT To find the size of the braking force required, the equation for work done can be used:

| **work done (kinetic energy) = force × distance (braking distance)** |

HT The table below shows the size of the braking force involved in two different braking situations:

Vehicle	Mass	Speed	Kinetic Energy	Braking Distance	Force
Car	1500kg	City ≈ 15m/s	168 750J	50m	3 375N
Lorry	7500kg	Motorway ≈ 30m/s	3 375 000J	50m	67 500N

The symbol ≈ means approximately equal to. Sometimes a single ~ is used to mean the same thing.

Quick Test

1. Explain what effect talking on a mobile phone might have on the stopping distance of a car.
2. HT Estimate the braking force needed to stop a 750kg motor cycle travelling at city speeds over a distance of 25m.
3. Describe a simple experiment to measure the reaction time of a person.

Key Words

braking distance
deceleration

Energy Stores and Transfers

You must be able to:

- Describe changes in the way energy is stored when a system changes
- Calculate the energy changes involved when a system changes
- Describe what is meant by internal energy
- Use specific heat capacity in calculations
- Describe how to investigate the specific heat capacity of materials.

Energy Stores and Systems

- A **system** is an object or group of objects.
- When a system changes, there are changes in the way **energy** is stored, e.g. a roller coaster transfers energy between gravitational and kinetic energy.
- Diagrams are a useful way to illustrate how the energy is redistributed when a system is changed.
- The Sankey diagram below shows that the light bulb redistributes the electrical energy as heat and light.

Calculating Energy Changes

- You need to be able to calculate the amount of energy associated with a moving object, an object raised above ground level and a stretched spring.
- The **kinetic energy** of a moving object can be calculated with the equation:

LEARN

$$\text{kinetic energy} = 0.5 \times \text{mass} \times (\text{speed})^2$$

$$E_k = \frac{1}{2}mv^2$$

- The **gravitational potential energy (GPE)** gained by raising an object above ground level can be calculated with the equation:

LEARN

$$\text{gravitational potential energy} = \text{mass} \times \text{gravitational field strength} \times \text{height}$$

$$E_p = mgh$$

- The amount of **elastic energy** stored in a stretched or compressed spring can be found with the equation:

$$\text{elastic potential energy} = 0.5 \times \text{spring constant} \times (\text{extension})^2$$

$$E_e = \frac{1}{2}ke^2$$

Specific Heat Capacity and Internal Energy

- **Internal energy** is the total kinetic and potential energy of all the particles that make up a system.
- Doing work on a system increases the energy stored in a system.

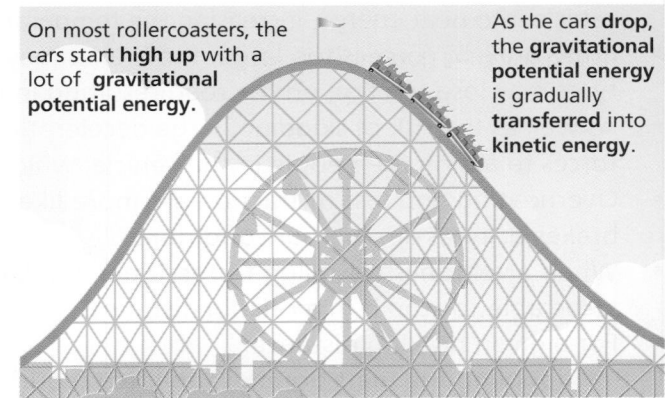

On most rollercoasters, the cars start **high up** with a lot of **gravitational potential energy.**

As the cars **drop,** the **gravitational potential energy** is gradually **transferred** into **kinetic energy.**

Sankey Diagram of an Incandescent Light Bulb

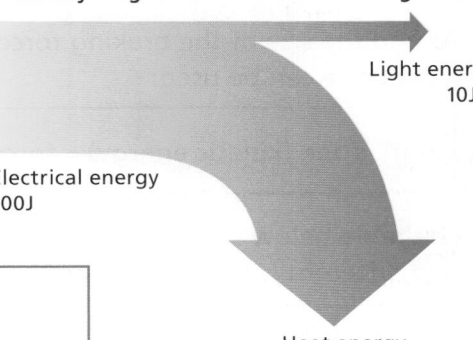

Light energy
10J

Electrical energy
100J

Heat energy
90J

Kinetic energy (E_k) is measured in joules (J).
Mass (m) is measured in kilograms (kg).
Speed (v) is measured in metres per second (m/s).

Gravitational potential energy (E_p) is measured in joules (J).
Mass (m) is measured in kilograms (kg).
Gravitational field strength (g) is measured in newtons per kilogram (N/kg).
Height (h) is measured in metres (m).

Elastic potential energy (E_e) is measured in joules (J).
Spring constant (k) is measured in newtons per metre (N/m).
Extension (e) in metres (m).

- Heating changes the energy stored in a system by increasing the energy of the particles within it.
- As the energy increases, this will either increase the temperature or produce a change of state (see pages 210–211).
- If the temperature increases, the increase depends on:
  – the mass of the substance heated
  – what the substance is
  – the energy input.
- The **specific heat capacity** of a substance is the amount of energy required to raise the temperature of one kilogram of the substance by one degree Celsius.

> **change in thermal energy =**
> **mass × specific heat capacity × temperature change**
>
> $$\Delta E = mc\Delta\theta$$

Calculate the increase in temperature when 2100J of energy is provided to 100g of water. The specific heat capacity of water is 4200 J/kg °C.

$$2100 = 0.1 \times 4200 \times \Delta\theta$$
$$\Delta\theta = \frac{2100}{(4200 \times 0.1)} = 5°C$$

Key Point

The amount of energy stored can also be found by the amount of work done, so you need to look at the information you are given. If you have not been given the information needed to use the elastic potential energy equation, but have been given force and distance, then use work done = force × distance

---

Change in thermal energy (ΔE) is measured in joules (J).
Mass (m) is measured in kilograms (kg).
Specific heat capacity (c) is measured in joules per kilogram per degree Celsius (J/kg °C).
Temperature change ($\Delta\theta$) is measured in degrees Celsius (°C).

Substitute in the given values into the equation $\Delta E = mc\Delta\theta$. Make sure they are in the correct units, i.e. 100g = 0.1kg.

Rearrange the equation to find the temperature change ($\Delta\theta$).

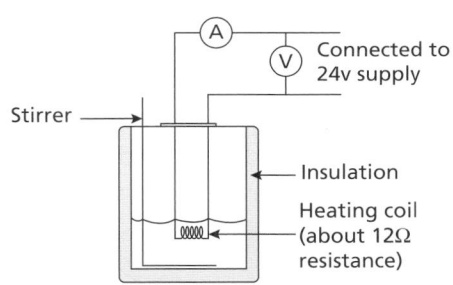

Stirrer

Connected to 24v supply

Insulation

Heating coil
(about 12Ω resistance)

REQUIRED PRACTICAL

Investigate the specific heat capacity of materials, linking the decrease of one energy store (or work done) to the increase in temperature and subsequent increase in thermal energy stored.

Sample Method	Considerations, Mistakes and Errors
1. Set up the apparatus as shown. 2. Measure the start temperature. 3. Switch on the electric heater for 5min. 4. Measure the end temperature. 5. Measure the voltage and current to find the power. 6. Repeat for different liquids. 7. Calculate the specific heat capacity. 8. Compare your results to another group's. If they get similar answers the experiment is **reproducible**.	• The energy provided by the heater is calculated as power × time. However, it could also be found using a joulemeter. • The specific heat capacity is calculated from the energy provided, the mass of the liquid and the temperature change. • If the temperature rise is too high, energy loss to the surroundings will affect the results.
Variables • The independent variable is the type of liquid. • The dependent variable is the temperature. • Control variables are the amount of liquid used and energy provided.	**Hazards and Risks** • The electric heater could be very hot so you must *not* touch it directly. • If the liquids become hot they could boil and spit, so safety goggles must be worn and the heater should not be left on for longer than is necessary.

Key Words

system
energy
kinetic energy
gravitational potential
  energy (GPE)
elastic energy
internal energy
specific heat capacity
reproducible

Quick Test

1. Use a Sankey diagram to illustrate the energy changes involved when a diesel powered crane is used to lift an object.
2. Work out the energy required to heat 1.5kg of water from 20°C to 100°C. (c = 4200J/kg °C)
3. Explain what is meant by 'reproducible' and why it is important.

Energy Transfers and Resources

You must be able to:

- Describe examples of energy transfers in a closed system
- Describe how thermal conductivity affects rate of cooling
- Distinguish between renewable and non-renewable energy resources.

Energy Transfers

- When looking at energy transfer, there are two key points:
  - Energy can be **transferred** usefully, stored or **dissipated** (spread out to the surroundings).
  - Energy cannot be created or destroyed.
- In a closed system the total energy never changes, but it can be transferred from one store to another.
- For example, when an electricity-powered lift raises the lift carriage:
  - it transfers electrical energy into gravitational potential energy
  - some energy is dissipated into the surroundings as heat and sound
  - this wasted energy is no longer available for useful transfers.
- Wasted energy is caused by unwanted energy transfers.
- These unwanted transfers can be reduced in several ways:
  - lubrication – reduces the friction that produces heat
  - tightening any loose parts – prevents unwanted vibration that wastes energy as sound
  - thermal insulation – reduces heat loss.
- A system isn't always a single device; it could be an entire building.
- A building wastes energy when it loses heat to the surroundings, causing it to cool down.
- The rate of cooling depends on the thickness and thermal **conductivity** of the walls.
- Thin walls with high thermal conductivity will conduct heat the quickest and the building will cool down rapidly.
- The table gives examples of energy transfers in different situations:

Situation	Energy Transfers
An object projected upwards	Kinetic energy (from the initial movement of the object) to gravitational potential energy (from its increase in height)
An object hitting an obstacle	Kinetic energy (from the moving object) to thermal energy and sound (when the object hits the obstacle)
A car accelerating	Chemical energy (in the fuel) to kinetic energy (the movement of the car)
A car braking	Kinetic energy (from the movement of the car) to thermal energy (from friction in the brakes)
Water boiling in a kettle	Electrical energy to thermal energy

> **Key Point**
>
> An anomalous result is one that doesn't fit the pattern. These:
>
> - should be looked at to try and determine the cause
> - should be ignored when plotting graphs
> - should not be included when calculating averages.

National and Global Energy Resources

- The main uses for energy resources are transport, electricity generation and heating.
- All energy resources fall into one or two categories:
  - **renewable** energy resources, which can be replenished
  - non-renewable energy resources, which will eventually run out.

Category	Energy Resource	Main Uses	Environmental Impacts, Ethics, Reliability and Other Information
Renewable	Biofuel	Transport and electricity generation	Large areas of land are needed for growing fuel crops. This can be at the expense of food crops in poorer countries.
Renewable	Wind	Electricity generation	Does not provide a constant source of energy. Turbines can be noisy / dangerous to birds. Some people think they think they ruin the appearance of the countryside.
Renewable	Water (hydro-electricity)	Electricity generation	Requires large areas of land to be flooded, altering ecosystems and displacing the people that live there.
Renewable	Geothermal	Electricity generation and heating	Only available in a limited number of places where hot rocks can be found close to the surface, e.g. Iceland.
Renewable	Tidal	Electricity generation	Variations in tides affect output. Have a high initial set-up cost. Can alter habitats / cause problems for shipping.
Renewable	Solar	Electricity generation and some heating	Depends on light intensity, so no power produced at night. High cost in relation to power output
Renewable	Water waves	Electricity generation	Output depends on waves, so can be unreliable. Can alter habitats.
Non-renewable	Nuclear fuel	Electricity generation and some military transport	Produces radioactive waste but no other emissions. Costly to build and decommission. Reliable output.
Non-renewable	Coal	Electricity generation, heating and some transport	Burning produces greenhouse gases (CO_2) and contributes to acid rain (SO_2). Reliable output.
Non-renewable	Oil	Transport and heating	Reliable output. Provides a compact source of energy for transport. Burning produces CO_2, NO_2 and SO_2. Serious environmental damage if spilt.
Non-renewable	Gas	Electricity generation, heating and some transport	Reliable output. Burning produces CO_2 but *not* SO_2.

Quick Test

1. A 20W light bulb transfers 15W usefully as light energy. What happens to the energy that is not transferred as light?
2. Describe **one** ethical consideration of hydroelectric power.

Review Questions

Alkanes

1. Which of the substances below is a saturated hydrocarbon?
   Tick **one** box.

   CH_4 ☐ C_2H_4 ☐ C_3H_6 ☐ C_5H_{10} ☐   [1]

2. Crude oil can be separated into different parts by fractional distillation.

   a) Explain how crude oil is separated by fractional distillation.   [3]

   b) Give **one** use for the diesel fraction of crude oil.   [1]

3. Propane, C_3H_8, is used as a fuel in camping stoves.

   a) Complete the equation below for the complete combustion of propane.

   $$C_3H_8 + \underline{\hspace{1cm}} O_2 \rightarrow \underline{\hspace{1cm}} CO_2 + \underline{\hspace{1cm}} H_2O$$ [3]

   b) Incomplete combustion of propane can produce carbon monoxide.

   Why is the production of carbon monoxide undesirable?   [1]

   > Total Marks _____ / 9

Cracking Hydrocarbons

1. A student has samples of propene and propane.

   Describe the test that the student should carry out to identify the two samples.   [4]

2. Name the type of bonds found in alkene molecules.
   Tick **one** box.

   Metallic and single covalent ☐   Single covalent and double covalent ☐

   Ionic and double covalent ☐   Double covalent and metallic ☐   [1]

   > Total Marks _____ / 5

Chemical Analysis

1 Chromatography can be used to separate mixtures of coloured substances.

The mixtures are separated because each component in the mixture distributes itself between the stationary phase and the mobile phase differently.

a) i) What is the stationary phase in paper chromatography? [1]

   ii) What is the mobile phase in paper chromatography? [1]

The R_f value can be used to identify the components in a soluble food colouring. **Figure 1** shows the chromatogram for a sample of food colouring.

b) Why is the line at the bottom of the absorbent paper drawn in pencil? [1]

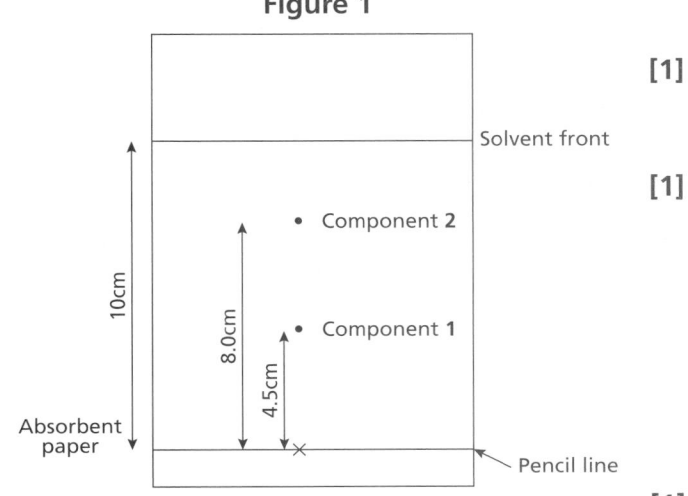

Figure 1

Solvent front

Component 2

Component 1

10cm · 8.0cm · 4.5cm

Absorbent paper

Pencil line

c) What is the R_f value for **Component 1**? [2]

d) How does this chromatogram show that the food colouring tested is a mixture and not a pure substance? [1]

Total Marks _____ / 6

The Earth's Atmosphere

1 a) How many years ago did algae evolve and start producing oxygen by photosynthesis?
   Tick **one** box.

   4.6 billion ☐   2.7 billion ☐

   4.6 million ☐   200 billion ☐   [1]

b) Complete the word equation to sum up what happens during photosynthesis.

   _____ + water → glucose + _____   [2]

Review Questions

2 The Earth's atmosphere has changed over time.

    a) How was the Earth's early atmosphere formed? **[1]**

    b) What was the main gas in the Earth's early atmosphere? **[1]**

    c) The level of nitrogen has increased over time.
Some nitrogen was produced when the gases in the early atmosphere reacted with oxygen.

       By what other means was nitrogen produced? **[1]**

    d) How has the formation of sedimentary rocks, like limestone, affected the Earth's atmosphere? **[2]**

3 Some of the carbon dioxide released into the atmosphere reacts with sea water.
As a result, there is a smaller increase in carbon dioxide levels than might be expected.

    a) Give **two** types of products that can be formed when carbon dioxide reacts with sea water. **[2]**

    b) How can algae in sea water reduce the levels of carbon dioxide in the atmosphere? **[1]**

> **Total Marks** .... / 11

Greenhouse Gases

1 Carbon dioxide is a greenhouse gas.

    a) Write down the chemical formula of carbon dioxide. **[1]**

    b) Explain how greenhouse gases increase the Earth's temperature. **[2]**

2 Why has the level of carbon dioxide in the atmosphere been increasing in recent times?
Tick **one** box.

Deforestation   ☐

More animal farming   ☐

Decomposition of rubbish in landfill sites   ☐

Photosynthesis   ☐ **[1]**

3 Carbon offsetting schemes allow companies and individuals to invest in environmental schemes.

    a) Which gas are these schemes designed to compensate for? **[1]**

b) Some carbon offsetting schemes involve the planting of trees.

How would this activity offset carbon? **[2]**

Earth's Resources

1 What can be used to sterilise water?
Tick **one** box.

Infrared radiation ☐   Carbon dioxide ☐

Ultraviolet radiation ☐   Salt ☐ **[1]**

2 Copper is a useful metal.

a) Why is copper a good metal for making wires? **[2]**

b) Why is copper a good metal for making saucepans? **[1]**

c) Copper can be extracted from its ore by heating with carbon.

What property of copper means that it can be extracted using this method? **[1]**

d) HT Copper can also be extracted by phytomining.

Describe how copper is extracted by phytomining. **[3]**

3 Copper can be extracted by phytomining or bioleaching.
These two methods of extraction are more environmentally friendly than traditional mining methods.

Explain why traditional methods of mining can cause environmental problems. **[3]**

Using Resources

1 a) What does the term LCA stand for? **[1]**

b) A drinks manufacturer compares the LCA for an aluminium can with the LCA for a plastic bottle.

Suggest **three** aspects they might look at before deciding which one to use. **[3]**

Practice Questions

Forces – An Introduction

1 The mass of an object is a scalar quantity, but the weight of an object is a vector quantity.

Explain what is meant by this statement and the link between mass and weight. **[4]**

2 **Figure 1** represents the forces acting on an object.

Figure 1

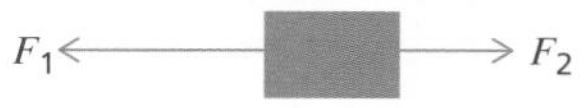

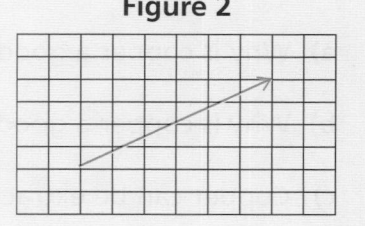

a) Compare the forces F_1 and F_2. **[2]**

b) The force F_2 increases until it is of equal magnitude to F_1.

What will be the magnitude of the resultant force? **[1]**

3 **HT** **Figure 2** represents the resultant force acting on an object.

Draw the horizontal and vertical components of the arrow onto **Figure 2**.

Figure 2

[1]

Total Marks _____ / 8

Forces in Action

1 A car travelling along a straight, level road at 30mph has 150kJ of kinetic energy. The driver applies the brakes and comes to a complete stop in 50m.

a) What happens to the temperature of the brakes during braking? **[1]**

b) How much work is done by the brakes to stop the car? **[1]**

c) Calculate the braking force applied by the brakes. **[2]**

2 A small boat floating on a lake has a weight of 2597N (g = 10N/kg).

a) Calculate the mass of the boat to the nearest kilogram. **[2]**

The boat is loaded with an additional 100kg. It sits lower on the water but still floats.

b) What is the new weight of the boat? **[2]**

c) How will the upwards force from the water compare to the downwards force from gravity? **[1]**

Total Marks _____ / 9

Forces and Motion

1 A hiker travels north for 2 miles, east for 1 mile and then south again for 2 miles.

　　a) What is the total distance travelled? [1]

　　b) What is the final displacement? [2]

2 The length of the race track at Silverstone is 3.65 miles.
A Formula One race is 52 laps of the race track.
The winner of the 2015 race completed the race in approximately 1 hour and 30 minutes.

　　a) Calculate the following:

　　　　i) The total distance travelled by one car during the race. [1]

　　　　ii) The displacement of the car at the end of the race, after completing 52 laps. [1]

　　　　iii) The average speed (in mph) of the winning car over the entire race. [2]

　　b) At some points in the race, the cars will be travelling at constant speed but their velocity will be changing.

　　　　Explain how this can be true and where on the track this might occur. [3]

3 **Figure 1** is a distance–time graph.

Figure 1

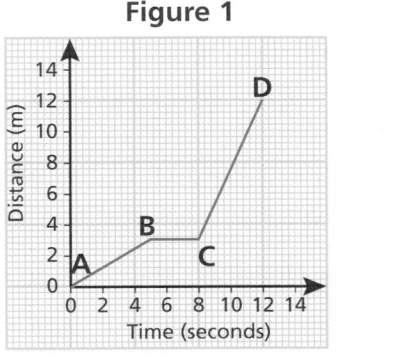

　　a) How long is the object stationary for in total? [1]

　　b) During which part of the journey is the object travelling at the greatest speed? [1]

　　c) What is the speed between points **A** and **B**? [2]

　　d) If the graph contained curved lines, what would the curved sections indicate? [1]

Total Marks .... / 15

Forces and Acceleration

1 A car travelling at 10m/s accelerates at a constant rate of 4m/s² over a distance of 100m.

　　Use the formula $v^2 - u^2 = 2as$ to work out the final velocity reached by the car. [4]

Practice Questions

2 **Table 1** shows the velocity of a remote-controlled car every two seconds for a 10 second period.

Table 1

Time (s)	Velocity (m/s)
0	0
2	4
4	8
6	8
8	8
10	0

a) Plot a velocity–time graph for the 10 second period. **[3]**

b) Using your graph, describe the motion of the car between 5 seconds and 7 seconds. **[1]**

c) Calculate the acceleration of the car during the first 4 seconds. **[2]**

d) From your graph, calculate the total distance travelled. **[3]**

Total Marks .... / 13

Terminal Velocity and Momentum

1 A falling object takes 0.4 seconds to accelerate from rest to a speed of 4m/s.

a) Assuming that no other forces act on the object, show that the acceleration due to gravity is $10m/s^2$. **[3]**

b) After 0.4 seconds the acceleration begins to reduce and eventually reaches zero.

   Explain why. **[3]**

2 A helicopter weighs 25 000N.

a) What is the size of the upwards force that must act on the helicopter for it to remain hovering in a stationary position? **[1]**

b) Use Newton's third law to explain how the helicopter produces the force required in part **a)**. **[3]**

Total Marks .... / 10

Stopping and Braking

1 Give **three** factors that would have a negative effect on the reaction time of a driver. **[3]**

2 In terms of energy, explain why the braking distance of a car travelling at 20mph will be greater than the braking distance of a car travelling at 15mph and why the faster car's brakes will get hotter. **[3]**

Total Marks .... / 6

Energy Stores and Transfers

1 The useful energy output from a petrol engine is kinetic energy. However, the engine wastes more energy as heat than it produces as kinetic energy. This and other transfers involved are shown on the Sankey diagram in **Figure 1**.

Figure 1

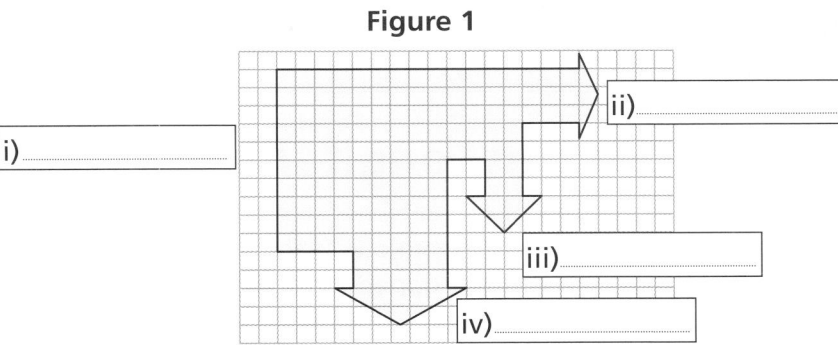

i)

ii)

iii)

iv)

a) Label the Sankey diagram in **Figure 1** using words from the box. **[4]**

Chemical energy	Kinetic energy	Heat energy	Sound energy

b) If the total input energy is 1000kJ, how much energy is converted to kinetic energy? **[2]**

2 A skateboarder of mass 60kg is practising on a half-pipe. She starts at the top, skates down one side and back up the other side.

a) Describe the energy transfers involved. **[6]**

b) The maximum speed the skateboarder reaches is 4m/s.

Calculate her kinetic energy at this point. **[3]**

c) The skater starts at a height of 2m.

Calculate how much gravitational energy she has at this point ($g = 10$N/kg). **[3]**

Total Marks _____ / 18

Energy Transfers and Resources

1 Coal is a non-renewable energy resource.

What is meant by 'non-renewable'? **[1]**

2 Explain the different factors that determine the amount of heat lost from a building. **[3]**

Total Marks _____ / 4

Waves and Wave Properties

You must be able to:

- Describe the difference between transverse and longitudinal waves
- Describe evidence that waves transfer energy not matter
- Describe waves in terms of amplitude, wavelength, frequency and period
- Explain how wave speed, frequency and wavelength are linked.

Transverse and Longitudinal Waves

- There are two types of wave: **transverse** and **longitudinal**.
- All waves transfer energy from one place to another.
- For example, if a stone is dropped into a pond, ripples travel outwards carrying the energy. The water does not travel outwards (otherwise it would leave a hole in the middle).
- The particles that make up a wave **oscillate** (vibrate) about a fixed point. In doing so, they pass the energy on to the next particles, which also oscillate, and so on.
- The energy moves along, but the matter remains.
- In a transverse wave, e.g. water wave, the oscillations are perpendicular (at right-angles) to the direction of energy transfer.
- This can be demonstrated by moving a rope or slinky spring up and down vertically – the wave then moves horizontally.
- In a longitudinal wave, e.g. sound wave, the oscillations are parallel to the direction of energy transfer.
- This can be demonstrated by moving a slinky spring moving back and forward horizontally – the wave also moves horizontally.

Properties of Waves

- All waves have a:
  - **frequency** – the number of waves passing a fixed point per second, measured in hertz (Hz)
  - **amplitude** – the maximum displacement that any particle achieves from its undisturbed position in metres (m)
  - **wavelength** – the distance from one point on a wave to the equivalent point on the next wave in metres (m)
  - **period** – the time taken for one complete oscillation in seconds (s).

Key Point

Waves transfer energy **not** matter.

Transverse Waves

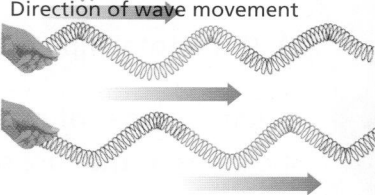

Hand movement up and down

Direction of wave movement

Longitudinal Waves

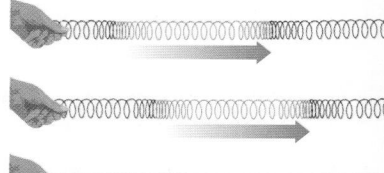

Hand movement in and out

Compression Rarefaction

Direction of wave movement

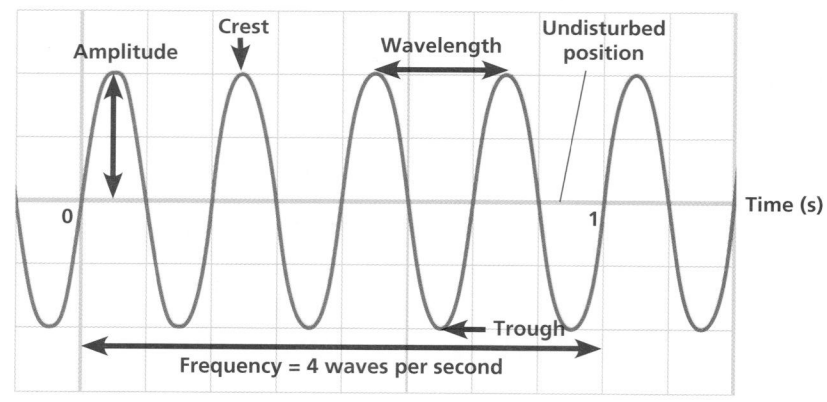

Amplitude · Crest · Wavelength · Undisturbed position · Time (s) · Trough

Frequency = 4 waves per second

- When observing waves (e.g. water waves):
  - amplitude is seen as the wave height
  - the period is the time taken for one complete wave to pass a fixed point.

$$\text{period} = \frac{1}{\text{frequency}}$$

$$T = \frac{1}{f}$$

Period (T) is measured in seconds (s). Frequency (f) is measured in hertz (Hz).

- Amplitude indicates the amount of energy a wave is carrying – the more energy, the higher the amplitude.

Wave Speed

- The speed of a wave is the speed at which the energy is transferred (or the wave moves).
- It is a measure of how far the wave moves in one second and can be found with 'the wave equation':

$$\text{wave speed} = \text{frequency} \times \text{wavelength}$$

$$v = f\lambda$$

Wave speed (v) is measured in metres per second (m/s). Frequency (f) is measured in hertz (Hz). Wavelength (λ) is measured in metres (m).

- Ripples on the surface of water are slow enough that their speed can be measured by direct observation and timing with a stopwatch.

REQUIRED PRACTICAL
Identify the suitability of apparatus to measure the frequency, wavelength and speed of waves in a ripple tank.

Sample Method	Considerations, Mistakes and Errors
1. Time how long it takes one wave to travel the length of the tank. Use this to calculate wave speed using $\text{speed} = \frac{\text{distance}}{\text{time}}$. 2. To find the frequency, count the number of waves passing a fixed point in a second. 3. Estimate the wavelength by using a ruler to measure the peak-to-peak distance as the waves travel. 4. Use a stroboscope to make the same measurements and compare the results.	• Using a stroboscope can significantly improve the accuracy of measurements. • By projecting a shadow of the waves onto a screen below the stroboscope, flash speed can be adjusted to make the waves appear stationary. This makes wavelength measurements much more accurate. • For high frequencies that are difficult to count, this can be used with the wave speed measurement to calculate the frequency using $f = \frac{v}{\lambda}$.
Variables	**Hazards and Risks**
• The key control variable is water depth. It important to ensure that the depth of the water is kept constant across the tank as, for a given frequency, the depth will affect the speed and wavelength.	• When using a stroboscope there is a risk to people with photo-sensitive epilepsy. It is important to check that there are no at risk people involved in the experiment or in the area.

Quick Test

1. In what direction are the oscillations in a transverse wave?
2. Work out the period of a wave with a frequency of 20Hz.
3. Describe a method for measuring wave speed.

Key Words

transverse
longitudinal
oscillate
frequency
amplitude
wavelength
period

Electromagnetic Waves

You must be able to:

- Recall the order of waves in the electromagnetic spectrum
- Construct ray diagrams to illustrate refraction of a wave at a boundary
- **HT** Use wave front diagrams to explain refraction.

Electromagnetic Waves

- **Electromagnetic (EM) waves** are transverse waves.
- All types of electromagnetic wave travel at the same velocity (the speed of light) in air or a vacuum.
- The electromagnetic spectrum extends from low frequency, low energy waves to high frequency, high energy waves.
- Human eyes are only capable of detecting visible light, i.e. a very limited range of electromagnetic waves.
- **HT** The wavelength of an electromagnetic wave affects how it is **absorbed**, **transmitted**, **reflected** or **refracted** by different substances. This affects its uses.

> ### Key Point
> All electromagnetic waves have the same velocity speed in air or a vacuum.

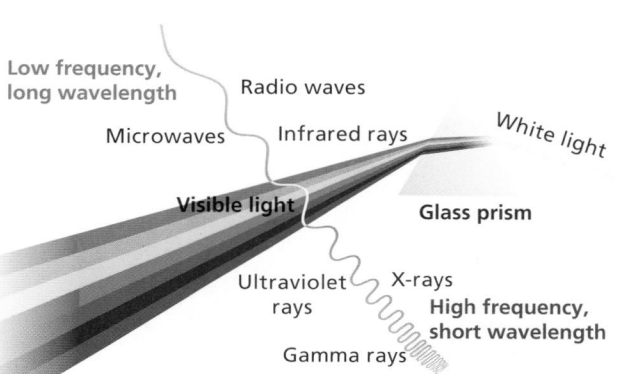

Low frequency, long wavelength — Radio waves — Microwaves — Infrared rays — White light — Visible light — Glass prism — Ultraviolet rays — X-rays — High frequency, short wavelength — Gamma rays

Refraction

- When a wave passes from one medium into another it can be refracted and change direction.
- The direction of refraction depends on:
  - the angle at which the wave hits the boundary
  - the materials involved.
- For light rays, the way in which a material affects refraction is called its **refractive index**.
- When light travels:
  - from a material with a low refractive index to one with a higher refractive index, it bends towards the normal
  - from a material with a high refractive index to one with a lower refractive index, it bends away from the normal.

> ### Key Point
> When slowing down, the wave is refracted towards the normal. When speeding up, it is refracted away.

HT Refraction is due to the difference in the wave speed in the different media.

- When a light wave enters, at an angle, a medium in which it travels slower:
  - the first part of the light wave to enter the medium slows down
  - the rest of the wave continues at the higher speed
  - this causes the wave to change direction, towards the normal.
- You need to be able to show this using a wave front diagram like the one below, which is for water waves.

Refraction

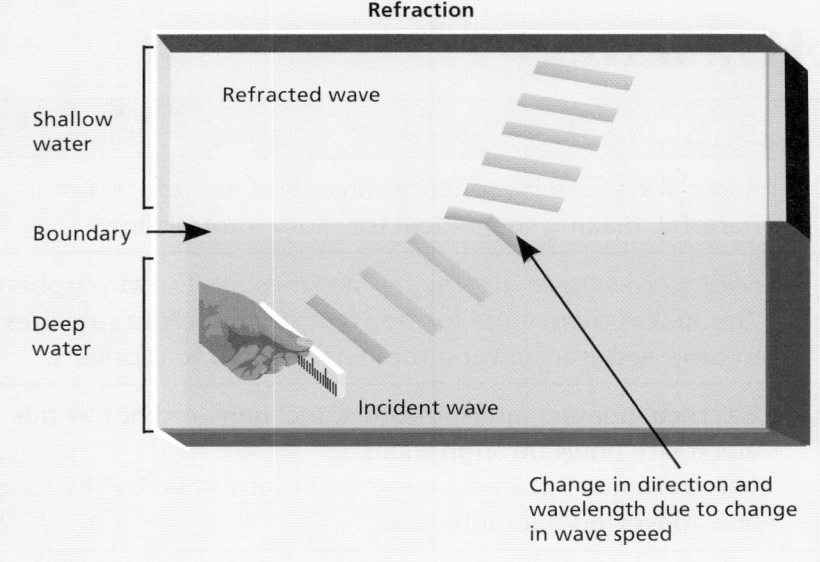

Change in direction and wavelength due to change in wave speed

Ray Diagrams

- When waves reach a boundary between one **medium** (material) and another, they can be refracted.
- Ray diagrams can be used to show what happens.
- When constructing a ray diagram:
  - Rays must be drawn with a ruler.
  - Each straight section of ray should have a single arrow drawn on it to indicate the direction of movement.
  - Where the ray meets the boundary, a 'normal' should be drawn at right-angles to the boundary.
  - All relevant angles should be labelled.
- Ray diagrams are used whenever a diagram is needed to represent how light moves from one place to another, e.g. reflection.

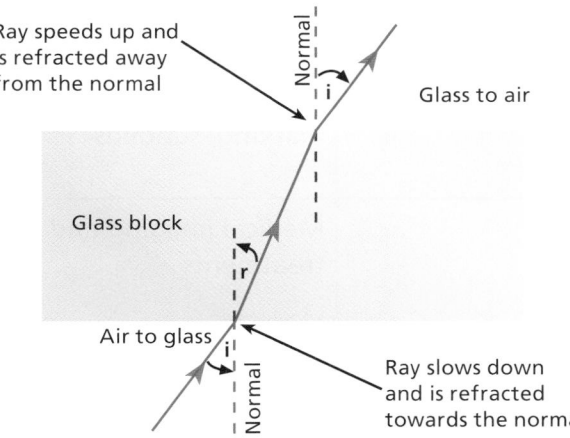

Ray speeds up and is refracted away from the normal

Glass to air

Glass block

Air to glass

Ray slows down and is refracted towards the normal

> **Quick Test**
>
> 1. List the types of wave in the electromagnetic spectrum in order, from long wavelength to short wavelength.
> 2. On a ray diagram what is the 'normal'?
> 3. **HT** A light wave travels from one medium to another. As it crosses the boundary, it speeds up. In which direction does the light ray turn?

> **Key Words**
>
> electromagnetic (EM) waves
> absorbed
> transmitted
> reflected
> refracted
> refractive index
> medium
> normal

The Electromagnetic Spectrum

You must be able to:

- Give examples of electromagnetic waves transferring energy
- Explain why different electromagnetic waves are suitable for particular applications
- Describe how electromagnetic waves can be produced
- Explain the risks and consequences of radiation exposure.

Uses and Applications of EM Waves

EM Wave	Uses	Explanations
Radio waves	Television, radio, Bluetooth	• Radio waves are low energy waves and, therefore, not harmful, making them ideal for radio transmission.
Microwaves	Satellite communications, cooking food	• Microwaves travel in straight lines through the atmosphere. • This makes them ideal for transmitting signals to satellites in orbit and transmitting them back down to receivers.
Infrared waves	Electrical heaters, cooking food, infrared cameras	• Electrical heaters, grills, toasters, etc. glow red hot as the electricity flows through them. • This transmits infrared energy that is absorbed by the food and converted back into heat.
Visible light	Fibre optic communications	• Visible light travels down optical fibres from one end to the other without being lost through the sides.
Ultraviolet waves	Energy efficient light bulbs, security marking, sunbeds	• In energy efficient light bulbs, UV waves are produced by the gas in the bulb when it is excited by the electric current. • These UV waves are absorbed by the coating on the bulb, which fluoresces giving off visible light.
X-rays	Medical imaging and treatments	• X-rays are able to penetrate soft tissue but not bone. • A photographic plate behind a person will show shadows where bones are.
Gamma rays	Sterilising food, treatment of tumours	• Gamma rays are the most energetic of all electromagnetic waves and can be used to destroy bacteria and tumours.

REQUIRED PRACTICAL

Investigate how the amount of infrared radiation absorbed or radiated by a surface depends on the nature of that surface.

Sample Method
1. Take four boiling tubes each painted a different colour: matt black, gloss black, white and silvered.
2. Pour hot water into each boiling tube.
3. Measure and record the start temperature of each tube.
4. Measure the temperature of each tube every minute for 10 min.
5. The tube that cools fastest, emits infrared energy quickest.

Considerations, Mistakes and Errors
- A common error in this experiment is not having the boiling tubes at the same temperature at the start – a hotter tube will cool quicker initially, which can affect results.
- Evaporation from the surface of the water can cause cooling too, which will affect the results. To minimise this, block the top of each tube with a bung or a plug of cotton wool.

Variables
- The independent variable is the colour of the boiling tube.
- The dependent variable is the temperature.
- Control variables include volume of water, start temperature and environmental conditions.

Hazards and Risks
- The main hazard is being burned when pouring the hot water and when handling the hot tubes. Using a test tube rack to hold the tubes minimises the need to touch the tubes and means hands can be kept clear when pouring the water into them.

HT Radio Signals

- Radio waves can be caused by oscillations in electrical circuits, i.e. an alternating current (see pages 194–195).
- The frequency of the radio wave produced matches the frequency of the electrical oscillation. This is how a radio signal is produced.
- When radio waves are absorbed by a conductor they may create an alternating current with the same frequency as the radio wave, this is how the signal is received.
- When this oscillation is induced in an electrical circuit it creates an electrical signal that matches the wave.

Hazards of EM Waves

- Changes in atoms and the nuclei of atoms can result in EM waves being generated or absorbed over a wide frequency range:
  - Electrons moving between energy levels as a result of heat or electrical excitation can generate waves, e.g. infrared waves, visible light, ultraviolet waves and X-rays.
  - Changes in the nucleus of an atom can generate waves, i.e. an unstable nucleus can give out excess energy as gamma rays.
- Ultraviolet waves, X-rays and gamma rays carry enough energy to have hazardous effects on the human body:
  - Ultraviolet waves can cause the skin to age prematurely and increase the risk of skin cancer.
  - X-rays and gamma rays are **ionising** radiation – they can damage cells by ionising atoms and, if absorbed by the nucleus of the cell, can cause gene mutations and cancer.

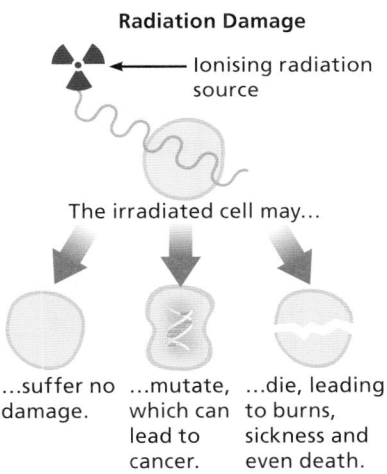

Radiation Damage

 ← Ionising radiation source

The irradiated cell may...

...suffer no damage.

...mutate, which can lead to cancer.

...die, leading to burns, sickness and even death.

> **Key Point**
>
> The risk of damage from EM waves depends on the type of radiation and the amount of exposure.
>
> Radiation dose is a measure of harm based on these two factors. It is measured in Sieverts (Sv).

Quick Test

1. Give **one** example of how electromagnetic waves transfer energy.
2. What property of microwaves makes them suitable for cooking?
3. Explain why UV, X-rays and gamma rays are the most hazardous electromagnetic radiations.

> **Key Words**
>
> microwaves
> infrared
> ultraviolet
> X-rays
> gamma rays
> ionising

An Introduction to Electricity

You must be able to:

- Draw and interpret circuit diagrams
- Calculate the charge that flows in a circuit
- Relate current, resistance and potential difference
- Explain how to investigate factors that affect the resistance of an electrical component.

Standard Circuit Symbols

- In diagrams of electrical circuits:
  - standard circuit symbols are used to represent the components
  - wires should be drawn as straight lines using a ruler.
- You need to know all of the circuit symbols in the table below:

Component	Symbol	Component	Symbol
Switch (open)		LED (light emitting diode)	
Switch (closed)		Bulb / lamp	
Cell		Fuse	
Battery		Voltmeter	
Diode		Ammeter	
Resistor		Thermistor	
Variable resistor		LDR (light dependent resistor)	

Electric Charge and Current

- Electric **current** is the flow of electrical **charge** – the greater the rate of flow, the higher the current.
- Current is measured in amperes (A), which is often abbreviated to amps, using an ammeter.
- Electric charge is measured in coulombs (C) and can be calculated with the equation:

> **LEARN**
>
> **charge flow = current × time**
>
> $$Q = It$$

- As the current in a single, closed loop of a circuit has nowhere else to go (i.e. no branches to travel down), the current is the same at all points in the loop.

> ### Key Point
>
> An ammeter is connected in series. A voltmeter is connected in parallel to the component.

> Charge flow (Q) is measured in coulombs (C).
> Current (I) is measured in amps (A).
> Time (t) is measured in seconds (s).

Resistance and Potential Difference

- The **resistance** of a component is the measure of how it resists the flow of charge.
- The higher the resistance:
  - the more difficult it is for charge to flow
  - the lower the current.
- Resistance is measured in ohms (Ω).
- **Potential difference** (or voltage) tells us the difference in electrical potential from one point in a circuit to another.
- Potential difference can be thought of as electrical push.
- The bigger the potential difference across a component:
  - the greater the flow of charge through the component
  - the bigger the current.
- Potential difference is measured in volts (V) using a voltmeter.
- Potential difference, current and resistance are linked by the equation:

LEARN

> **potential difference = current × resistance**
>
> $$V = IR$$

Key Point

Increasing the resistance reduces the current.

Increasing the voltage increases the current.

Potential difference (V) is measured in volts (V).
Current (I) is measured in amps (A).
Resistance (R) is measured in ohms (Ω).

REQUIRED PRACTICAL

Investigate the factors that affect the resistance of electrical circuits.

Sample Method

This example looks at how length affects the resistance of a wire:

1. Set up the standard test circuit as shown.
2. Pre-test the circuit and adjust the supply voltage to ensure that there is a measurable difference in readings taken at the shortest and longest lengths.
3. Record the voltage and current at a range of lengths, using crocodile clips to grip the wire at different points.
4. Use the variable resistor to keep the current through the wire the same at each length.
5. Use the voltage and current measurements to calculate the resistance.

Considerations, Mistakes and Errors

- Adjusting the supply voltage to ensure as wide a range of results as possible is important, as measurements could be limited by the **precision** of the measuring equipment.
- The range of measurements to be tested should always include at least five measurements at reasonable intervals. This allows for patterns to be seen without missing what happens in between, but also without taking large numbers of unnecessary measurements.

Variables

- The independent variable is the length of the wire.
- The dependent variable is the voltage.
- The control variable is the current (which is kept the same, because if it was too high it would cause the wire to get hot and change its resistance).

Hazards and Risks

- Current flowing through the wire can cause it to get very hot.
- To avoid being burned by the wire:
  - a low supply voltage should be used, such as the cell in the diagram
  - adjust the variable resistor to keep the current low.

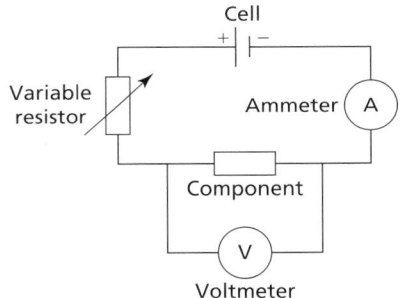

Cell

Variable resistor

Ammeter (A)

Component

V

Voltmeter

Quick Test

1. A 600C charge flows in a circuit in 1 minute. What current is flowing?
2. Draw a standard test circuit for an investigation to measure the resistance of a diode.
3. A bulb has a 10V potential difference across it. A 2A current flows. What is the resistance of the bulb?

Key Words

current
charge
resistance
potential difference
precision

Circuits and Resistance

You must be able to:

- Interpret potential difference–current graphs
- Describe the shape of potential difference–current graphs for various components
- Use circuit diagrams to construct circuits.

Resistors and Other Components

- Potential difference–current graphs ($V–I$ graphs) are used to show the relationship between the potential difference (voltage) and current for any component.
- A straight line through the origin indicates that the voltage and current are directly proportional, i.e. the resistance is constant.
- A steep gradient indicates low resistance, as a large current will flow for a small potential difference.
- A shallow gradient indicates a high resistance, as a large potential difference is needed to produce a small current.
- For some resistors the value of R is not constant but changes as the current changes, this results in a non linear graph

> **Key Point**
>
> A $V–I$ graph shows the relationship between voltage and current. It, therefore, can be used to determine the resistance.

REQUIRED PRACTICAL
Investigate the $V–I$ characteristics of a filament lamp, a diode and a resistor at constant temperature.

Sample Method	**Considerations, Mistakes and Errors**
1. Set up the standard test circuit as shown. 2. Use the variable resistor to adjust the potential difference across the test component. 3. Measure the voltage and current for a range of voltage values. 4. Repeat the experiment at least three times to be able to calculate a mean. 5. Repeat for the other components to be tested.	• Before taking measurements, check the voltage and current with the supply turned off. This will allow zero errors to be identified. • A common error is simply reading the supply voltage as the voltage across the component. At low component resistances, the wires will take a sizeable share of the this voltage, resulting in a lower voltage across the component. This is why a voltmeter is used to measure the voltage across the component.
Variables	**Hazards and Risks**
• The independent variable is the potential difference across the component (set by the variable resistor). • The dependent variable is the current through the component, measured by the ammeter.	• The main risk is that the filament lamp will get hotter as the current increases and could cause burns. If it overheats, the bulb will 'blow' and must be allowed to cool down before attempting to unscrew and replace it.

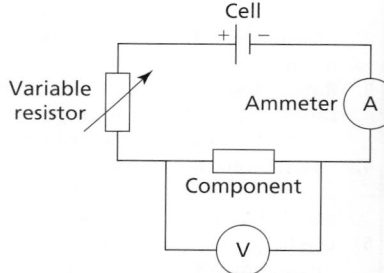

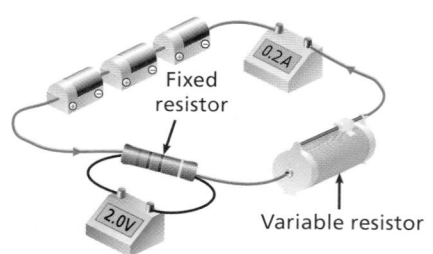

Resistors
- An **ohmic conductor** is a resistor in which the current is directly proportional to the potential difference at a constant temperature.
- This means that the resistance remains constant as the current changes.
- It is indicated by a linear (straight line) graph.

Filament Lamps
- As the current through a filament lamp increases, its temperature increases.
- This causes the resistance to increase as the current increases.
- It is indicated by a curved graph.

Diodes
- The current through a **diode** will only flow in one direction.
- The diode has a very high resistance in the reverse direction.
- This is indicated by a horizontal line along the *x*-axis, which shows that no current flows.

Thermistors
- The resistance of a thermistor decreases as the temperature increases.
- This makes them useful in circuits where temperature control or response is required.
- For example, a thermistor could be used in a circuit for a thermostat that turns a heater off at a particular temperature or an indicator light that turns on when a system is overheating.

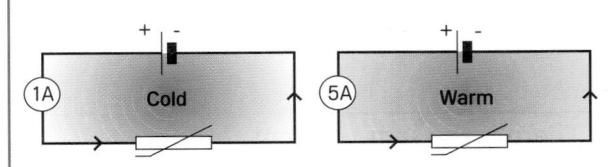

Light Dependent Resistors (LDRs)
- The resistance of an LDR decreases as light intensity increases.
- This makes them useful where automatic light control or detection is needed, e.g. in dusk till dawn garden lights / street lights and in cameras / phones to determine if a flash is needed.

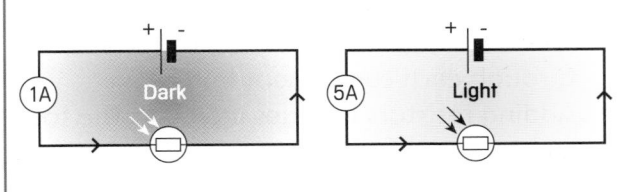

Quick Test
1. A *V–I* graph is plotted for a component. When the potential difference is negative, no current flows. What component has been tested?
2. A *V–I* graph is steep for high temperatures and shallow for low temperatures. What component does it represent?

Key Words
ohmic conductor
diode

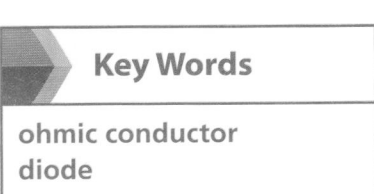

Circuits and Power

You must be able to:

- Explain the difference between series and parallel circuits
- Explain the effect of adding resistors in series and parallel.
- Explain what is meant by 'power' using examples
- Explain how the power transfer in any circuit device is related to the potential difference across it and the current through it, and to the energy changes over time.

Series and Parallel Circuits

- Electrical components can be connected in **series** or **parallel**.
- Some electrical circuits contain series and parallel parts.

Series Circuits	Parallel Circuits
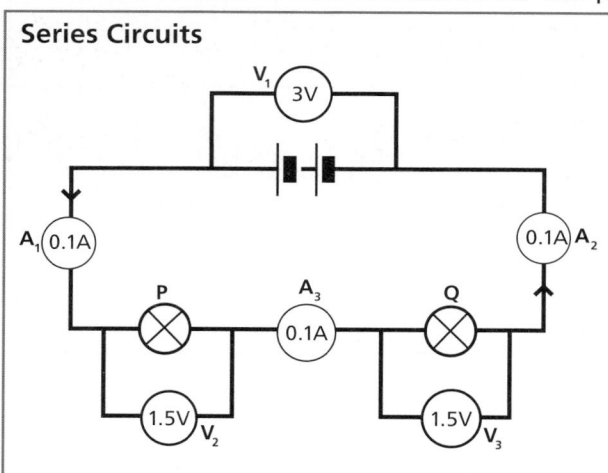	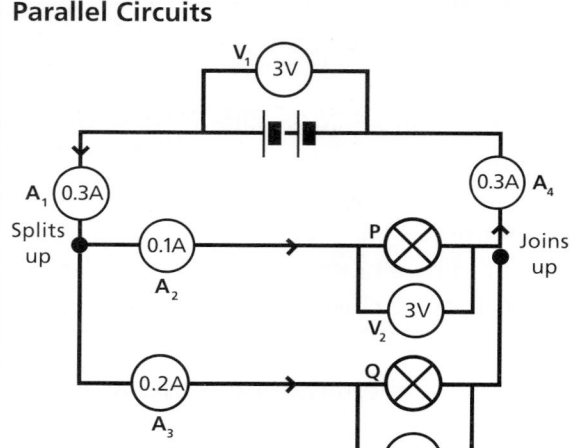
<ul><li>There is the same current through each component.</li><li>The total potential difference of the power supply is shared between the components.</li><li>The total resistance of two components is the sum of the resistance of each component. This is because the current has to travel through each component in turn.</li><li>Adding resistors in series increases the total resistance (R) in ohms (Ω): $R_{total} = R_1 + R_2$</li></ul>	<ul><li>The potential difference across each component is the same.</li><li>The total current drawn from the power supply is the sum of the currents through the separate components.</li><li>The total resistance of two resistors is less than the resistance of the smallest individual resistor. This is because, in parallel, there are more paths for the current to take – it can take one or the other, allowing it to flow more easily.</li><li>Adding resistors in parallel reduces the total resistance.</li></ul>

- You need to be able to calculate the currents, potential differences and resistances in d.c. series circuits.

A circuit containing two resistors in series has a 12V supply. R_1 is a 4Ω resistor and has a voltage of 8V across it.

a) Work out the voltage across R_2.

$$V_{total} = V_1 + V_2$$
$$12 = 8 + V_2$$
$$V_2 = 12 - 8 = 4V$$

Remember, in a series circuit, the power supply voltage is shared.

Substitute in the given values.

Rearrange to find the voltage across the second resistor.

b) Calculate the current that flows in the circuit.

$$V = IR$$

$$I = \frac{V}{R} = \frac{8}{4} = 2A$$ ← The current flow is the same through every component, so it can be calculated using the known voltage and resistance of R_1.

c) Work out the resistance of R_2.

$$V = IR$$ ← Use the values for voltage and current calculated in parts **a)** and **b)**.

$$R = \frac{V}{I} = \frac{4}{2} = 2\Omega$$

Power in Circuits

- The **power** of a device depends on the potential difference across it and the current flowing through it.
- A device with a higher potential difference or current will use more energy per second than one with a lower potential difference or current, i.e. it will be more powerful.

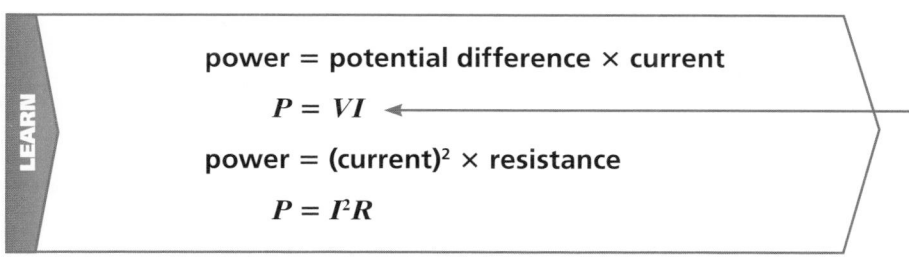

power = potential difference × current

$$P = VI$$ ←

power = (current)² × resistance

$$P = I^2R$$

> Power (P) is measured in watts (W).
> Potential difference (V) is measured in volts (V).
> Current (I) is measured in amps (A).
> Resistance (R) is measured in ohms (Ω).

The heating element in a kettle produces an output power of 2300W when a potential difference of 230V is applied across it.

a) Calculate the current flowing across the element.

$$P = VI$$

$$I = \frac{P}{V}$$

$$I = \frac{2300}{230} = 10A$$

b) Work out the resistance of the element. ←

$$P = I^2R$$

$$R = \frac{P}{I^2}$$

$$R = \frac{2300}{10^2} = \frac{2300}{100} = 23\Omega$$

> You could also use $V = IR$ here:
> $$R = \frac{V}{I} = \frac{230}{10} = 23\Omega$$
> This is a useful way to double check your answers.

Key Point

> You could also use equivalent resistance to answer part **c)**:
>
> From the 12V supply, R_1 took 8V and R_2 took 4V. R_1 must be twice as difficult for current to flow through, i.e. it must have twice the resistance. Therefore, $R_2 = 2\Omega$

Quick Test

1. A series circuit with two resistors has a 6V supply. R_1 has a 5V potential difference across it. What can be said about the resistance of R_2 and what voltage will it have across it?
2. A torch bulb has a resistance of 6Ω. A current of 2A flows through it. Calculate the total power used by the bulb.
3. An oven has a 230V supply and a 30A current. Work out the power of the oven.

Key Words

series
parallel
power

Domestic Uses of Electricity

You must be able to:

- Explain the difference between direct and alternating potential difference
- Describe a three-core cable
- Explain why a live wire may be dangerous even when a switch in the circuit is open
- Calculate the power of a device
- HT Understand how efficiency can be increased.

Direct and Alternating Potential Difference

- A **direct current (d.c.)** supply:
  - has a direct potential difference, i.e. one that is always positive or always negative – this makes the current direction constant
  - is the type of current that is supplied by cells and batteries.
- An **alternating current (a.c.)** supply:
  - has an alternating potential difference, i.e. one that alternates from positive to negative – this makes the current direction alternate
  - is the type of current used in mains electricity.

Mains Electricity

- Mains electricity in the UK is 230V and changes direction 50 times a second, i.e. it has a frequency of 50Hz.
- The mains supply uses three-core cable, i.e. the cable contains three wires.
- Each wire carries a different **electrical potential** and is colour-coded:
  - live wire (brown) – 230V potential
  - neutral wire (blue) – at or close to the 0V earth potential
  - earth wire (green and yellow stripes) – 0V potential.
- During operation:
  - the potential difference causes current to flow through the live and the neutral wires
  - the live wire carries the alternating potential from the supply
  - the neutral wire completes the circuit
  - current will only flow in the earth wire if there is a fault connecting it to a non-zero potential.
- The earth wire is a safety wire, which stops the exterior of an appliance becoming live.

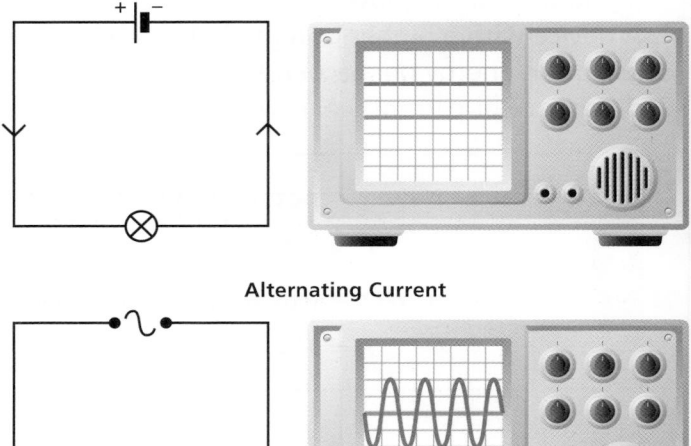

Direct Current

Alternating Current

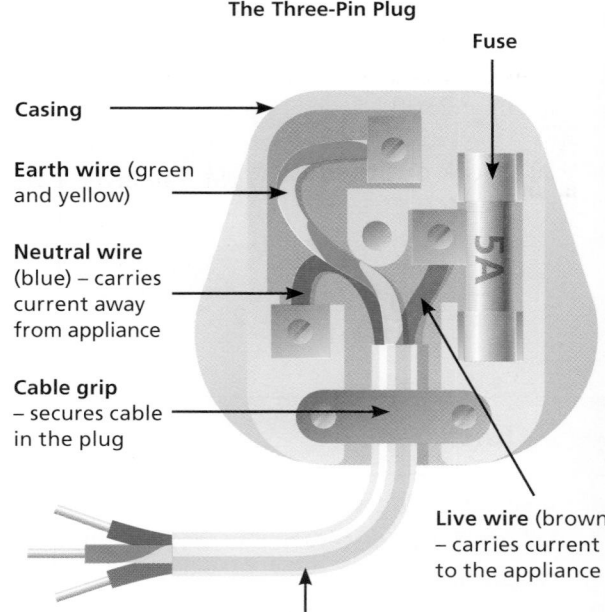

The Three-Pin Plug

Fuse

Casing

Earth wire (green and yellow)

Neutral wire (blue) – carries current away from appliance

Cable grip – secures cable in the plug

Live wire (brown) – carries current to the appliance

Cable

Dangers of Mains Electricity

- Mains electricity can be very dangerous – an electric shock from a mains supply can easily be fatal.

- Touching the live wire can create a large potential difference across the body and result in a large current flowing through the body.
- The live wire can be dangerous even if a switch in the circuit is open.
- For example, a television might be switched off (so no current flows), but still plugged in and switched on at the wall:
  - The live wire between the wall and the switch on the television is still at an alternating potential.
  - All it needs is a path for the electricity to flow through.
  - This path could be provided by a damaged cable exposing the live wire.
  - If someone then touches the live wire, creating a potential difference from the live to the earth and causing current to flow, they will get an electric shock.

> **Key Point**
>
> Any connection between live and earth can be dangerous, because it creates a potential difference, causing a current to flow.

Power and Efficiency

- **Power** is the rate at which energy is transferred or work is done:

LEARN

$$\text{power} = \frac{\text{energy transferred}}{\text{time}} \quad \text{or} \quad \text{power} = \frac{\text{work done}}{\text{time}}$$

$$P = \frac{E}{t} \qquad\qquad P = \frac{W}{t}$$

> Power (P) is measured in watts (W).
> Energy transferred (E) is measured in joules (J).
> Work done (W) is measured in joules (J).
> Time (t) is measured in seconds (s).

- An energy transfer of 1J per second is equal to 1W of power.
- You must be able to demonstrate the meaning of power by comparing two things that use the same amount of energy but over different times.
- For example, if two kettles are used to bring the same amount of water to the boil and one takes less time, it is because it has a higher power.
- In an energy transfer, **efficiency** is the ratio of useful energy out to total energy in:
  - An efficiency of 0.5 or 50% means that half the energy is useful, but half is wasted.
  - An efficiency of 0.75 or 75% means that three-quarters of the energy is useful, but a quarter is wasted.

LEARN

$$\text{efficiency} = \frac{\text{useful energy transfer}}{\text{total energy transfer}}$$

$$\text{efficiency} = \frac{\text{useful power output}}{\text{total power input}}$$

> **Key Point**
>
> Power is the rate at which energy is transferred.

HT To increase the efficiency of an energy transfer, the amount of wasted energy needs to be reduced.

> **Quick Test**
>
> 1. Describe the difference between alternating current and direct current.
> 2. What is the voltage of the UK mains electricity supply?
> 3. A 2.5V bulb has a current of 1.2A flowing through it. What is the power of the bulb and how much energy would it use in 10s?
> 4. How much energy is transferred by a 2kW kettle in 50s?

> **Key Words**
>
> direct current (d.c.)
> alternating current (a.c.)
> electrical potential
> power
> efficiency

Electrical Energy in Devices

You must be able to:

- Describe how different domestic appliances transfer energy
- Calculate the energy transferred by a device
- Describe the structure of the national grid
- Explain why the national grid is efficient.

Energy Transfers in Appliances

- Whenever a charge flows, it has to overcome the resistance of the circuit. This requires energy, therefore:
  - work is done when charge flows
  - the amount of work done depends on the amount of charge that flows and the potential difference.
- The amount of energy transferred can also be found from the power of the appliance and how long it is used for, e.g. a 20W bulb uses 20J of energy in every second.

LEARN

energy transferred = power × time

$$E = Pt$$

energy transferred = charge flow × potential difference

$$E = QV$$

Energy transferred (E) is measured in joules (J).
Power (P) is measured in watts (W).
Time (t) is measured in seconds (s).
Charge flow (Q) is measured in coulombs (C).
Potential difference (V) is measured in volts (V).

A 2kW heater is on for one hour. How much energy does it use?

1 hour = 60 minutes × 60 seconds = 3600 seconds
2kW = 2 × 1000W = 2000W

$E = Pt$
= 2000 × 3600
= 7 200 000J

Start by converting the values into watts and seconds.

The given values are for power and time, so use the first equation.

Substitute in the values.

- Electrical appliances are designed to cause energy transfers.
- The type and amount of energy transferred between stores depends on the appliance.

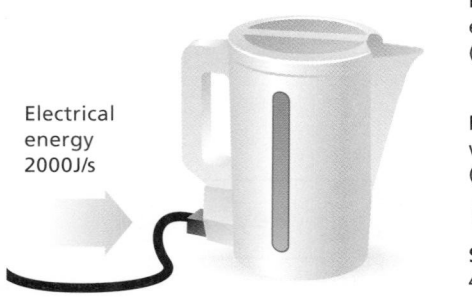

Electrical energy 2000J/s

Heat energy (for element) 160J/s (wasted)

Heat energy (to water) 1800J/s (useful)

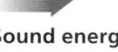

Sound energy 40J/s (wasted)

Key Point

The energy transferred by an appliance depends on the power and the time it is on for.

The National Grid

- The **National Grid** is a system of cables and **transformers** linking power stations to homes and businesses.

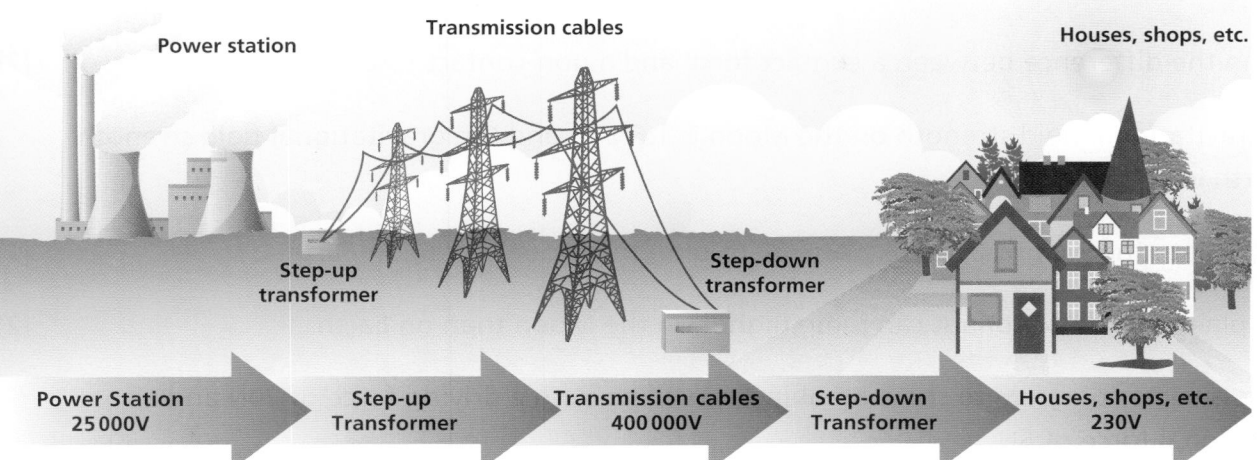

- Each component of the grid has a particular function.
- **Power station**:
  - The power station transfers the energy supply into electrical energy.
  - Using a smaller number of large power stations is more efficient than building many small, local power stations, because large stations can be made more efficient.
  - This is because most power plants use steam turbines, which are more efficient at higher steam temperatures, and the bigger the plant, the bigger the boiler, so the higher the steam temperature.
- **Step up transformers**:
  - The transformers increase the potential difference from the power station to the transmission cables.
  - This reduces the current and, therefore, reduces the heating effect caused by current flowing in the transmission cables.
  - Reducing the heating effect, reduces energy loss so makes the transmission more efficient.
- **Transmission cables**:
  - Transmission cables transfer the electricity.
- **Step down transformers**:
  - The transformers reduce the potential difference from the transmission cables to a much lower value for domestic use.

 Key Point

Electrical power is produced and transferred efficiently to consumers using the National Grid.

Quick Test

1. A bulb transfers 30J of energy in 10 seconds. During this time a charge of 12C is transferred. What is the voltage of the bulb?
2. Use a Sankey diagram to represent the energy transfers that take place in a washing machine that is 50% efficient.
3. A washing machine intensive cycle takes two hours. If the power is 500W, how much energy is transferred?
4. Give **two** ways in which the National Grid is designed to be an efficient way to transfer energy.

 Key Words

National Grid
transformer

Review Questions

Forces – An Introduction

1 Explain the difference between a contact force and a non-contact force. **[2]**

2 The gravitational field strength on the Moon is 1.6N/kg and the gravitational field strength of Earth is 10N/kg.

    a) Calculate the weight of a 70kg astronaut on **i)** the Moon and **ii)** on Earth. **[2]**

    b) Explain why an astronaut can jump higher on the Moon than on Earth. **[2]**

3 HT Use a vector diagram to show an object travelling with a driving force of 20N and a frictional force of 5N.

    Your diagram should show:
- The driving force
- The frictional force
- The resultant force
- The size of the resultant force. **[4]**

> **Total Marks** _____ / 10

Forces in Action

1 In a tall building, the height between floors is 3.5m.
The lift car that carries people between floors weighs 1200N.

Calculate the work done by the engines when the lift car is raised up five floors. **[3]**

2 **Figure 1** shows a flight of stairs.

Calculate the work done by a 40kg child climbing the stairs (g = 10N/kg). **[3]**

Figure 1

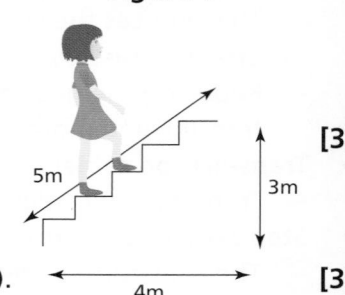

5m, 3m, 4m

3 Explain why at least two forces are needed to stretch a spring. **[2]**

4 Using springs as an example, explain what is meant by the 'limit of proportionality'. **[2]**

5 A student carries out an experiment to investigate the extension of a rubber band.

The results are plotted in a graph of force over extension.
The line produced is curved and gets steeper as the force increases.

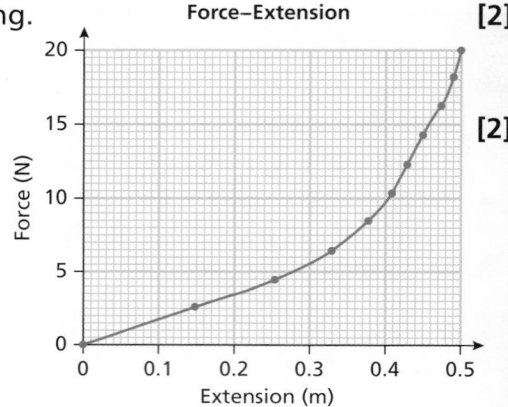

Force–Extension

Force (N) vs Extension (m)

What conclusions can be drawn from the graph? **[3]**

Total Marks _____ / 13

Forces and Motion

1 Describe the difference between speed and velocity. **[2]**

2 A hiker travels south for three miles, west for two miles and then north for one mile.

a) What is the total distance travelled by the hiker? **[1]**

b) HT Use a scale vector diagram to show the final displacement of the hiker. **[2]**

3 A sound wave travelling in the ocean takes 3 seconds to travel 4.5km.

Calculate the speed of sound in water in metres per second (m/s). **[2]**

4 On a distance–time graph, what do the following features represent?

a) A straight line with a very steep gradient. **[1]**

b) A horizontal line. **[1]**

c) A curved line. **[1]**

5 HT Using an example, explain how an object can travel at constant speed but with a changing velocity. **[3]**

6 HT What name is given to the tendency of objects at rest to remain at rest and for moving objects to keep moving? **[1]**

Total Marks _____ / 14

Forces and Acceleration

1 Write down whether each of the following statements is **true** or **false**.

a) If the resultant force on an object is doubled, its acceleration will double. **[1]**

b) If the force on an object is constant and the mass of the object increases, the acceleration will also increase. **[1]**

c) Newton's second law states that for every force there is an equal and opposite reaction force. **[1]**

Review Questions

2 A car travelling at 20m/s decelerates at a constant rate of 2m/s².
Use the formula $v^2 - u^2 = 2as$ to calculate how far the car travels before coming to rest. **[3]**

3 State what the following features would represent on a velocity–time graph.

a) A straight line with a positive gradient.

b) A straight line with a negative gradient.

c) A horizontal line above the x-axis. **[2]**

d) A horizontal line below the x-axis. **[2]**

e) HT The area under the graph. **[1]**

Total Marks _____ / 11

Terminal Velocity and Momentum

1 Use Newton's third law and the idea of equal and opposite forces to explain why a hovercraft moves forward when the propeller spins. **[3]**

2 HT Write down the formula that links momentum, mass and velocity. **[1]**

3 HT A 240kg cannon fires a 1kg cannonball with a velocity of 120m/s.

Work out the velocity at which the cannon recoils backwards when it is fired. **[4]**

Total Marks _____ / 8

Stopping and Braking

1 Give **three** factors that would have a negative effect on the braking distance of a vehicle. **[3]**

2 HT Two students, Lisa and Jack, are discussing the braking of vehicles.
Jack thinks that if the braking force is the same but a car is going twice as fast, it will take twice the distance to stop.

Is he correct? You must explain your answer. **[3]**

3 Explain what is meant by 'reaction time' and how this applies to thinking and stopping distances. **[3]**

4 A car travelling at 12m/s has a braking distance of 14m.

If the driver's reaction time is 0.5 seconds, what is the total stopping distance? **[3]**

Total Marks _____ / 12

Energy Stores and Transfers

1 On a rollercoaster, the carts are winched to the top of the ride, 16.2m above the start point. Use 10N/kg for gravitational field strength.

a) The carts and riders on the rollercoaster have a combined mass of 2200kg.

Calculate the work done by the motor in winching the carts to the top of the ride (assume the winching process is 100% efficient). **[3]**

b) The bottom of the first drop of the rollercoaster is at the same height as the start point.

Assuming no energy is lost as the carts move down the track, state how much kinetic energy the carts will have when they reach the bottom. **[1]**

c) Use your answer to part **b)** to work out the velocity of the carts at the bottom of the first drop. **[2]**

2 For this question you will need to refer to the Physics Equations on page 273.
A catapult has a spring constant of 100N/m and is stretched 60cm before being fired.

a) Calculate how much energy has been stored in the catapult. **[3]**

b) Use your answer to part **a)** to work out the velocity of a 57.6g ball bearing fired from the catapult. Assume all of the energy from the catapult goes into the ball bearing. **[3]**

c) When actually measured, it is found that the velocity of the ball bearing is much slower than the calculated value. Suggest why this might be. **[2]**

Total Marks _____ / 14

Energy Transfers and Resources

1 Complete **Table 1** to show the main energy transfers that occur in the situations given.

Table 1

Input Energy	Situation	Useful Output Energy
	Walking uphill	
	Sliding down a slide	
	Firing a catapult	
	Fireworks / sparklers	

[4]

Total Marks _____ / 4

Practice Questions

Waves and Wave Properties

1. Give **one** example to show that waves do not transfer matter. [2]

2. Complete the sentences to describe the process by which energy is transferred by a wave. Use answers from the box.

oscillate	fixed	particles	energy

   The _____ that make up a wave _____ about a _____ point.

   When doing this, they pass _____ onto adjacent particles and start them oscillating. [4]

3. **Figure 1** shows a wave.

   Figure 1

   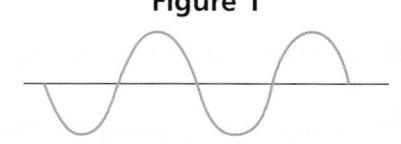

   a) Add labels to **Figure 1** to show the wavelength and amplitude. [2]

   b) How many complete waves are shown in **Figure 1**? [1]

   c) The waves shown have a period of 2 seconds.

   How long would the waves pictured take to pass a fixed point? [1]

4. A low frequency sound wave has a time period of 0.05s and travels at a speed 330m/s.

   $$\text{period} = \frac{1}{\text{frequency}}$$

   a) Calculate the frequency of the wave. [2]

   b) Use the wave equation to calculate the wavelength of the wave. [2]

5. Describe a method that can be used to measure the speed of sound in air. [4]

   Total Marks _____ / 18

Electromagnetic Waves

1. **Figure 1** shows a ray of light entering a glass block.

   Complete **Figure 1** to show:
   - The path of the light ray as it passes through the block.
   - The normal where the ray enters and exits the block.

   Figure 1

   [4]

2 HT What happens to a wave as it passes from one medium to another?
Circle **one** answer.

Its frequency changes   Nothing happens   Its speed changes   [1]

3 How does the speed of light in a vacuum differ from the speed of light in glass?   [1]

Total Marks _____ / 6

The Electromagnetic Spectrum

1 a) Mobile phones send signals using electromagnetic waves.

Which type of electromagnetic wave is used?   [1]

b) A scientist being interviewed for television says:

I think that using mobile phones too much increases the risk of developing a brain tumour.

Is this a conclusion, a fact or an opinion?   [1]

c) A study on the link between mobile phones and brain tumours observed 420 000 adults over a 10-year period.
The study found no correlation between the amount of mobile phone use and brain tumours.

i) Which of the following conclusions is most accurate?
Tick **one** box.

Short to medium-term use by adults does not cause brain tumours. ☐

Mobile phones are completely safe to use. ☐

Children should not use mobile phones. ☐

It is not safe for adults to use mobile phones. ☐   [1]

ii) Would you expect mobile phone use to result in cancer?
Use your knowledge of electromagnetic waves and their hazards to explain your answer.   [3]

Total Marks _____ / 6

An Introduction to Electricity

1 Some students investigate the factors that affect the resistance of a light dependent resistor. One student uses a circuit containing a variable resistor connected to a 12V supply and adjusts the resistor to set the output voltages.

Another student uses a lab pack with outputs 2V, 4V, 6V, 8V, 10V and 12V to set their voltages.

a) What advantage could using a variable resistor have over using a lab pack with a range of fixed outputs? **[2]**

b) During this experiment, the voltage and current across the resistor are measured.

   Why is it important to keep the current low? **[2]**

c) **Table 1** shows the results of the experiment.

Table 1

Light Level	Voltage	Current
Total darkness	10V	0.2A
Low light	6V	0.3A
Maximum brightness	0.4V	0.4A

Use the results to calculate the maximum change in resistance of a light dependent resistor. **[2]**

Total Marks _____ / 6

Circuits and Resistance

1 **Figure 1** shows the $V–I$ graph for an ohmic conductor.

a) What can be deduced from **Figure 1** about the resistance of an ohmic conductor? **[1]**

b) Add a second line to **Figure 1** to represent a different ohmic conductor with a higher resistance than the original. **[2]**

Figure 1

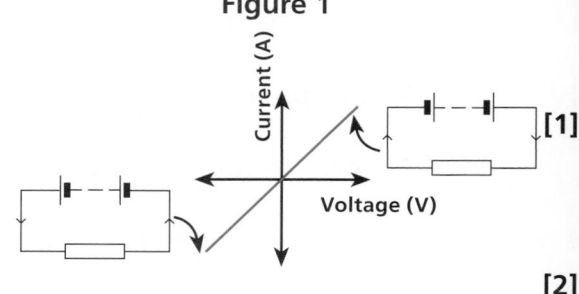

2 A circuit is constructed with a battery and LED. The current through the LED is found to be 1.2A.

What would the current through the LED be if the battery was reversed? **[1]**

Total Marks _____ / 4

Circuits and Power

1 **a)** Draw a series circuit with a 6V battery and three bulbs. [2]

b) Each bulb in your series circuit is identical. Work out the potential difference across each bulb. [1]

c) The current through the battery is 2A. Work out the current through each bulb. [1]

d) Use your answers to parts **a)** and **b)** to calculate the resistance of each bulb. [3]

e) What is the total resistance of the circuit? [1]

Total Marks .... / 8

Domestic Uses of Electricity

1 Name the **three** wires in a UK mains cable and, for each one, give its colour. [3]

2 Explain why touching a live wire will cause a large current to flow through the person touching it. You must refer to electrical potential and potential difference in your answer. [4]

3 What is the function of a neutral wire in a circuit? [1]

4 In a plug, at what potential is the earth wire? [1]

Total Marks .... / 9

Electrical Energy in Devices

1 **Table 1** gives the power rating and daily usage for a number of different appliances.

Table 1

Appliance	Power Rating	Total Hours Used Per Day
Kettle	2000W	1
Tumble Dryer	1800W	3
Television	500W	6
Oven	5000W	1

a) Which appliance uses the least energy per hour? [1]

b) Work out which appliance uses the most energy during one day. [2]

c) Calculate how many joules of energy are used by the oven per day. [4]

Total Marks .... / 7

Magnetism and Electromagnetism

You must be able to:

- Describe the difference between permanent and induced magnets
- Describe what happens when magnets interact
- Explain magnetic fields and magnetic field patterns
- Describe how the magnetic effect of a current can be demonstrated
- Explain why a solenoid can increase the magnetic effect of the current.

Magnetic Poles and Fields

- There are two types of magnetic **pole**: a north (seeking) pole and a south (seeking) pole.
- The poles of a magnet are the places where the magnetic forces are strongest.
- Unlike (opposite) poles attract – the north pole of a magnet will attract the south pole of another magnet.
- Like poles repel – a north pole will repel a north pole and a south pole will repel a south pole.
- The region around a magnet, where a force acts on another magnet or magnetic material (e.g. iron, steel or nickel cobalt), is called the magnetic field.
- The strength of the field depends on the distance from the magnet – it is strongest at the poles
- **Permanent magnets** produce their own magnetic field.
- **Induced magnets** become a magnet when placed in a magnetic field. When removed from the field, they lose their magnetism quickly.
- The force between a permanent magnet and a magnetic material or an induced magnet is always one of attraction.
- The arrows on field lines always run from north to south and show the direction of the force that would act on a north pole placed at that point.
- The density of the field lines is called the **flux density** and indicates the strength of the field at that point – the closer together the lines, the higher the flux density.
- The higher the flux density, the stronger the field and the greater the force that would be felt by another magnet.

> **Key Point**
>
> Like poles repel. Unlike poles attract.

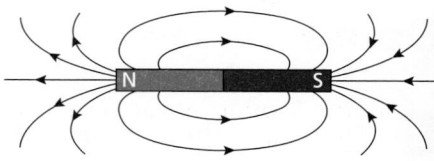

Plotting Fields

- A magnetic compass contains a small bar magnet.
- The compass needle aligns with the Earth's magnetic field and always points to the magnetic north.
- This provides evidence that the Earth's core must be magnetic.
- A magnetic compass can be used to plot the field around a bar magnet:
  1. Place the bar magnet on a piece of paper.
  2. Place the compass at one end of the magnet.
  3. On the paper, mark where the point of the compass needle is.
  4. Move the compass so the tail of the needle is at the point that has just been marked.

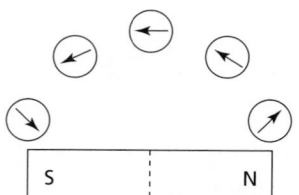

⑤ On the paper, mark a new point where the needle is.
⑥ Repeat and connect the marks until the full field is plotted.

Electromagnetism and Solenoids

- Whenever a current flows in a conducting wire, a magnetic field is produced around the wire.
- The direction of the field lines depends on the direction of the current and can be found with the right hand grip method:
  - Grip the wire in your right hand, with the thumb pointing in the direction of the current.
  - The fingers curled around the wire will point in the direction that the field lines should be drawn
- The strength of the field depends on the size of the current and the distance from the wire.
- This effect can be seen by placing a magnetic compass at different points along the wire and turning the power supply on an off.
- A solenoid is formed when a wire is looped into a cylindrical coil.
- Shaping the wire into a solenoid increases the strength of the magnetic field, creating a strong uniform field inside the solenoid.
- To increase the field strength further, an iron core can be added. This creates an electromagnet.
- The solenoid increases the magnetic field strength because:
  - it concentrates a longer piece of wire into a smaller area
  - the looped shape means that the magnetic field lines around the wire are all in the same direction.
- The magnetic field around a solenoid has a similar shape to that around a bar magnet.
- The north pole of a solenoid can be found with the right hand grip method:
  - Hold the solenoid in your right hand with your fingers following the direction the current flows.
  - Your thumb will point to the north pole of the solenoid.

Electromagnetic Devices

- Many devices use electromagnets.
- You need to be able to interpret diagrams to determine how they work. For example, here is an electric bell:
  ① When the switch is pushed, the electromagnet is magnetised.
  ② The electromagnet attracts the armature.
  ③ The hammer strikes the gong and breaks the circuit.
  ④ The armature springs back, completing the circuit again and remagnetising the electromagnet.
  ⑤ The cycle repeats for as long as the button remains pushed.

Right Hand Grip Method

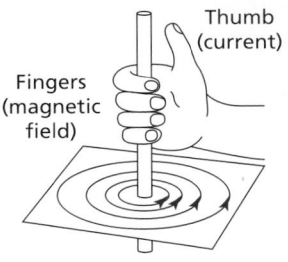

Thumb (current)

Fingers (magnetic field)

A Solenoid

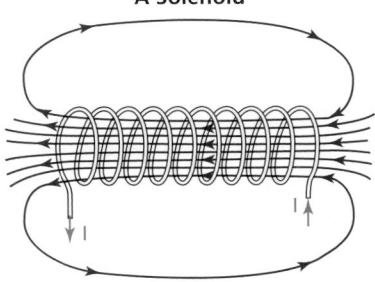

Right Hand Grip Method

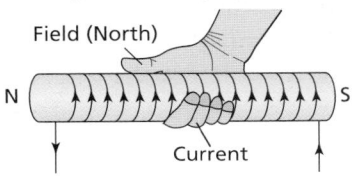

Field (North)

N   S

Current

Electric Bell

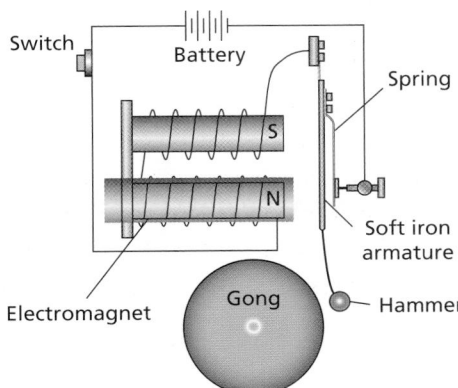

Switch

Battery

Spring

S

N

Soft iron armature

Electromagnet

Gong

Hammer

Quick Test

1. The field lines around a magnet are drawn and the lines are very close together. What does this tell you about the strength of the magnet?
2. Sketch the field lines around a current carrying straight wire.
3. How could the strength of an electromagnet be increased?

Key Words

pole
permanent magnet
induced magnet
flux density
solenoid
uniform
electromagnet

The Motor Effect

You must be able to:

- **HT** Recall and use Fleming's left hand rule
- **HT** Calculate the force felt by a current carrying conductor in a magnetic field
- **HT** Explain how rotation is caused in an electric motor.

Fleming's Left Hand Rule

- When a current carrying conductor is placed in a magnetic field it experiences a force. This is called the **motor effect**.
- The motor effect is caused by the field created by the current interacting with the magnetic field.
- The force can be increased by increasing either:
  - the size of the current
  - the length of conductor in the magnetic field
  - the **flux density** (see page 206).

Creating a Force

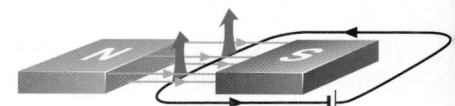

> force =
> magnetic flux density × current × length (of wire within the field)
> $F = BIl$

Force (F) is measured in newtons (N).
Magnetic flux density (B) is measured in tesla (T).
Current (I) is measured in amps (A).
Length (l) is measured in metres (m).

A 10cm length of wire with a 4A current flowing passes through a magnetic field. What magnetic flux density is needed to create a 2N force on the wire?

$$F = BIl$$

$$B = \frac{F}{Il}$$

$$B = \frac{2}{4 \times 0.1} = \frac{2}{0.4} = 5T$$

Rearrange the equation to make B the subject.

Convert the length from cm to m.

- Reversing the direction of either the current or the magnetic field will cause the direction of the force to reverse.

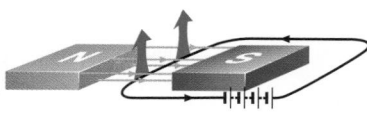

Several cells (bigger voltage) produce a big current and, therefore, a large force.

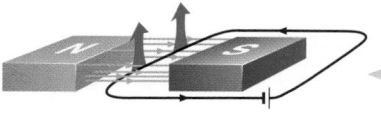

A large force can also be produced with stronger magnets, giving a higher flux density.

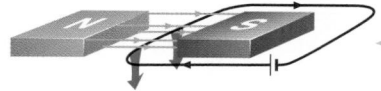

Reversing the cell reverses the current and so reverses the forces.

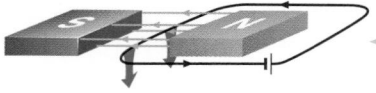

Reversing the magnets reverses the field and so reverses the direction.

- The direction of the force on the conductor can be found using Fleming's left hand rule:
  1. Hold the left hand so that the thumb, first finger and second finger are all at right-angles to one another.
  2. Align the first finger, so that it points in the direction of the magnetic field, from north to south.
  3. Rotate the wrist so that the second finger points along the wire in the direction the current is flowing.
  4. The thumb will be pointing in the direction of the force, i.e. the direction the conductor would move.
- The left hand rule can be used to find the direction of either the field, current or movement as long as the other two are known.

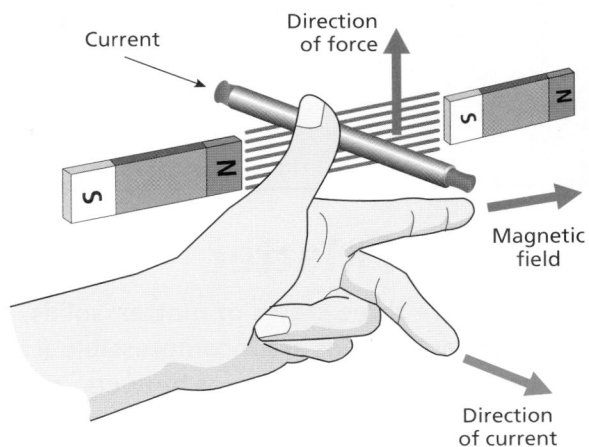

Fleming's Left Hand Rule

Electric Motors

- A current carrying coil in a magnetic field will rotate.
- This is because the current going up one side of the coil is in the opposite direction to the current coming back down the other side, so one side moves up and the other moves down.
- This is the basis of an electric motor, which is designed so that the coil rotates continuously.

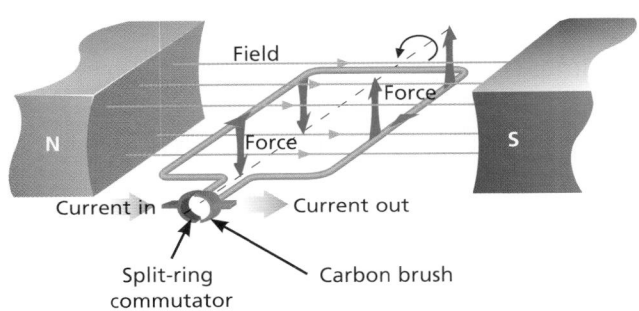

- The brush contacts at the **commutator** ensure that the current direction in the coil is always in the same direction.
- This ensures that the motor continues rotating and does not simply stop in the upright position.
- Fleming's left hand rule can be used on one side of the coil to work out which direction the motor will rotate.
- Increasing the current or the magnetic field will make the motor rotate faster.
- Reversing the current or the magnetic field will make the motor rotate in the opposite direction.

HT **Quick Test**

1. How can the speed of a motor be increased?
2. Why will a motor not work without a commutator?
3. A high power electric motor has a magnetic flux density of 2T and a current of 10A. If the total length of wire in the coil is 8m, calculate the force on the motor.

Particle Model of Matter

You must be able to:

- Draw simple diagrams to model the different states
- Describe how to investigate the density of objects
- Explain what is meant by 'internal energy' and how this links to temperature and changes of state
- Use the particle model to explain pressure in gases.

States of Matter

- There are three states of matter: solids, liquids and gases.
- Solids and liquids are incompressible (cannot be squashed) because there are no gaps between the particles in them.
- Solids contain particles in a fixed pattern and have a fixed size and shape.
- Liquids have a fixed size but contain particles that are free to move, allowing them to change shape to fit their container.
- Gases have large gaps between the particles, making them compressible and enabling them to change size and shape.

Gas

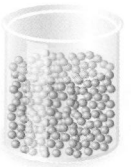
Liquid

Solid

Temperature decreases

Density

- The density of a material is its mass per unit volume.

$$\text{density} = \frac{\text{mass}}{\text{volume}}$$

$$\rho = \frac{m}{V}$$

Density (ρ) is measured in kilograms per metre cubed (kg/m³).

Mass (m) is measured in kilograms (kg).

Volume (V) is measured in metres cubed (m³).

REQUIRED PRACTICAL	
Investigate the density of regular and irregular solids and liquids.	
Sample Method 1. Set the equipment up as shown. 2. Record the height of the water in the measuring cylinder and the mass of the solid / liquid being tested. 3. Add the solid / liquid being tested to the measuring cylinder. 4. Record the new height in the measuring cylinder. 5. Subtracting the original height from the new height gives the volume of the solid / liquid being tested. 6. Now the density can be calculated.	**Considerations, Mistakes and Errors** • If a solid that is less dense than water is tested, the volume measurement will be incorrect because the solid will not be fully submerged. • When reading from the measuring cylinder, the reading should be taken from the bottom of the meniscus. • The temperature of the water must be exactly the same throughout all tests, as an increase in temperature could cause the material or water to change volume slightly through expansion.
Variables • The independent variable is the material being tested. • The dependent variables are the volume and mass. • The control variable is the temperature.	**Hazards and risks** • There are very few hazards, unless the materials being tested are hazardous or react with water. • The main hazard could be a slip hazard if water is spilt.

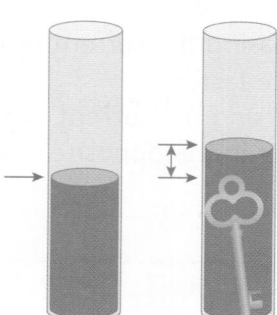

Change of State and Specific Latent Heat

- As energy is added to a system, its temperature will increase.
- At the melting or boiling point, the energy being added causes the substance to change state not temperature.
- A change of state is reversible.
- It is a physical change that alters the internal energy but not the temperature or mass. It is not a chemical change.

> **energy for a change of state = mass × specific latent heat**
>
> $$E = mL$$

Energy (E) is measured in joules (J).
Mass (m) is measured in kilograms (kg).
Specific latent heat (L) is measured in joules per kilogram (J/kg).

- The **latent heat of fusion** is the energy needed for a substance to change from solid to liquid (melt).
- The **latent heat of vaporisation** is the energy needed for a substance to change from liquid to gas (evaporate).
- The horizontal parts of heating and cooling graphs indicate where energy is being used to change state.

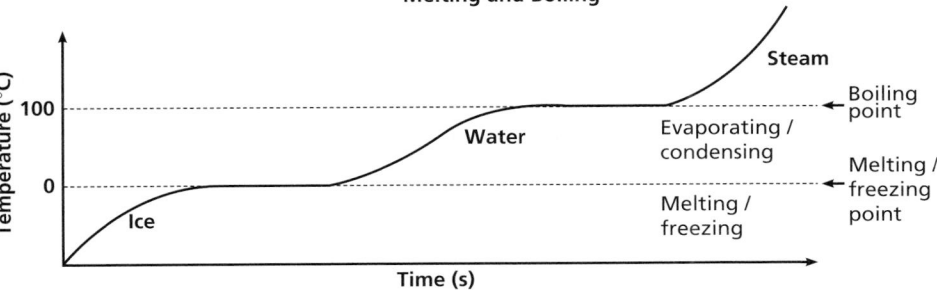

Melting and Boiling

- At certain pressures, some substances have a boiling point that is the same or lower than their melting point.
- These substances change from a solid to a gas without having a liquid phase in the middle.
- This is called **sublimation**.
- The reverse process, changing from a gas to a solid, is called desublimation.

Particle Motion and Pressure in Gases

- The particles of a gas are in constant random motion and its temperature is related to the average kinetic energy of the particles.
- When the particles of a gas collide with the walls of their container they exert a force on the wall, which is felt as pressure.
- If the volume is kept constant, increasing the temperature increases the speed of the particles. This increases the frequency and force with which the particles hit the walls and increases the pressure.

Quick Test

1. Why are solids and liquids incompressible?
2. How could you find the density of an irregular solid?
3. Why would an increase in temperature affect the pressure of a gas?

Atoms and Isotopes

You must be able to:

- Recall the approximate size of an atom
- Describe the structure of an atom
- Explain what an isotope is
- Describe how isotopes of the same element are different
- Describe the plum pudding model of the atom
- Describe evidence that has led to the nuclear model of the atom.

The Structure of the Atom

- **Atoms** are very small with a radius of around 1×10^{-10} metres.
- Atoms contain a positively charged **nucleus** made up of **protons** and **neutrons**, which is surrounded by negatively charged **electrons**.
- Protons have an electrical charge of +1 and electrons have a charge of −1.
- Atoms have equal numbers of electrons and protons, so they have no overall electrical charge.
- The nucleus of the atom contains most of the mass, but its radius is less than $\frac{1}{10\,000}$ of the radius of the atom.
- The electrons are arranged at different distances from the nucleus (in different energy levels).
- The energy level of an electron may change when the atom emits or absorbs electromagnetic radiation:
  - Absorbing electromagnetic radiation moves electrons to a higher energy level, further from the nucleus.
  - Electromagnetic radiation is emitted when an electron drops to a lower energy level.
- An atom that loses one of its outer electrons becomes a positive **ion**.
- If it gains an extra electron it becomes a negative ion.

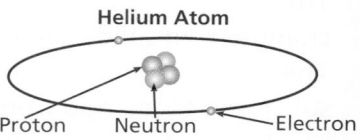

Helium Atom

Proton Neutron Electron

Isotopes

- All atoms of a particular **element** have the same number of protons.
- The number of protons in an atom of an element is called its **atomic number**.
- The total number of protons and neutrons in an atom is called its mass number.
- Atoms are represented in the following way:
- Atoms of the same element can have different numbers of neutrons.
- These atoms are called **isotopes**.
- For example, carbon has two common isotopes:
  - carbon-12, which contains 6 protons and 6 neutrons
  - carbon-14, which contains 6 protons and 8 neutrons.
- These isotopes are represented with the following symbols:

$$^{14}_{6}C \qquad ^{12}_{6}C$$

- Note that in both examples, the atomic number is 6 but the mass number is different.

Mass number → 4
Atomic number → 2 **He** ← Element symbol

Key Point

All isotopes of an element contain the same number of protons, otherwise they would be different elements. It is the number of neutrons that is different.

The Plum Pudding Model

- The model of the atom has changed over the years.
- Atoms were once thought to be tiny spheres that could not be divided.
- The discovery of the electron by J. J. Thompson, in 1897, led to the plum pudding model of the atom, which depicts the atom as a ball of positive charge with electrons embedded in it, like plums in a pudding.

Plum Pudding Model

Rutherford, Geiger and Marsden

- In 1905, Geiger and Marsden carried out Rutherford's experiment: bombarding thin gold foil with alpha particles.
- If the plum pudding model was correct, the heavy, positively charged alpha particles would have passed straight through.
- Most particles did pass through, but not all.
- Some of the alpha particles were deflected, so they must have come close to a concentration of charge, unlike the spread out charge described by the plum pudding model.
- Some alpha particles were reflected back, so:
  - they must have been repelled by the same charge that the alpha particles carried
  - the repelling charge must have been much heavier than the alpha particle, or the alpha particle would have passed through.
- The conclusion was that:
  - the mass of the atom was concentrated in a central nucleus, which was positively charged
  - the electrons surround this nucleus.

Rutherford, Geiger and Marsden Experiment

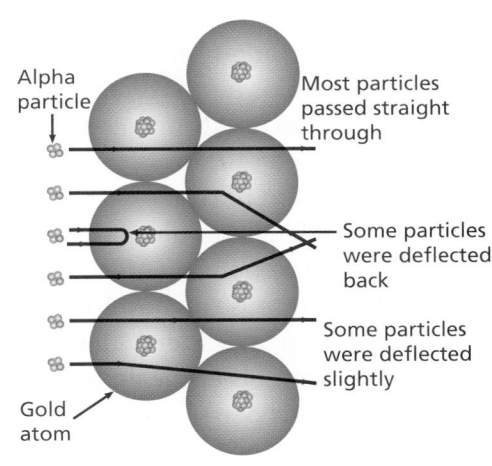

Alpha particle

Most particles passed straight through

Some particles were deflected back

Some particles were deflected slightly

Gold atom

Further Developments

- Niels Bohr adapted this nuclear model by suggesting that electrons orbit the nucleus at specific distances.
- Later experiments led to the idea that the positive charge of the nucleus can be divided into a whole number of smaller particles, each with the same amount of positive charge.
- These particles were given the name 'proton'.
- In 1932, 20 years after the nuclear model was accepted, James Chadwick carried out a number of experiments.
- These provided evidence that within the nucleus there was another particle, which was called the 'neutron', leading to further refinement of the nuclear model for the structure of the atom.
- The structure of the atom is an example of a theory which, through experiment and peer review, has changed and developed over time.

> **Key Point**
>
> New experimental evidence can lead to a scientific model being changed or replaced over time.

> **Key Words**
>
> atom
> nucleus
> proton
> neutron
> electron
> ion
> element
> atomic number
> isotopes

> **Quick Test**
>
> 1. What charge does a proton carry?
> 2. How is an isotope different from an element?
> 3. What does the mass number tell us about an element?
> 4. a) What was the most surprising result from the Rutherford and Marsden experiment?
>   b) What did this result mean?

Nuclear Radiation

You must be able to:

- Explain what is meant by count-rate
- Describe what alpha, beta and gamma radiation are
- Describe the properties of alpha, beta and gamma radiation
- Explain the difference between irradiation and contamination.

Nuclear Decay and Radiation

- Some atomic nuclei are **unstable** and give out radiation in order to become more stable.
- The type of radiation emitted depends on why the nucleus is unstable and is a random process (it is not possible to predict exactly when an atom will decay).
- The **activity** of a **radioactive** source is the rate at which it decays. It is measured in becquerels (Bq).
- One becquerel is equivalent to one decay per second, i.e. 1Bq = 1 decay per second.
- The count-rate is the number of decays recorded each second by a detector (e.g. a Geiger-Muller tube).
- One becquerel is equivalent to one count per second, i.e. 1Bq = 1 count per second.

Alpha, Beta and Gamma Decay

- There are three main types of nuclear radiation: **alpha** (α), **beta** (β) and **gamma** (γ).

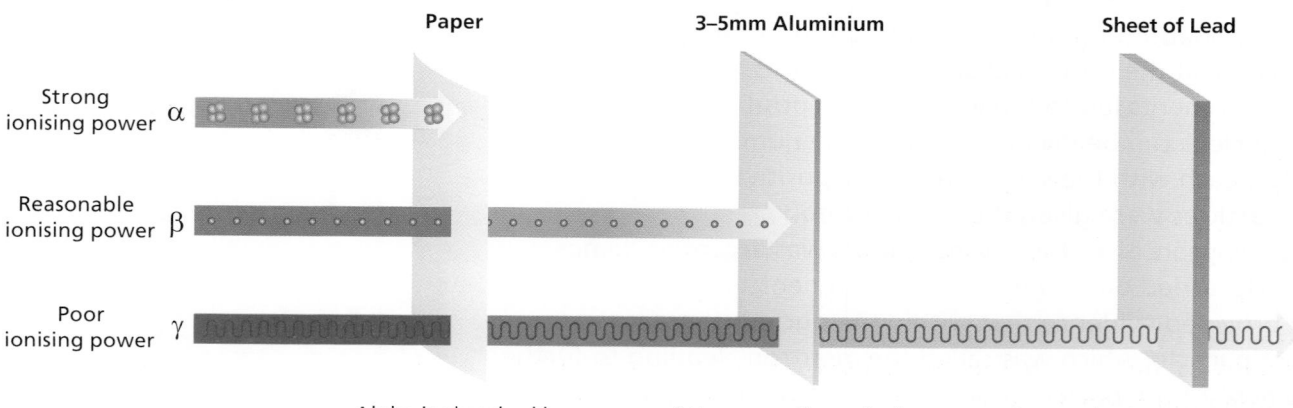

Alpha is absorbed by a few centimetres of air or a thin sheet of paper.	Beta passes through air and paper but is absorbed by a few millimetres of aluminium.	Gamma is very penetrating and needs many centimetres of lead or many metres of concrete to absorb most of it.

- Each type of radiation consists of different particles or electromagnetic radiation.
- All three types of nuclear radiation are ionising – they are capable of knocking electrons from atoms.
- Alpha is the most ionising and gamma is the least ionising.

Radiation Type	Components	Hazards
alpha (α)	• Two neutrons and two protons (the same as a helium nucleus). • Ejected from the nucleus.	• Highly likely to be absorbed and cause damage if passing through living cells.
beta (β)	• A high-speed electron. • Ejected from the nucleus as a neutron turns into a proton.	• Likely to cause damage if absorbed by living cells. • Can penetrate the body to inner organs.
gamma (γ)	• Electromagnetic radiation. • Emitted from the nucleus.	• Likely to pass through living cells without being absorbed and causing ionisation.

- A neutron (*n*) is the fourth type of nuclear radiation, which can be emitted during radioactive decay.

Radioactive Contamination

- Radioactive contamination is the unwanted presence of materials containing radioactive atoms on other materials.
- The hazard from the contamination is due to the decay of the contaminating atoms.
- The type of radiation emitted affects the hazard.
- Irradiation:
  - is the process of exposing an object to nuclear radiation
  - can be deliberate or accidental
  - does not cause the object to become radioactive.
- When using radioactive sources, it is important to protect against unwanted irradiation by:
  - using sources of the lowest activity possible for the shortest amount of time possible
  - wearing appropriate protective clothing such as a lead apron
  - not handling sources with bare hands.
- You need to be able to compare the hazards associated with contamination and irradiation, e.g.
  - Food contaminated with an alpha source would be more hazardous than food contaminated with a gamma source, because alpha radiation is more strongly ionising.
  - An area contaminated with an alpha source would not be dangerous, unless it was entered, due to the low penetration of alpha radiation. However, if it was contaminated with a source that emitted gamma radiation, this would irradiate people nearby.

Key Point

A contaminated object continues to give out radiation until decontaminated.

An irradiated object does not become radioactive.

Key Words

unstable
activity
radioactive
alpha
beta
gamma
contamination
irradiation

Quick Test

1. Which type of radiation is the most penetrative?
2. Alpha radiation is more ionising than beta radiation. However, it could be argued that beta radiation is more dangerous. Explain why.
3. How is radioactive contamination different from irradiation?

Half-Life

You must be able to:

- Work out the half-life of a radioactive isotope from given information
- Use balanced nuclear equations to illustrate radioactive decay.

Half-Life

- The random nature of radioactive decay makes it impossible to predict which nucleus will decay next.
- However, with a large enough number of nuclei, it is possible to predict how many will decay in a certain time period.
- The **half-life** of a radioactive isotope is:
  - the average time it takes for half of the nuclei to decay
  - the time it takes for the count rate, or activity, of a sample containing the isotope to fall to 50% of its original value.
- The graph below shows how the count rate of a sample of Iodine-128 changes over time.

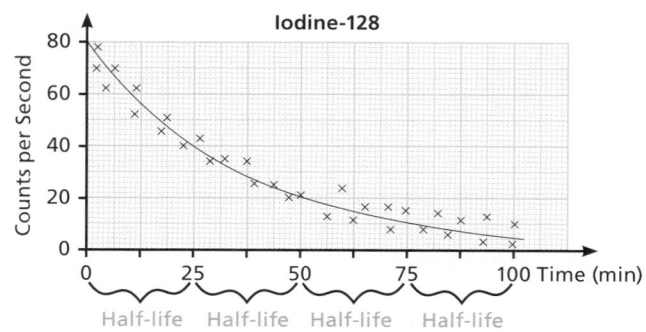

- The count rate takes 25 minutes to fall from 80 to 40, so its half-life is 25 minutes

> **Key Point**
>
> Half-life is a measure of how long it takes for half of the radioactive atoms to decay.

HT If the half-life is known, then the activity of a source after a certain amount of time can be calculated.

> The half-life of a radioactive isotope is 2 years and the initial activity is 800Bq. What will be the activity after 6 years?
>
> 2 years = $\frac{1}{2}$ count rate
>
> 4 years = $\frac{1}{4}$ count rate
>
> 6 years = $\frac{1}{8}$ count rate
>
> final count rate = $\frac{1}{8}$ × initial activity
>
> $\qquad = \frac{1}{8} \times 800 = 100Bq$

If the half-life is 2 years, then in 6 years' time, 3 half-lives will have passed.

This is the count rate, or activity, after 6 years.

Nuclear Equations

- Nuclear equations are used to represent radioactive decay:
  - An alpha particle is represented by the symbol ^4_2He.
  - A beta particle by the symbol $^0_{-1}e$.
- When an alpha particle is emitted:
  - the mass number of the element is reduced by 4
  - the atomic is number reduced by 2.
- This is because 2 protons and 2 neutrons are emitted from the nucleus, e.g.

$$^{219}_{86}Radon \rightarrow {}^{215}_{84}Polonium + {}^4_2He$$

- With beta decay:
  - the mass number does not change
  - the atomic number is increased by 1.
- This is because a neutron turns into a proton and an electron, and the electron is emitted as the beta particle, e.g.

$$^{14}_{6}Carbon \rightarrow {}^{14}_{7}Nitrogen + {}^0_{-1}e$$

- The emission of a gamma ray does not cause a change in the mass or the charge of the nucleus.
- You need to be able to be able to write balanced decay equations for alpha and beta decay:
  - The mass numbers on the right-hand side must add up to the same number as those on the left.
  - The atomic numbers on the right must have the same total as those on the left.

$$^{219}_{86}Radon \rightarrow {}^{215}_{84}Polonium + {}^4_2He \longleftarrow$$ 219 = 215 + 4 and 86 = 84 + 2

$$^{14}_{6}Carbon \rightarrow {}^{14}_{7}Nitrogen + {}^0_{-1}e \longleftarrow$$ 14 = 14 + 0 and 6 = 7 − 1

> **Key Point**
>
> An alpha particle is represented by the symbol ^4_2He.
>
> A beta particle by the symbol $^0_{-1}e$.

Quick Test

1. Radioactive decay is a random event. Explain what this means.
2. The count rate of a radioactive source takes 4 months to fall from 1000Bq to 250Bq. What is the half-life?
3. **HT** A radioactive source has a half-life of 3 years. What fraction of the original isotope would remain after 15 years?

> **Key Words**
>
> half-life

Review Questions

Waves and Wave Properties

1 **a)** Waves can be made on a spring by holding one end and moving it up and down, as shown in **Figure 1**.

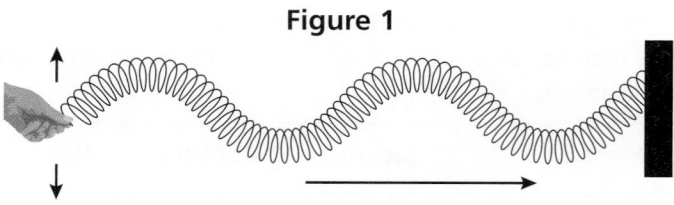

Figure 1

    i) What type of wave is shown in **Figure 1**? [1]

    ii) How could the person producing the wave increase the frequency of the wave? [1]

    iii) How could the person producing the wave increase the amplitude of the wave? [1]

b) Waves can also be made by moving a spring backwards and forwards, as shown in **Figure 2**.

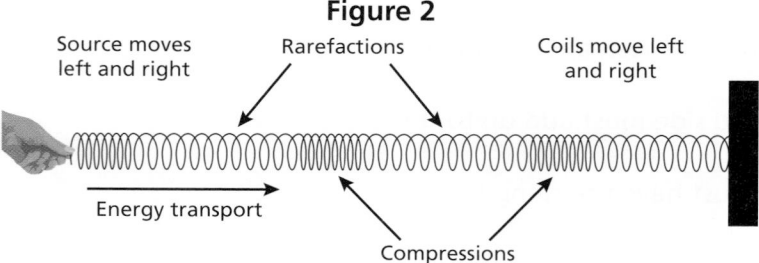

Figure 2

Source moves left and right   Rarefactions   Coils move left and right

Energy transport

Compressions

    i) What type of wave is shown in this diagram? [1]

    ii) Give an example of a wave that travels in this way. [1]

> Total Marks _____ / 5

Electromagnetic Waves

1 Use the wave model to explain why light is refracted as it travels from one medium to another. [4]

2 A student carries out an investigation into how the angle of incidence affects refraction by a glass block.

Give **one** variable that should be controlled and explain why it is important. [3]

> Total Marks _____ / 7

The Electromagnetic Spectrum

1 When electromagnetic waves are absorbed, they transfer energy to the object absorbing them.

Describe how this enables the following processes to occur.
You must name the type of electromagnetic wave that is carrying the energy in each case.

a) Plants are able to grow. [2]

b) Bread is toasted when put into a grill. [2]

c) Television signals are received by an aerial. [2]

2 Complete the sentences using words from the box.

bacteria	high	most	low	least	viruses

Gamma rays have a very _____ frequency and are the _____ energetic

of all electromagnetic waves. They can be used to destroy _____ . [3]

3 X-rays are an ionising radiation.
They can be used to treat cancer. However, exposure to X-rays can also cause cancer.

Explain what is meant by 'ionising radiation' and how it can cause cancer. [4]

4 What type of current is produced when an electromagnetic wave is absorbed by a conductor? [1]

Total Marks _____ / 14

An Introduction to Electricity

1 A bulb has a current of 0.1A flowing through it.

a) Calculate the amount of charge that is transferred if the bulb is turned on for 1 hour. [3]

b) The bulb is changed for a different type of bulb with twice the resistance.

How would the voltage have to change for the same current to flow? [1]

2 A light emitting diode has a voltage of 2V across it and transfers 6C of charge per minute.

   a) Calculate the current that flows through the LED. [1]

   b) Use your answer to part **a)** to calculate the resistance of the LED. [1]

> **Total Marks** .... / 6

Circuits and Resistance

1 The current–potential difference graphs for three electrical devices are shown in **Figure 1**.

Figure 1

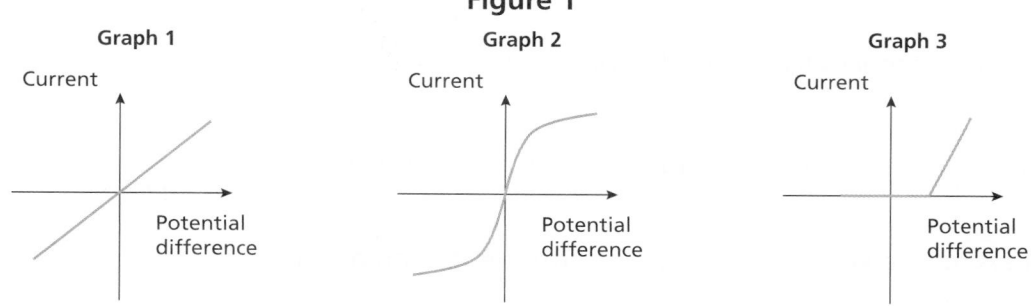

Graph 1   Graph 2   Graph 3

   a) Which graph corresponds to each of the components listed below?

     i) A diode   **ii)** A resistor at constant temperature   **iii)** A filament lamp [3]

   b) Which graph shows a constant resistance? [1]

   c) Which graph shows the resistance increasing as the current increases? [1]

> **Total Marks** .... / 5

Circuits and Power

1 a) Draw a parallel circuit with a 2V battery and two bulbs. [3]

   b) Each bulb in the circuit is identical. Work out the potential difference across each bulb. [1]

   c) The current through the battery is 2A. Work out the current through each bulb. [1]

   d) Use your answers to parts **a)** and **b)** to calculate the resistance of each bulb. [2]

> **Total Marks** .... / 7

Domestic Uses of Electricity

1 Give the voltage and frequency of the UK mains electricity supply. [2]

2 **Figure 1** shows a three-pin plug.

Add the following labels to **Figure 1**.

Figure 1

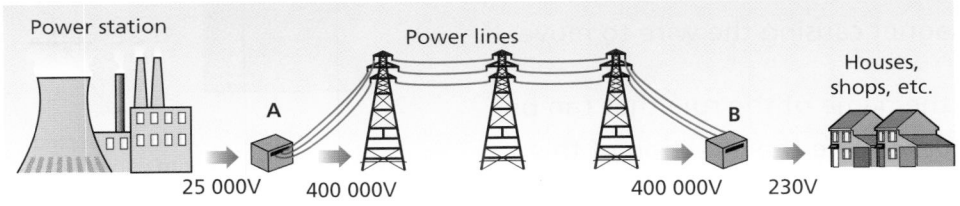

a) Earth Wire [1]

b) Live Wire [1]

c) Neutral Wire [1]

d) Fuse [1]

e) Cable Grip [1]

Total Marks _____ / 7

Electrical Energy in Devices

1 Write down the formula that links power, potential difference and current. [1]

2 **Figure 1** shows the national grid, which is used to distribute electricity from power stations to customers.

Figure 1

Power station   Power lines   Houses, shops, etc.

25 000V   400 000V   400 000V   230V

a) What name is given to **i)** part **A** and **ii)** part **B**? [2]

b) Part **A** increases the voltage. What effect does this have on the current and why is this important? [2]

c) Describe what part **B** does. [2]

Total Marks _____ / 7

Practice Questions

Magnetism and Electromagnetism

1. Draw the magnetic field lines around a bar magnet. **[3]**

2. Explain what is meant by 'induced magnetism'. **[1]**

3. Some students are carrying out an experiment with plotting compasses and electric circuits. They notice that the compass needle moves when the circuit is switched on, indicating that the current produces a magnetic field.

   a) Give **two** ways in which the size of this magnetic field could be increased. **[2]**

   b) How could the direction of the magnetic field be reversed? **[1]**

   c) **Figure 1** shows the magnetic field lines around the wire.

      Draw an arrow on the wire in **Figure 1** to show the direction of the current through the wire. **[1]**

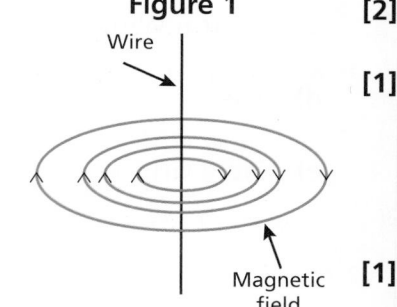

Figure 1

Wire

Magnetic field

Total Marks _____ / 8

The Motor Effect

1. **Figure 1** shows a conducting wire carrying a current placed in a permanent external magnetic field.

   The field from the wire interacts with the field from the magnet causing the wire to move.

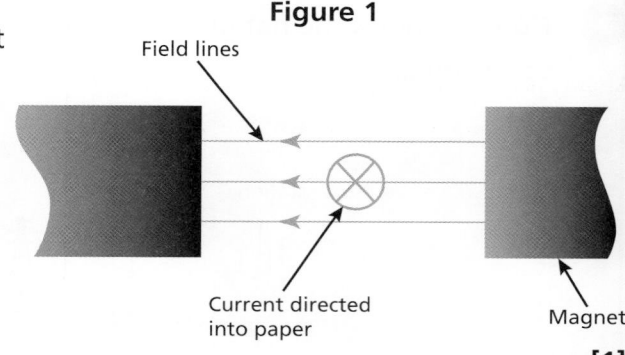

Figure 1

Field lines

Current directed into paper

Magnet

   a) What is the name of the rule that can be used to determine the direction of this movement? **[1]**

   b) Use this rule to determine the direction in which the wire in **Figure 1** moves. **[1]**

   c) Give **one** way in which the size of the force can be increased. **[1]**

   d) Give **one** way in which the direction of the force can be reversed. **[1]**

Total Marks _____ / 4

Particle Model of Matter

1 Solids and liquids are described as incompressible.

What does this mean? **[1]**

2 An experiment is carried out to measure the density of an irregular-shaped object.
The object is submerged in water to calculate its volume.

Explain why this method will not work if the object is less dense than water. **[2]**

3 A pan of boiling water is left on the hob.
The hob remains on and continually transfers energy to the water.
However, the temperature of the water does not increase.

Explain why this happens in terms of particle motion and energy. **[2]**

4 A burn from steam at 100°C causes worse injuries than a burn from hot water that is also at 100°C.

Explain why this is using the idea of specific latent heat. **[2]**

5 A sealed container of fixed volume is filled with air.
The container is heated and after a period of time the lid of the container pops off.

Use the ideas of pressure and particle motion to explain why this happens. **[3]**

6 **Table 1** contains information about the mass and volume of different objects.
The density of water is 1g/cm³.

Table 1

	Mass	Volume	Will it Float? Y/N
Object A	60g	40cm³	
Object B	60g	80cm³	
Object C	30g	80cm³	

Calculate the density of each object and determine whether it will float or sink in water. **[3]**

Total Marks .... / 13

Practice Questions

Atoms and Isotopes

1. **Figure 1** shows the structure of a helium atom.

   a) Add labels to **Figure 1** to identify the protons, neutrons and electrons. **[3]**

   b) What charge does a proton carry? **[1]**

   c) Write down the typical radius of an atom. **[1]**

   d) What would have to happen for the atom to become a negative ion? **[1]**

   e) Use **Figure 1** to determine the mass number and the atomic number of helium. **[2]**

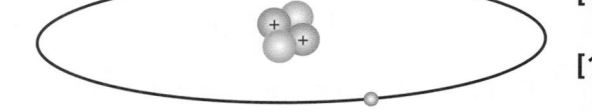

Figure 1

2. Chlorine has two common isotopes: Chlorine-35 and Chlorine-37.
   These are represented by the following symbols:

   $^{35}_{17}Cl$ $^{37}_{17}Cl$

   a) How many protons does a chlorine atom contain? **[1]**

   b) How many electrons does a neutral atom of chlorine contain? **[1]**

   c) Compare the two isotopes of chlorine in terms of the numbers of neutrons in the nucleus. **[2]**

> **Total Marks** .... / 12

Nuclear Radiation

1. Draw **one** line from each type of radiation to the correct description.

Alpha	Weakly ionising – likely to pass straight through living cells without being absorbed.
Beta	Moderately ionising – can pass through the skin and damage organs inside the body.
Gamma	Highest likelihood of being absorbed and causing damage when passing through living cells.

[2]

2 **Figure 1** shows three different materials.

Figure 1

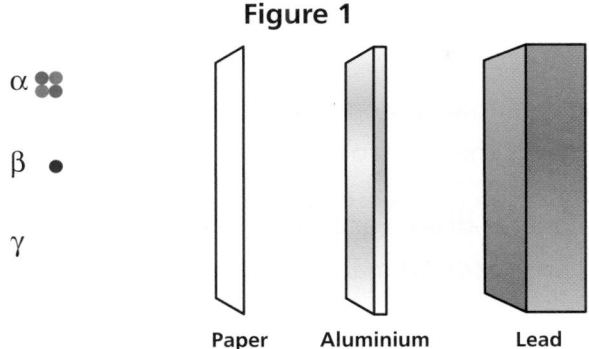

Paper   Aluminium   Lead

On **Figure 1**, draw a line from each type of radiation to show which of the materials it can penetrate. **[3]**

> **Total Marks** .... / 5

Half-Life

1 A small sample of a radioactive material is collected and tested in a laboratory.
When first tested the count-rate of the sample is 1000 counts per second.
After 4 minutes, the count-rate falls to 250 counts per second.

a) Work out the half-life of the sample tested. **[2]**

b) After a few hours, the count-rate remains constant at 8 counts per minute.

What is causing the count-rate to remain constant? **[1]**

c) A second source was found to have an activity of 150Bq.
After several months, this activity was still measured to be 150Bq.

What conclusion can be drawn about the half-life of this source? **[1]**

> **Total Marks** .... / 4

Review Questions

Magnetism and Electromagnetism

1 On **Figure 1**:

Figure 1

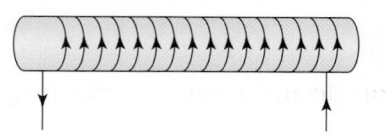

- Mark which end of the solenoid is the north pole.
- Draw the field lines around the electromagnet.

[3]

2 A student looks at the field lines drawn around two different magnets.
They notice that the field lines around magnet A are much closer together than those around magnet B.

What does this indicate?

[1]

> **Total Marks** .... / 4

The Motor Effect

1 **Figure 1** shows a conducting wire in a magnetic field.

Figure 1

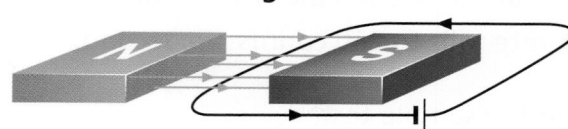

a) Use Fleming's left hand rule to work out the direction the wire will move when the current flows. Mark this on **Figure 1** with an arrow.

[1]

b) The poles of the magnets are reversed.

How will this affect the direction the wire moves?

[1]

2 The force experienced by a current carrying wire in a magnetic field depends on the magnetic flux density.

a) What is meant by flux density?

[2]

b) What unit is magnetic flux density measured in?

[1]

3 HT A 4cm length of wire passes through a magnetic field.
It experiences a force of 0.2N when a current of 4A flows through it.

Calculate the strength of the magnetic field.
Refer to the Physics Equations on page 273.

[3]

> **Total Marks** .... / 8

Particle Model of Matter

1 All matter is made of particles.

Sketch the particle arrangement for the different states of matter: solid, liquid and gas.
You do not need to draw more than nine particles in each diagram. **[3]**

2 **a)** Explain what happens as a solid is heated to become a liquid in terms of particle
movement and energy. **[3]**

b) Explain what happens as a liquid is heated to become a gas in terms of particle
movement and energy. **[3]**

3 Describe a method that can be used to find the density of an irregular-shaped object. **[4]**

4 Energy is provided to a sealed system containing ice.
Figure 1 shows how the temperature of the system changes with time.

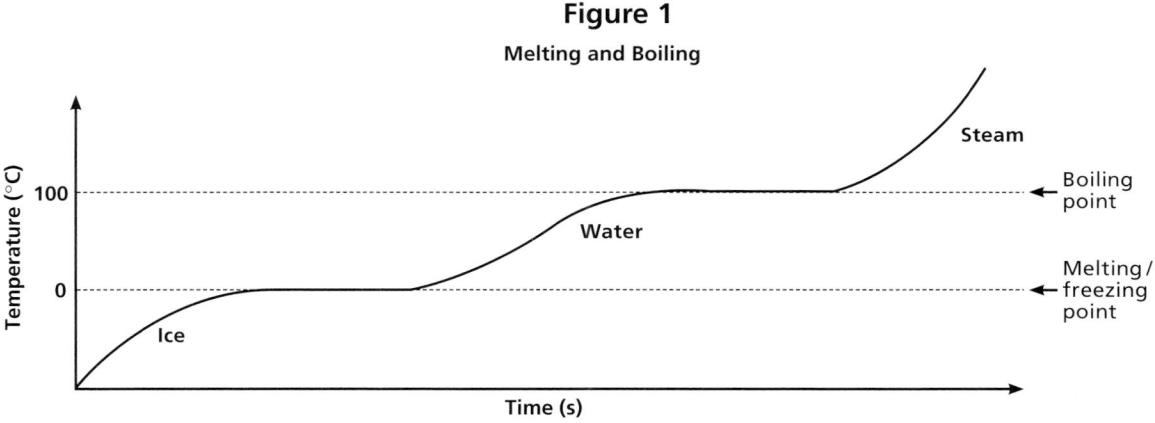

Figure 1
Melting and Boiling

a) Explain the shape of the graph. **[6]**

b) The water becomes a gas and the temperature of the system continues to rise.

What effect will this have on the pressure of the gas? **[1]**

c) The specific latent heat of fusion of water is 334 000 J/kg.

If there was 500 g of ice in the system at the start, calculate how much energy is needed to
completely melt the ice once it has reached 0°C. **[2]**

Total Marks _____ / 22

Review Questions

Atoms and Isotopes

1 In 1905, Rutherford and Marsden carried out an experiment to test the plum pudding model of the atom.

During the experiment a thin gold foil was bombarded with alpha particles and the path of the alpha particles recorded.

Figure 1

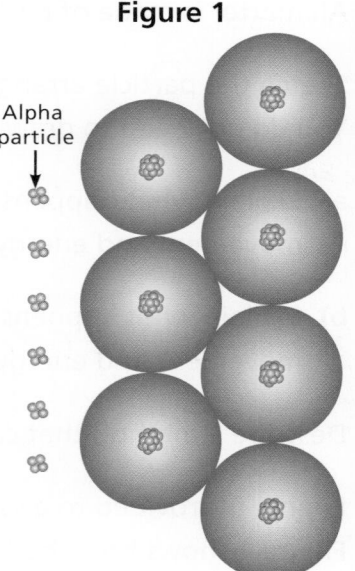

a) On **Figure 1**, draw three possible paths taken by the alpha particles as they strike the gold foil. [3]

b) What conclusion did Rutherford make from the path of the alpha particles? [2]

c) What adaptation did Niels Bohr make to the nuclear model proposed by Rutherford. [2]

2 Uranium is a very heavy atom. It has a number of different isotopes:

- uranium-233 $^{233}_{92}U$
- uranium-235 $^{235}_{92}U$
- uranium-238 $^{238}_{92}U$

a) How many protons does an atom of uranium contain? [1]

b) Which of the isotopes of uranium is the lightest? [1]

c) Compare the nuclei of uranium-235 and uranium-238. [2]

d) When uranium forms ions, it can form U^{3+} and U^{4+} ions.

How many electrons does a U^{3+} ion have? [2]

Total Marks _____ / 13

Nuclear Radiation

1 Draw **one** line from each type of radiation to the correct description.

Alpha		An electron emitted from the nucleus

Beta		High-frequency electromagnetic radiation

Gamma		A helium nucleus (2 protons and 2 neutrons)

[2]

2 Explain why it is not possible to predict exactly when a particular radioactive nucleus will decay. **[2]**

Total Marks _____ / 4

Half-Life

1 HT Carbon dating is used to determine the age of dead organic materials.
The process uses radioactive carbon-14, which has a half-life of around 6000 years.

a) Scientists are studying a mammoth found frozen in ice.
They find that a sample of carbon from the remains has a count-rate of
200 counts per minute.
The scientists work out that when the mammoth died, the count-rate
would have been 800 counts per minute.

Use this information to calculate how long ago the mammoth died. **[3]**

b) A second set of remains is found that are 18 000 years older than the first set
of remains.

If the same size sample is tested, calculate the count-rate of the second sample. **[3]**

c) Carbon dating can only be used to calculate the age of carbon samples up to
60 000 years old.

Explain why. **[2]**

Total Marks _____ / 8

Mixed Exam-Style Biology Questions

1. **Figure 1** shows a cheek cell.

Figure 1

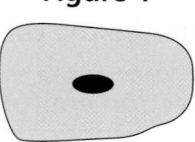

a) Put a ring around the most likely width of this cell.

0.002mm   0.02 mm   2mm   20mm   [1]

b) Write down **three** structures that are found in plant cells but **not** in animal cells.   [3]

c) Use words from the box to complete **Table 1**.

| chloroplast   cell membrane   nucleus   vacuole |

Table 1

Sub-Cellular Structure	Function
	stores cell sap
	allows gases and water to enter and leave the cell
	controls what the cell does

[3]

2. **Figure 2** shows the villi that line the small intestines in a healthy person.

Figure 2

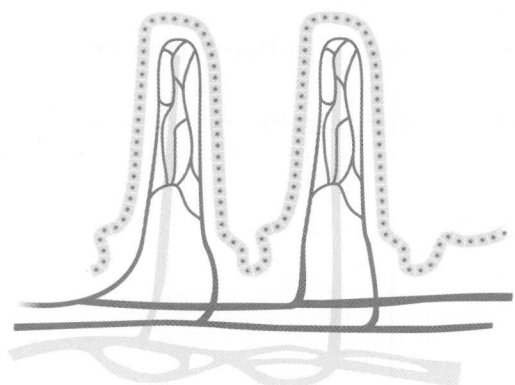

a) Describe **two** features of the villi shown in **Figure 2** that help the small intestine to function efficiently. **[2]**

b) The villi that line the small intestine are also covered in microvilli.
Each cell that has microvilli is packed with lots of mitochondria.

   i) Explain the advantage of microvilli in the absorption of digested food molecules. **[2]**

   ii) Explain why it is necessary for these cells to be packed with lots of mitochondria. **[2]**

c) The villi of a person with coeliac disease are damaged and have a much smaller surface area compared with the villi of a healthy person.

   What effect will this damage have on the function of the small intestine? **[1]**

3 A sports scientist investigated the amount of lactic acid in the leg muscle of a runner.

The runner ran for 30 minutes and then rested.

Table 2 shows the results.

Table 2

Time After Start of Exercise (Minutes)	0	10	20	30	40	50	60	70	80	90
Lactic Acid (Arbitrary Units)	0	1	6	13	8	6	4	3	1	0

a) What was the percentage increase in lactic acid between 20 and 30 minutes? **[2]**

b) How long after the exercise finished did it take for the lactic acid to be completely removed from the muscle? **[1]**

c) HT Explain what happens to the lactic acid build-up after exercise. **[2]**

d) The lactic acid of a second athlete is investigated in the same way.

   Why is it important to keep variables the same in the investigation? **[1]**

e) It takes 20 minutes for the lactic acid levels in the second athlete's muscle to return normal.

   Which athlete is the fittest? **[1]**

4 This article about squirrels was published in a natural history magazine.

Red Versus Grey?

There are two main species of squirrel living in Britain: the native red squirrel (*Sciurus vulgaris*) and the grey squirrel (*Sciurus carolinensis*), which was introduced from America.

The number of reds has declined in Britain over the last sixty years and they are now rare.

There have been a number of studies to try to find out why the number of reds has declined.

The main reason is probably due to competition. The reds are lighter animals and so spend more time up in the trees. They prefer coniferous woodland, feeding on the seeds from pine cones, high up in the trees. The larger greys spend more time on the ground and are better adapted to eating acorns and other seeds that are found in the more common, deciduous woodlands.

The reds lack the enzymes to digest acorns. Therefore, they are soon out-competed in deciduous woodlands.

a) What information in the article shows that the red and grey squirrels are closely related? **[1]**

b) The reason for the decline of the red squirrels is thought to be competition.

What are the squirrels competing for? **[1]**

c) Why are the two types of squirrel **not** competing for mates? **[1]**

d) Some conservationists want to put up boxes that will give out poisoned bait to any squirrels that are heavier than a certain mass.

Suggest **one** advantage and **one** disadvantage of this conservation method. **[2]**

e) Conservationists hope that a new strain of the red squirrel will evolve, which can digest acorns.

Explain how natural selection could bring this about. **[4]**

5 HT A plant is receiving plenty of light but the rate of photosynthesis stops increasing.

What other **environmental** factors might be responsible?
Tick **one** box.

Concentration of carbon dioxide or the amount of oxygen ☐

Concentration of carbon dioxide or the temperature ☐

Concentration of chlorophyll or the temperature ☐

Concentration of glucose or the amount of oxygen ☐ **[1]**

6 **Figure 3** shows the inheritance of cystic fibrosis in a family.

Figure 3

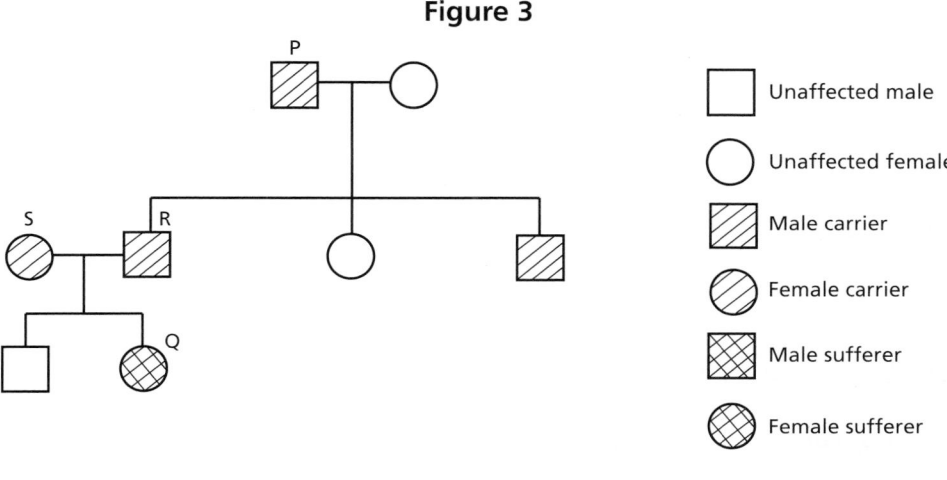

a) Using **C** for the non-cystic fibrosis allele and **c** for cystic fibrosis allele, write down the genotype of:

   i) Person **P**. **[1]**

   ii) Person **Q**. **[1]**

b) What is the phenotype of person **Q**? **[1]**

c) Draw a genetic diagram and work out the percentage probability of a carrier and a sufferer having a baby that has cystic fibrosis. **[4]**

7 **Figure 4** shows an evolutionary tree for some present-day vertebrates.

Figure 4

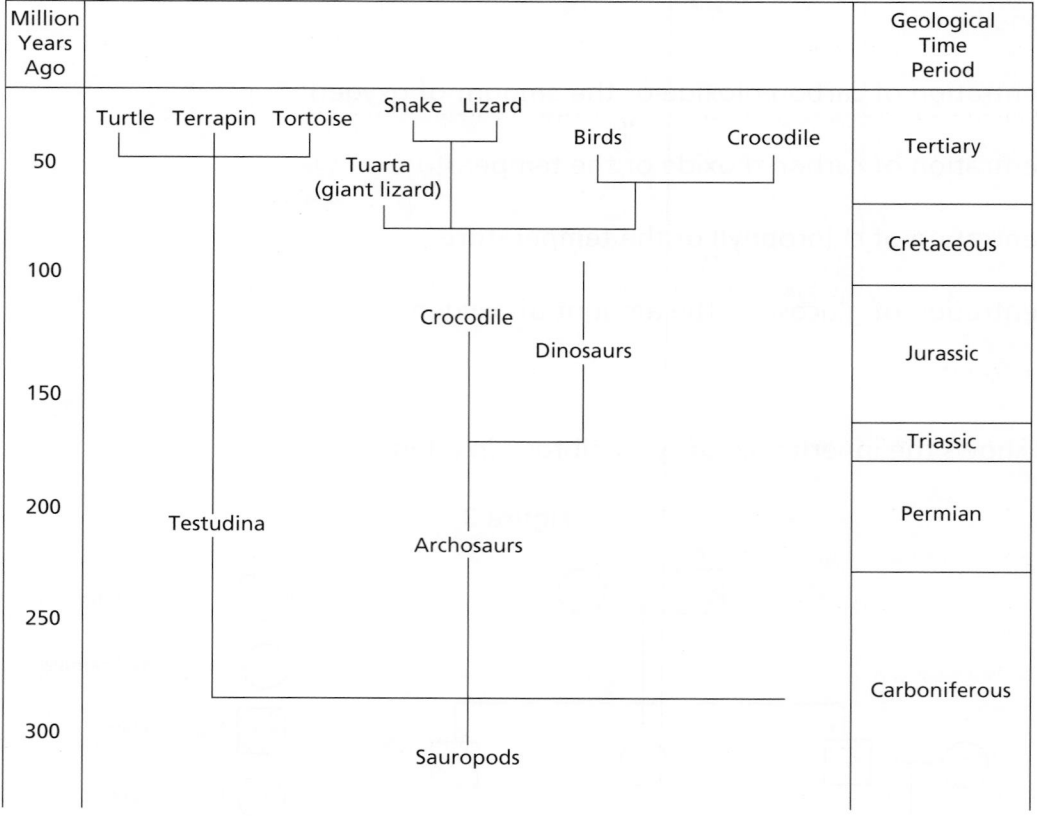

a) How many millions of years ago did the *Testudina* appear? [1]

b) In what geological time period did the dinosaurs become extinct? [1]

c) How do scientists know that dinosaurs once lived on Earth? [1]

d) What group of animals alive today is most closely related to the snake? [1]

e) Which ancestor is shared by dinosaurs, crocodiles and the giant lizard, but is not an ancestor of tortoises? [1]

8 The MMR vaccination is usually given to children when they are 13 months old.

It protects them from three diseases: measles, mumps and rubella.

a) Explain how a vaccination like MMR can protect a person from getting a disease. **[4]**

b) MMR contains three different vaccinations in one injection.
About 10 days after the injection, the child might develop a fever and a measles-like rash.
After three weeks, the child might get a very mild form of mumps.
After six weeks, a rash of small spots like rubella may develop.
However, most children suffer no symptoms after vaccinations.

   i) People may feel ill after having a vaccination.

   Explain why this is. **[2]**

   ii) Some parents do not want their children to have the MMR vaccine.
   They would like their children to visit the doctor three different times for three
   separate vaccinations.

   Suggest why they might feel this way. **[2]**

   iii) Suggest why the government might want children to have one vaccination rather
   than three separate injections. **[1]**

c) Doctors are worried that the number of children being vaccinated might fall.

Explain why it is important that the percentage of children vaccinated stays above
a certain level. **[2]**

9 **Figure 5** shows a red-eyed tree frog.
These frogs are bright green with red eyes, blue stripes and
orange feet.
They live in trees in the rainforest and feed on small insects.

Figure 5

Explain how each of the following adaptations helps the frog
to survive:

a) It has sticky pads on its 'fingers' and toes. **[1]**

b) When sleeping, it hides its bright colours by closing its eyes and tucking its feet beneath its body. **[1]**

c) It has a long, sticky tongue. **[1]**

10 Duckweed is a small floating plant found in ponds.
It reproduces quite quickly to produce large populations.

Helen decides to investigate duckweed by growing some in a beaker of water.
She counts the number of duckweed plants at regular intervals.

Table 3 shows her results.

Table 3

Day	1	4	8	10	15	19	20	23	26	30
Number of Plants	1	2	3	4	15	28	28	28	29	29

a) Plot a graph of the results to show how the population changed over time.

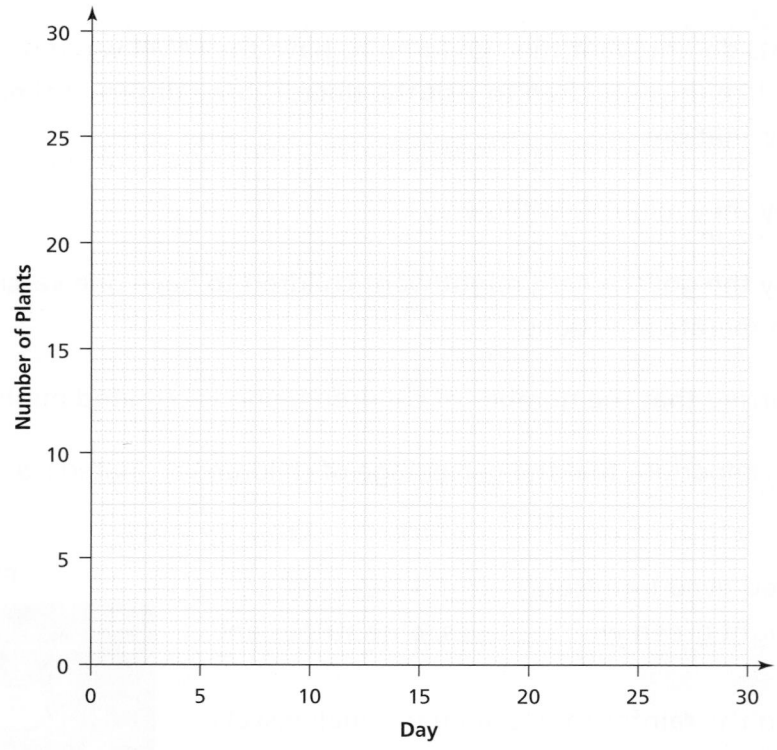

[3]

b) Describe what is happening to the population between day one and day ten. [1]

c) Suggest **two** factors that could have prevented the population from continuing
to increase. [2]

11 There are a number of ways in which we can help the environment.

a) Draw one line from each method of helping the environment to the corresponding result.

Method	Result
Increasing the amount of metal recycled	Less acid rain is produced
Reducing acid emissions	Fewer quarries are dug to provide raw materials
Using fewer pesticides and fertilisers	Fewer forests are cut down
Increasing the amount of paper recycled	Less pollution of rivers flowing through farmland

[3]

b) In the past, bags of compost from garden centres contained lots of peat.

Explain how this can damage the environment. [2]

c) Sustainable food production is another way of helping the environment.

Explain what is meant by 'sustainable food production'. [2]

12 a) How many chromosomes does a human body cell contain? [1]

b) How may chromosomes do human sex cells contain? Tick **one** box.

Half the number of chromosomes in a normal body cell. ☐

The same number of chromosomes in a normal body cell. ☐

Half the number of chromosomes in a sperm cell. ☐

Twice the number of chromosomes in a normal body cell. ☐ [1]

c) Explain what determines the sex of an individual. [2]

d) The probability of having a baby boy is 50%.

Draw a genetic diagram to show why this is. [3]

Total Marks .... / 77

Mixed Exam-Style Chemistry Questions

1. **Figure 1** shows a potassium atom.

Figure 1

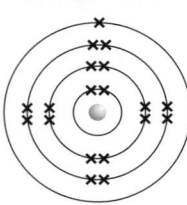

   a) Write down the electron configuration of this potassium atom. [1]

   b) Explain why potassium belongs to Group 1 of the periodic table. [2]

   c) Sodium has an atomic number of 11.

   Draw a diagram to show how the electrons are arranged in a sodium atom. [1]

   d) Explain why the chemical reactions of potassium and sodium are similar. [1]

2. What is an 'ion'? [2]

3. An ion of lithium has the symbol $^{7}_{3}Li^{+}$.

   Give the number of protons, electrons and neutrons in this ion of lithium. [3]

4. Tin is extracted from its ore, tin oxide, by heating the tin oxide with carbon.

   a) Complete the word equation for the extraction of tin.

   $$\text{tin oxide} + \text{carbon} \rightarrow \underline{\hspace{3cm}} + \underline{\hspace{3cm}}$$ [2]

   b) In terms of oxidation and reduction, explain what happens to tin oxide and carbon in this reaction. [2]

5. Which salt is produced when sulfuric acid is neutralised by sodium hydroxide?
   Tick **one** box.

Sulfur sulfate	☐	Sodium sulfate	☐
Sodium chloride	☐	Sulfur chloride	☐	[1]

6. Graphene has some special properties, which could make it very useful in the future.

   Tick each of the properties of graphene.

Property	Does graphene have this property?
Very strong	
Low melting point	
Good electrical conductor	
Good thermal conductor	
Opaque	
Nearly transparent	
Excellent electrical insulator	
Fragile	

[4]

7 Potassium fluoride is an ionic compound.

a) Suggest why potassium fluoride has a high melting point. [2]

b) Does potassium fluoride conduct electricity when solid?
You must give a reason for your answer. [1]

c) Does potassium fluoride conduct electricity when it is molten?
You must give a reason for your answer. [1]

d) Does potassium fluoride conduct electricity when dissolved in water to form an
aqueous solution?
You must give a reason for your answer. [1]

8 A calcium atom can be represented by: $^{41}_{20}Ca$

a) How many protons does this atom of calcium have? [1]

b) How many neutrons does this atom of calcium have? [1]

c) How many electrons does this atom of calcium have? [1]

9 Ammonia, NH_3, is produced from nitrogen and hydrogen.

Relative atomic masses (A_r): N = 14, H = 1

a) Calculate the relative molecular mass of ammonia, NH_3. [2]

b) HT Calculate the mass of 1.00 mole of ammonia, NH_3. [1]

c) HT Calculate the mass of 0.5 moles of ammonia, NH_3. [1]

10 Balance the following symbol equations.

a) _____$Ca + O_2 \rightarrow$ _____CaO [2]

b) _____$Na + Br_2 \rightarrow$ _____$NaBr$ [2]

c) $H_2 + Br_2 \rightarrow$ _____HBr [1]

d) _____$H_2 + N_2 \rightarrow$ _____NH_3 [2]

e) _____$K + I_2 \rightarrow$ _____KI [2]

11 What technique could be used to extract a sample of pure water from a solution of salt and water?
Tick **one** box.

Crystallisation ☐ Filtering ☐

Distillation ☐ Chromatography ☐ [1]

12 **Figure 2** shows a lithium atom.

Figure 2

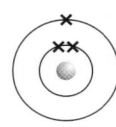

a) Write down the electron configuration of this lithium atom. [1]

b) Explain why lithium belongs to Group 1 of the periodic table. [1]

13 Everything is made of matter. There are three states of matter.

a) Complete **Figure 3** to show how the particles are arranged in solids, liquids and gases.
Use the symbol 'o' to represent a particle.

Figure 3

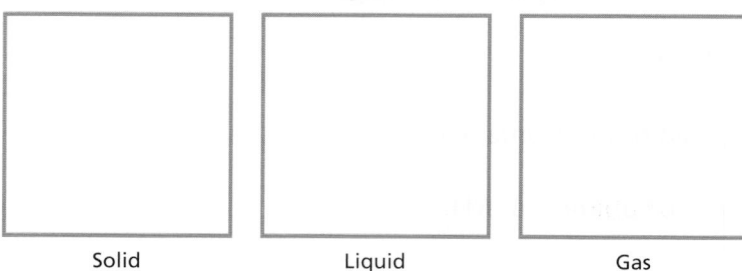

Solid   Liquid   Gas [3]

b) Explain how the particles are moving in a gas. **[2]**

c) Complete the sentence below.

When substances boil, they change from the _____ to the _____ state. **[2]**

d) Why does water have a much lower boiling point than iron?

Your answer should include:
- The type of bonding and the structure of each substance.
- The strength of these forces of attraction. **[6]**

14 **a)** HT In terms of electron transfer, explain why bromide ions have a 1− charge. **[2]**

b) HT In terms of electron transfer, explain why calcium ions have a 2+ charge. **[2]**

15 **a)** What is the chemical symbol for silver? **[1]**

b) Name the type of bonding in silver. **[1]**

c) Pure silver is too soft for many uses.

Why is pure silver soft? **[2]**

d) Silver is sometimes made into an alloy.

What is an alloy? **[1]**

16 Balanced symbol equations can be used to sum up what happens in chemical reactions.

Balance these symbol equations:

a) $Mg + \text{_____} HCl \rightarrow MgCl_2 + H_2$ **[1]**

b) $\text{_____} HgO \rightarrow \text{_____} Hg + O_2$ **[2]**

17 Name the type of bond found in alkane molecules.
Tick **one** box.

Double covalent ☐ Single covalent ☐

Giant ionic ☐ Giant metallic ☐ **[1]**

18 **Figure 4** shows the outer electrons of a magnesium atom and an oxygen atom.

Figure 4

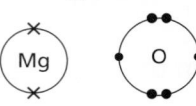

a) i)   Which group of the periodic table does magnesium belong to? [1]

   ii) Which group of the periodic table does oxygen belong to? [1]

   iii) Draw a diagram to show the electronic structure of a magnesium ion and an oxide ion. Show the outer shell of electrons only. [2]

b) The compound magnesium oxide is a solid at room temperature.

   What type of structure does magnesium oxide have?
   Tick **one** box.

   Metallic ☐   Giant covalent ☐

   Simple molecular ☐   Giant ionic ☐ [1]

c) Explain why magnesium oxide does not conduct electricity when solid but does when molten. [2]

19 Ethane, C_2H_6, is a simple molecule.

a) What sort of bonding is found in ethane molecules? [1]

b) **Figure 5** shows the outer electrons of a carbon atom and a hydrogen atom.

Figure 5

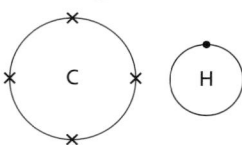

   Complete **Figure 6** to show the electron arrangement in an ethane molecule.

Figure 6

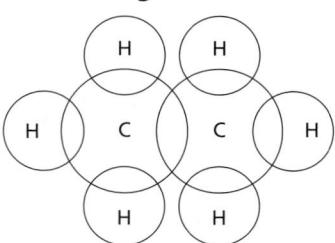

[1]

20 Ethane, C_2H_6, is a fuel.

a) By what process is ethane obtained from crude oil? [1]

b) Complete the equation below for the complete combustion of ethane.

$$2C_2H_6 + \underline{\hspace{1cm}} O_2 \rightarrow \underline{\hspace{1cm}} CO_2 + \underline{\hspace{1cm}} H_2O$$ [3]

c) Incomplete combustion of ethane can produce carbon monoxide.

Why is the production of carbon monoxide a concern? [1]

21 HT The Avogadro constant has a value of 6.02×10^{23}.

a) How many atoms are present in 39g of potassium? [1]

b) How many atoms are present in 16g of sulfur? [1]

22 Which of the substances below is a saturated hydrocarbon?
Tick **one** box.

C_3H_6 ☐ C_2H_4 ☐ C_3H_8 ☐ C_4H_8 ☐ [1]

23 Lead bromide is an ionic compound. It can be separated by electrolysis.

a) i) Name the element formed at the positive electrode during the electrolysis of
molten lead bromide. [1]

ii) Name the element formed at the negative electrode during the electrolysis of
molten lead bromide. [1]

b) HT Complete the half equations to show the reactions that take place at each electrode.

i) At the anode: $2Br^- \rightarrow \underline{\hspace{1cm}} + 2e^-$ [1]

ii) At the cathode: $Pb^{2+} + \underline{\hspace{1cm}} \rightarrow Pb$ [1]

24 Collision theory can be used to explain the rate of a chemical reaction.

a) What **two** things must happen for a chemical reaction to take place? [2]

b) Explain, in terms of collision theory, why increasing the pressure of a reaction that
involves gases increases the rate of a chemical reaction. [2]

25 During the electrolysis of molten lead iodide, lead and iodine are produced.

At the anode: $2I^- \rightarrow I_2 + 2e^-$
At the cathode: $Pb^{2+} + 2e^- \rightarrow Pb$

a) Describe what happens to the lead ions during the electrolysis of lead iodide. **[2]**

b) During the electrolysis of lead iodide, oxidation and reduction take place.

In terms of oxidation and reduction, explain what happens to the iodide ions. **[2]**

c) Why does the lead iodide have to be molten? **[1]**

26 Metals such as lead and copper are extracted from their ores.

a) What is an ore? **[1]**

b) Lead can be extracted by smelting.

Describe what happens during smelting. **[2]**

c) Why can lead be extracted by smelting? **[1]**

d) **HT** Copper can be extracted from its ore by phytomining.

What is phytomining? **[1]**

e) Aluminium is more reactive than iron, but it is used to make drinks cans and saucepans.

Why can aluminium be used in these ways? **[2]**

27 Which of the following changes would increase the rate of a chemical change?
Tick **one** box.

Carrying out the reaction in the dark ☐

Reducing the temperature by 10°C ☐

Increasing the surface area of solids ☐

Decreasing the concentration ☐ **[1]**

28 Chromatography can be used to separate mixtures of coloured substances.

The R_f value can be used to identify the components.

Figure 7 shows a chromatogram of a colouring ingredient used in fruit-flavoured drinks.

Figure 7

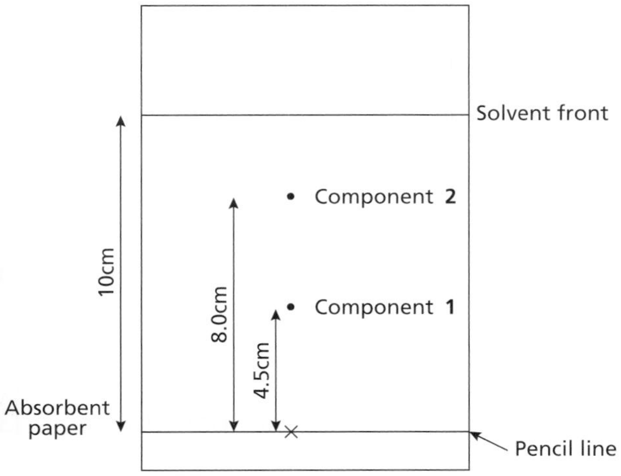

a) Why is the line at the bottom of the absorbent paper drawn in pencil? **[1]**

b) What is the R_f value for **Component 2**? **[2]**

c) Explain how this chromatogram shows that the colouring ingredient is a mixture. **[1]**

29 A solution of sodium sulfate contains sodium, Na^+, and sulfate, SO_4^{2-}, ions.

a) What is the formula of sodium sulfate? **[1]**

b) Which group of the periodic table does sulfur belong to?

You must explain your answer. **[2]**

30 When magnesium carbonate is added to a beaker of hydrochloric acid, a chemical reaction takes place.

$$MgCO_3(s) + 2HCl(aq) \rightarrow MgCl_2(aq) + H_2O(l) + CO_2(g)$$

a) HT Hydrochloric acid is a strong acid.

What is a strong acid? **[2]**

b) During the reaction, the mass of the reaction mixture and beaker goes down.

Why does the mass go down? **[2]**

Total Marks .... / 118

Mixed Exam-Style Physics Questions

1 A skydiver steps out of an aeroplane.
After 10 seconds, she is falling at a steady speed of 50m/s.

a) What name is given to this steady speed?   [1]

The skydiver opens her parachute.

After another 5 seconds, she is once again falling at a steady speed. This speed is now only 10m/s.

b) Calculate the skydiver's average acceleration during the time from when she opens her parachute until she reaches 10m/s.   [3]

c) Explain as fully as you can:

   i) Why the skydiver eventually reaches a steady speed (with or without her parachute).   [3]

   ii) Why the skydiver's steady speed is lower when her parachute is open.   [2]

d) The skydiver and her equipment have a total mass of 75kg.

   Calculate the weight of the skydiver.   [2]

2 The graph in **Figure 1** shows changes in the velocity of a racing car.

Figure 1

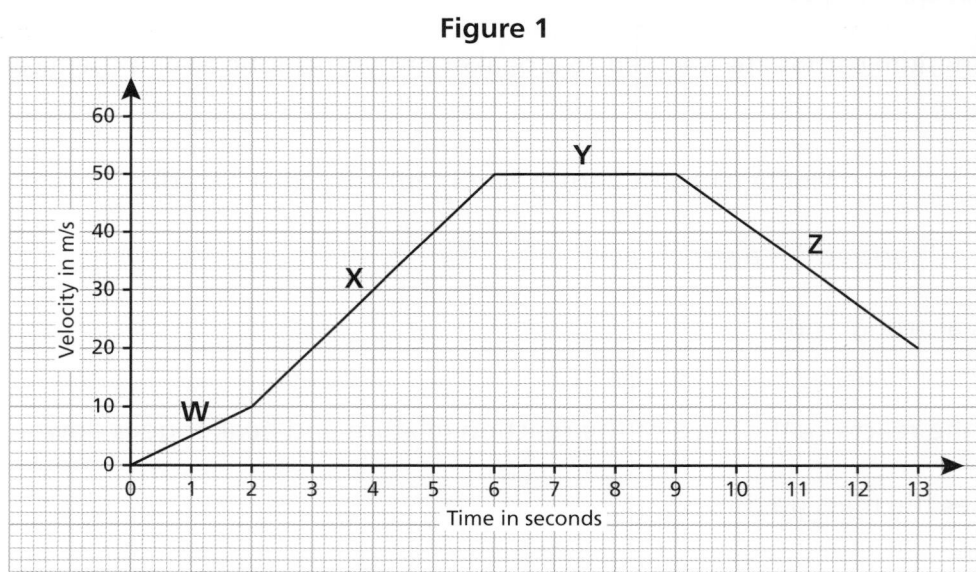

a) Describe the motion of the car in the section marked **Z**.   [1]

b) Calculate the acceleration of the racing car during the period labelled **X**.   [4]

c) Calculate the distance travelled by the car during the period marked **Y**.   [3]

3 When a car driver has to react and apply the brakes quickly, the car travels some distance before stopping.

Part of this distance is called the 'thinking distance'.

This is the distance travelled by a car while the driver reacts to a dangerous situation.

Table 1 shows the thinking distance (m) for various speeds (km/h).

Table 1

Thinking Distance (m)	0	3	6	9	12	15
Speed (km/h)	0	16	32	48	64	80

a) Describe how thinking distance changes with speed. **[2]**

b) A driver drank two pints of lager.
Some time later, the thinking time of the driver was measured as 1.0s.

   i) Calculate the thinking distance for this driver if the driver was travelling at 9m/s. **[1]**

   ii) A speed of 9m/s is the same as 32km/h.

   Use the table to find the thinking distance for a driver who has not had a drink travelling at 32km/h. **[1]**

   iii) What has been the effect of the drink on the thinking distance of the driver? **[1]**

4 To get a bobsleigh moving quickly, the crew need to do work to overcome friction.
They push the sleigh hard for a few metres and then jump in.

a) Write down the formula used to calculate work done. **[1]**

b) If the crew exert a force of 400N over 5m, calculate how much work have they done on the sled. **[2]**

c) As the crew do work on the sled they transfer energy to it.

   What type of energy does the sled gain? **[1]**

d) The energy the crew use to get the sled moving comes from their food.

   What type of energy store is this? **[1]**

5 The Sankey diagrams in **Figure 2** show what happens to each 100 joules of energy stored in coal when it is burned on an open fire and in a stove.

Figure 2

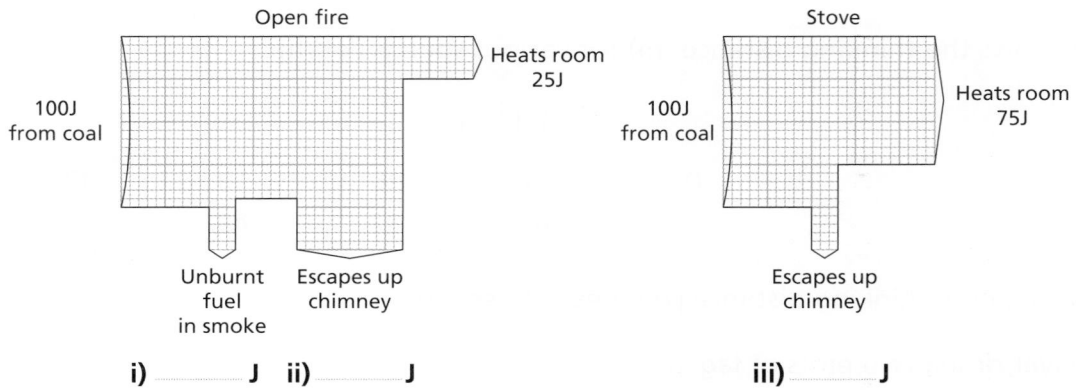

i) .... J ii) .... J   iii) .... J

a) Add the **three** missing numbers to the diagrams in **Figure 2**. **[3]**

b) Work out the efficiency of the open fire. **[1]**

c) Work out the efficiency of the stove. **[1]**

6 A cyclist accelerates from a standstill at a set of traffic lights.
The driving force of the back tyre on the ground is 250N.

a) How much work is done by this force when the cyclist travels five metres? **[2]**

b) Describe what happens to the energy transferred by this force. **[1]**

7 A man's car will not start, so two friends help by pushing it. The car has a mass of 800kg.
By pushing as hard as they can for 12 seconds, the two friends make the car reach a speed of three metres per second.

a) Calculate the acceleration given to the car. **[2]**

b) Whilst pushing the car, the two friends together do a total of 2400 joules of work.

Calculate their total power. **[2]**

c) Another motorist has the same problem.
The two friends push his car along the same stretch of road with the same force as before.
It takes them 18 seconds to get the second car up to a speed of three metres per second.

Calculate the mass of the second car. **[4]**

8 **Table 2** shows the main sources of energy used for electricity generation in Britain in 2015.

Table 2

Coal	28.2%
Oil and other	2.6%
Gas	30.2%
Nuclear	22.2%
Renewables	16.7%

a) How does the amount of electricity obtained from nuclear sources compare with the amount obtained from renewables?   **[1]**

b) Hydroelectricity is a renewable source.

   Name **two** other renewable energy sources.   **[2]**

c) Give **one** advantage and **one** disadvantage of nuclear power.   **[2]**

d) A hydroelectric dam stores water at a high level.
   When the water is released, it flows down pipes to the power station where it turns turbines.

   Complete the diagram to show the **useful** energy transfers that occur as the water flows down the pipes **to** the power station.

   _____ energy ⟶ _____ energy   **[2]**

e) The electricity generated in a power station is transmitted over long distances.
   Before this happens, its voltage is increased by using a step-up transformer.

   Describe **one** advantage and **one** disadvantage of transmitting electricity at high voltage.   **[2]**

9 A forklift truck was used to stack boxes onto a trailer.
It lifted a box weighing 1900N through 4.5m in a time of 9 seconds.

a) Calculate the work done on the box.   **[2]**

b) Calculate the useful power output of the motor.   **[2]**

c) The engine and motor combined are 50% efficient.

   Calculate the energy used by the forklift truck in lifting the box.   **[2]**

10 An inventor develops a solar-powered bike.

A battery is connected to solar cells, which charge it up.

There is a switch on the handlebars.

When the switch is closed, the battery drives a motor attached to the front wheel.

a) Complete the sentences using words from the box.

Words may be used once, more than once, or not at all.

chemical	electrical	heat (thermal)	kinetic	light
gravitational potential		sound	elastic potential	

i) The solar cells transfer _____ energy to _____ energy. **[2]**

ii) When the battery is being charged up _____ energy is transferred to

_____ energy. **[2]**

iii) The motor is designed to transfer _____ energy to _____ energy. **[2]**

b) The cyclist stops pedalling for 10 seconds.

During this time the motor transfers 1500 joules of energy.

i) Calculate the power of the motor. **[2]**

ii) Name **one** form of wasted energy that is produced when the motor is running. **[1]**

11 **Figure 3** shows part of the National Grid.

At **X**, the transformer increases the voltage to a very high value.

At **Y**, the voltage is reduced to 230V for use by consumers.

Figure 3

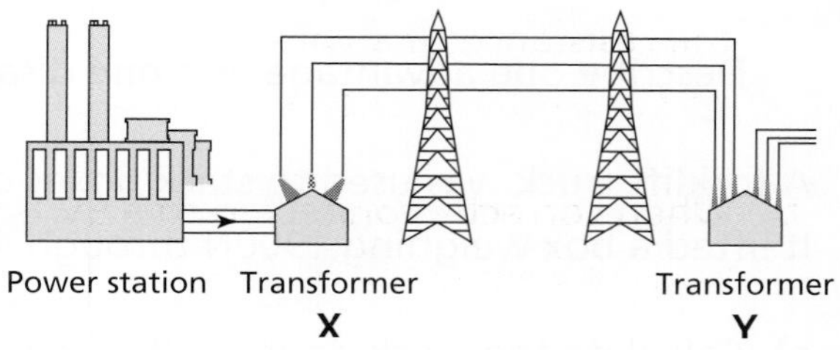

Power station   Transformer **X**   Transformer **Y**

a) What happens to the current as the voltage is increased at **X**? **[1]**

b) Why is electrical energy transmitted at very high voltages? **[1]**

c) What name is given to the type of transformer at **Y**? **[1]**

12 **Figure 4** shows a circuit.

Figure 4

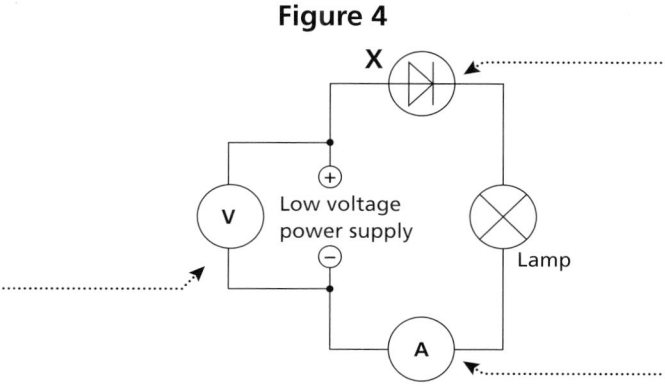

a) Add the missing labels to **Figure 4**. [3]

b) The circuit in **Figure 4** is tested with a range of different voltages.
The graph in **Figure 5** shows the results.

Figure 5

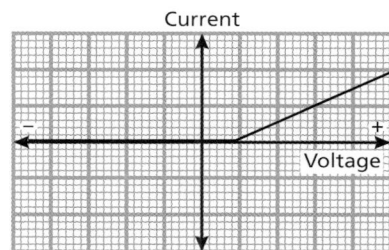

Describe what happens to the current through component **X** as the voltage changes.
You must refer to the resistance of the component. [4]

13 a) Draw a circuit diagram to show how a battery, ammeter and voltmeter can be used to find
the resistance of a wire.
You must use the correct circuit symbols. [3]

b) When correctly connected to a 9V battery, the wire in the circuit has a current of 0.30A
flowing through it.

i) Write down the equation that links current, resistance and voltage. [1]

ii) Calculate the resistance of the wire. [2]

iii) When the wire is heated, the current drops to 0.26A.

How has the resistance of the wire changed? [1]

14 The circuit diagram in **Figure 6** includes a component labelled **X**.

a) Calculate the potential difference across the 8Ω resistor. **[2]**

b) What is the potential difference across component **X**? **[1]**

c) What is the current through component **X**? **[1]**

Figure 6

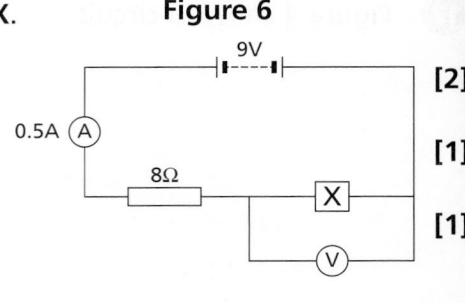

15 The circuit in **Figure 7** has four identical ammeters.

Figure 7

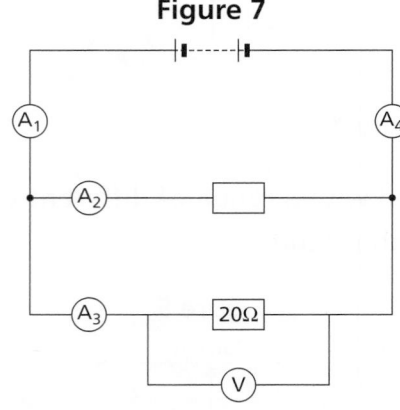

Table 3 gives the current through two of the ammeters.

Table 3

Ammeter	Current in amps
A_1	
A_2	0.2
A_3	0.3
A_4	

a) Complete the table to show the current through A_1 and A_4. **[2]**

b) Calculate the reading on the voltmeter.
You must show your working. **[2]**

c) Give the potential difference of the power supply. **[1]**

16 Microwave ovens can be used to heat many types of food.

a) Describe in detail how microwave ovens heat food. **[3]**

b) Microwaves used in ovens have a frequency of 2450 million Hz. Their wavelength is 0.122m.

Calculate the speed of microwaves. Show clearly how you work out your answer. **[3]**

17 **Figure 8** shows a beam of light striking a Perspex block.

Figure 8

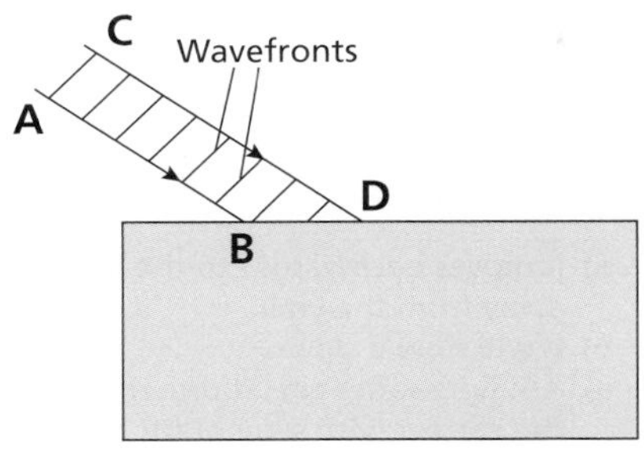

a) Continue the paths of the rays **AB** and **CD** inside the Perspex block on **Figure 8**. [2]

b) Draw the wavefronts of the beam of light in the Perspex. [2]

c) Explain why the beam behaves in the way you have shown. [2]

Figure 9

18 The arrows in **Figure 9** represent the size and direction of the forces on a space shuttle, fuel tank and booster rockets one second after launch.

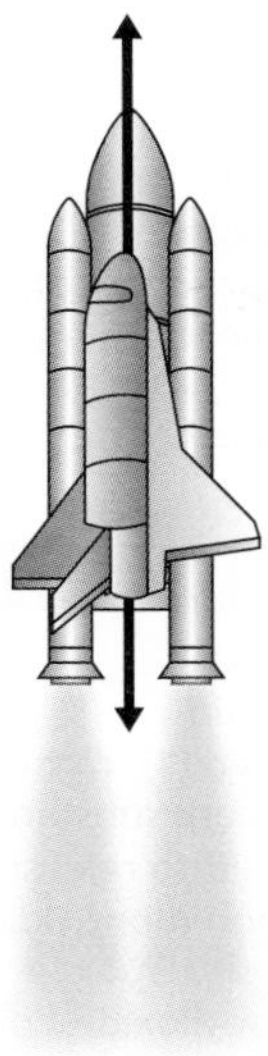

a) Name the force illustrated by the arrow pointing downwards. [1]

b) Describe the upward motion of the space shuttle one second after launch. [1]

c) By the time it moves out of the Earth's atmosphere, the total weight of the space shuttle, fuel tank and booster rockets has decreased and so has the air resistance.

How does this change the motion of the space shuttle? [2]
Assume the thrust force does not change.

d) The space shuttle takes nine minutes to reach its orbital velocity of 8100m/s.

i) Write down the equation that links acceleration, change in velocity and time taken. [1]

ii) Calculate the average acceleration of the space shuttle during the first nine minutes of its flight in m/s².
To gain full marks you must show how you worked out your answer. [3]

iii) How is the velocity of an object different from the speed of an object? [2]

Total Marks _____ / 117

Answers

Pages 10–11 Biology Review Questions

1. Cells [1]
2. a) Cell membrane [1]; cytoplasm [1]
   b) **Any two of:** cell wall [1]; chloroplast [1]; vacuole [1]
3. a) cells [1]
   b) Tail [1]
   c) Testis [1]
   d) Fertilisation [1]
4. Carbon monoxide: binds with haemoglobin / less oxygen carried in blood [1]; Nicotine: addictive / raises blood pressure [1]; Tar: causes lung cancer / bronchitis [1]
5. Photosynthesis [1]
6. (Aerobic) respiration [1]
7. **Any two of:** supports the body [1]; acts as a framework that enables muscles to move the body [1]; protects the organs [1]; makes red blood cells [1]
8. a) i) C [1]
      ii) D [1]
      iii) A [1]
      iv) D [1]
   b) An organism that feeds on / in another living organism [1]
9. a) Bacteria / fungi / virus / protista [1]
   b) To prevent harmful microorganisms from entering the medic's body [1]
10. a) i) Oxygen [1]
       ii) Carbon dioxide [1]
    b) To allow gases to pass across [1]
    c) To increase the rate at which gases can leave or enter the blood [1]

Pages 12–13 Chemistry Review Questions

1. a) i) Solid [1]
      ii) Liquid [1]
   b) oxygen [1]; carbon dioxide [1]
   c) Fuels are substances that can be burned to release energy. [1]
   d) Carbon monoxide [1]
   e) S [1]
   f) Sulfur dioxide [1]
2. a) 3 [1]
   b) Potassium [1]
   c) 5 [1]
3. a) A = alkaline [1]; B = alkaline [1]; C = neutral [1]; D = acidic [1]
   b) To clean it / stop contamination [1]
   c) Lime / alkali [1] (Accept a named alkali)
4. a) Crystallisation / by evaporating the water [1]
   b) Filtration [1]
   c) Three correctly drawn lines [2] (1 mark for one correct line)
      Mixture – Contains two or more elements or compounds, which are not chemically joined.
      Compound – Contains atoms of two or more elements, which are chemically joined.
      Element – Contains only one type of atom.

5. a) To mix it / make it react faster [1]
   b) Wear goggles / do not touch the copper sulfate solution [1]
   c) Colour change / temperature change [1]
   d) Copper [1]
   e) iron + copper sulfate → iron sulfate [1]; + copper [1]

Pages 14–15 Physics Review Questions

1. a) It moves backwards / to the left / away from the wall [1]
   b) It will slow it down [1]
2. a) A simple series circuit drawn with the correct with symbol for battery [1]; motor [1]; and switch [1]

   b) The battery [1]
   c) It will be slower [1]
3. Both bulbs will go out [1]
4. Both bulbs will stay the same [1]
5. a) There is a limited amount [1]; so it will run out [1]
   b) **Any two of:** coal [1]; gas [1]; nuclear [1]
6. chemical [1]; electrical [1]; light [1]; thermal [1]
7. 30N [1]
8. a) B [1]
   b) C is bigger than A [1]
   c) B and D [1]
   d) D is bigger than B [1]
9. a) It is reflected [1]
   b) Black absorbs light / the cyclist does not reflect any light [1]; so no light from the cyclist enters the driver's eyes [1]; and they appear black against a black background [1]

Pages 16–33 Revise Questions

Page 17 Quick Test
1. Nucleus
2. On the ribosomes
3. Cell wall; chloroplasts; permanent vacuole
4. To support the cell
5. In loops / plasmids in the cytoplasm, floating free

Page 19 Quick Test
1. Liver cell, nucleus, bacterium, ribosome
2. 5μm
3. $\frac{150\,000}{50}$ = 2000 times

Page 21 Quick Test
1. 46 (23 pairs)
2. So that each new cell gets the full amount of chromosomes
3. A cell that is not specialised and can divide to form various types of cells
4. To produce new cells so the root and shoot can grow

Page 23 Quick Test
1. The perfume diffuses faster
2. Energy from respiration
3. Thin walls / rich blood supply

Page 25 Quick Test
1. To become specialised and more efficient
2. To provide energy for the sperm to swim
3. A tissue
4. To cover the body
5. An organ

Page 27 Quick Test
1. A protein
2. **Any two of:** temperature; pH; concentration of enzyme; concentration of substrate
3. In the stomach, pancreas and small intestine
4. Lipase
5. In the liver

Page 29 Quick Test
1. Platelets
2. No nucleus; biconcave disc; contain haemoglobin
3. Artery
4. Right atrium
5. Valves

Page 31 Quick Test
1. a) Communicable
   b) Non-communicable
   c) Non-communicable
2. The UV light can trigger cancer
3. Being overweight
4. Layers of fatty material build-up in the arteries, reducing the blood flow so that not enough oxygen and glucose (needed for respiration) can reach the heart muscle

Page 33 Quick Test
1. Palisade mesophyll
2. Water and minerals
3. **Any two of:** low light; high humidity; low temperature; still air
4. To stop the plant losing water when carbon dioxide is not needed

Pages 34–39 Practice Questions

Page 34 Cell Structure
1. 1 mark for each correct row [3]

Type of Cell	Nucleus	Cytoplasm	Cell Membrane	Cell Wall
Plant Cell	✓	✓	✓	✓
Bacterial Cell	✗	✓	✓	✓
Animal Cell	✓	✓	✓	✗

Page 34 Investigating Cells

1. a) $\frac{50}{0.1}$ [1]; = 500 times [1]

    b) i) E   [1]
       ii) A   [1]
       iii) C   [1]
       iv) B   [1]
       v) A   [1]
       vi) D   [1]
       vii) C   [1]
       viii) B   [1]

Page 35 Cell Division

1. a) Cells that are unspecialised / undifferentiated [1]; and can form any type of cell [1]
    b) Stem cells could be used to replace damaged cells [1]; and treat diseases, such as Parkinson's disease / diabetes. [1]; Embryonic stem cells are usually extracted from embryos [1]; which results in the destruction of the embryos [1]; some people have ethical / religious objections to this [1]

Page 35 Transport In and Out of Cells

1. a) particles / molecules [1]; high [1]; low [1]; greater / faster [1]
    b) Active transport moves molecules from low to high concentration [1]; it requires a transfer of energy from respiration [1]
2. a) Concentration of sugar solution [1]

    > The independent variable is the one that is deliberately changed.

    b) **Any one of:** all the chips from the same potato [1]; same size chip [1]; length of time in the solution [1]; volume of solution [1]
    c) 0.0mol/dm³ [1]
    d) The potato chip took in water [1]; by osmosis [1]; because the concentration inside the cell was greater than the solution outside [1]
    e) Repeat the experiment for each concentration several times and calculate the mean (average) [1]
3. 1 mark for each correct row [4]

	Osmosis	Diffusion	Active Transport
Can cause a substance to enter a cell	✓	✓	✓
Needs energy from respiration	✗	✗	✓
Can move a substance against a concentration gradient	✗	✗	✓
Is responsible for oxygen moving into the red blood cells in the lungs	✗	✓	✗

Page 37 Levels of Organisation

1. Four correctly drawn lines [3] (2 marks for two correct lines and 1 mark for one correct line)
    A cell that is hollow and forms tubes – to transport water
    A cell that has a flagellum – to swim
    A cell that is full of protein fibres – to contract
    A cell that has a long projection with branched endings – to carry nerve impulses

Page 37 Digestion

1. a) Liver [1]
    b) Gall bladder [1]
    c) Small intestine / duodenum [1]
    d) To give the best pH for enzymes in the small intestine to work [1]
2. a) Protease [1]
    b) Amino acids [1]

Page 38 Blood and the Circulation

1. a) A = trachea [1]; B = bronchus [1]; C = bronchiole [1]
    b) i) Carbon dioxide [1]
       ii) Oxygen [1]
2. lungs [1]; vein [1]; body [1]; deoxygenated [1]; vena cava [1]

> Remember: arteries always take blood away from the heart ('A' for away and 'A' for artery).

Page 38 Non-Communicable Diseases

1. Four correctly drawn lines [3] (2 marks for two correct lines and 1 mark for one correct line)
    Type 2 diabetes – obesity
    liver damage – excess alcohol intake
    lung cancer – smoking
    skin cancer – UV light
2. a) $89.5 - 38.9 = 50.6$, $\frac{50.6}{89.5} \times 100$ [1]; = 56.5% [1]
    b) There are fewer alveoli [1]; so less surface area [1]; therefore, slower / less diffusion [1]
    c) There is less oxygen for respiration [1]; therefore, less energy available [1]

Page 39 Transport in Plants

1. absorbing carbon dioxide [1]; giving off water vapour [1]; giving off oxygen [1]
2. a) Xylem [1]
    b) Phloem [1]

Pages 40–55 Revise Questions

Page 41 Quick Test

1. Use of a handkerchief / tissue
2. So that microorganisms from the uncooked meat do not contaminate the cooked meat
3. It is not a barrier method of contraception
4. To prevent the fungal spores spreading to healthy leaves

Page 43 Quick Test

1. With acid
2. The process that a white blood cell uses to engulf pathogens
3. Antibodies
4. To stimulate antibody production without making the person too ill

Page 45 Quick Test

1. Antibiotics only kill bacteria (HIV is caused by a virus)
2. Resistant
3. Because willow bark contains aspirin
4. Neither the patient nor the doctor know which medication the patient is taking

Page 47 Quick Test

1. Chlorophyll
2. From the air
3. **Any three of:** temperature; carbon dioxide concentration; light availability; chlorophyll concentration
4. Nitrate ions (nitrogen)

Page 49 Quick Test

1. Aerobic
2. Respiration rate increases to generate more heat
3. Lactic acid
4. The liver
5. Lactic acid builds up and causes cramp

Page 51 Quick Test

1. Sensory neurone
2. A synapse
3. The brain and spinal cord

Page 53 Quick Test

1. In the neck
2. In the pancreas
3. It stimulates them to store glucose as glycogen
4. By injecting insulin

Page 55 Quick Test

1. Testosterone
2. Pituitary gland
3. LH
4. Outside the body / in a Petri dish

Pages 56–59 Review Questions

Page 56 Cell Structure

1. cytoplasm [1]; cell wall [1]; free within the cell [1]; plasmids [1]

Page 56 Investigating Cells

1. a) 20µm [1]
    b) There are chloroplasts present [1]
    c) **Any one of:** no cell wall [1]; moves around due to flagellum [1]
    d) The chromosomes must divide / DNA must be replicated [1]

Answers

Page 57 Cell Division
1. a) $\frac{15}{0.03}$ [1]; = 500 times [1]
    b) Cell A [1]; sets of chromosomes have moved apart [1]

Page 57 Transport In and Out of Cells
1. a) $\frac{-0.6}{2.7} \times 100$ [1]; = −22.2% [1]
    b) No change in mass [1]; which means, no net water uptake or loss / no net osmosis [1]; so, the concentration of the sugar solution is the same as the cell contents [1]
    c) To remove any liquid [1]; so that the mass of the chip is not affected [1]

Page 58 Levels of Organisation
1. Four correctly drawn lines [3] (2 marks for two correct lines and 1 mark for one correct line)
    glandular – can produce enzymes and hormones
    nervous – can carry electrical impulses
    muscular – can contract to bring about movement
    epithelial – a lining / covering tissue

Page 58 Digestion
1. protease – proteins – amino acids [1]; amylase – starch – maltose [1]; lipase – fats – glycerol and fatty acids [1]

Page 58 Blood and the Circulation
1. a) Red blood cell / erythrocyte [1]
    b) So that it can fit in more haemoglobin [1]
2. a) i) double [1]
       ii) twice [1]
       iii) arteries [1]
    b) tissues [1]; lungs [1]; glucose / food [1]

Page 59 Non-Communicable Diseases
1. a) Build-up of fatty deposits / blockages [1]; in the arteries [1]

> Remember to say that that cholesterol build-up happens in the arteries, not veins or just blood vessels.

    b) Less blood reaches heart muscle [1]; less oxygen / glucose available [1]; muscle cells cannot contract / beat [1]; possible heart attack [1]

Page 59 Transport in Plants
1. epidermis [1]; mesophyll [1]; gaps [1]

Pages 60–63 Practice Questions

Page 60 Pathogens and Disease
1. Four correctly drawn lines [3] (2 marks for two correct lines and 1 mark for one correct line)

bacterium – salmonella
virus – measles
protist – malaria
fungus – rose black spot

Page 60 Human Defences Against Disease
1. dead [1]; weakened [1]; antibodies [1]; immune [1] (first two marks can be given in any order)

> Do not use the term 'resistant' here. Resistance is when an organism is born with the ability not to get a disease.

Page 60 Treating Diseases
1. a) i) A chemical made by microorganisms [1]; that kills bacteria / stops bacteria reproducing [1]
       ii) Viruses [1]
    b) i) MRSA [1]
       ii) Antibiotics are only available by prescription [1]; only prescribed when necessary [1]; patients are told that they must finish the dose [1]

Pages 61 Photosynthesis
1. a) water [1]; oxygen [1]
    b) Light [1]; chlorophyll [1]
    c) Soil [1]

Page 61 Respiration and Exercise
1. a) Accurately plotted points [1]; joined by a smooth line [1]

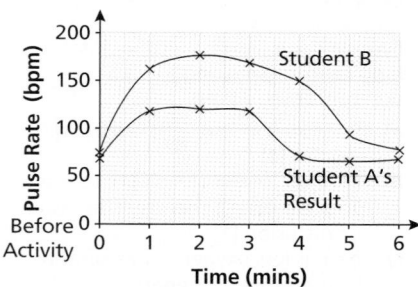

    b) Student A [1]; **Any one of:** their pulse rate did not rise so high [1]; their pulse rate returned to normal quicker [1]
    c) Fatigue is caused by a lactic acid build-up [1]; produced by anaerobic respiration [1]; because insufficient oxygen is available [1]

Page 62 Homeostasis and the Nervous System
1. a) stimulus, receptor, sensory neurone, relay neurone, motor neurone, effector, response [3] (2 marks for three or four stages in the correct place; 1 mark for two stages in the correct place)
    b) Between the sensory and relay neurone [1]; and between the relay and motor neurone [1]

2. a) i) C [1]
       ii) A [1]
       iii) B [1]
    b) A motor neurone [1]

Page 63 Hormones and Homeostasis
1. a) 93mg / 100cm³ of blood [1]
    b) 280µg / 100cm³ of blood [1]
    c) 26 minutes after eating the glucose [1]
    d) Increased glucose levels cause release of more insulin [1]; causing greater uptake of glucose into cells [1]; so more glucose is converted to glycogen [1]
    e) The patient's glucose levels would have been higher [1]; and would have stayed higher for longer [1]

Page 63 Hormones and Reproduction
1. a) Ovaries [1]
    b) Possible side effects [1] (Accept named side effects)

Pages 64–79 Revise Questions

Page 65 Quick Test
1. Long shoots that are produced in asexual reproduction
2. Four
3. 39
4. DNA

Page 67 Quick Test
1. Alleles
2. Heterozygous
3. Recessive
4. X and Y

Page 69 Quick Test
1. Darwin
2. When the parasites reproduce, mutations occur; some mutated parasites might be resistant to the drugs traditionally used; these parasites survive and reproduce, so a resistant population develops.
3. **Any one of:** soft-bodied organisms did not fossilise; fossils have been destroyed

Page 71 Quick Test
1. **Any one of:** high milk yield; fat content of milk; high-quality beef (any other sensible answer)
2. An enzyme
3. Worries about the effects on wild plant populations / human health

Page 73 Quick Test
1. Order
2. Three
3. Its genus is *Felis* and species is *(Felis) catus*
4. **Any two of:** changes to the environment over long periods of time; new predators; new diseases; more successful competitors; a single catastrophic event

Page 75 Quick Test

1. Tropical forest / oak woodland (Accept any other sensible answer)
2. Light; minerals; water; space
3. Organisms that can survive in extreme conditions
4. A square frame used for estimating populations
5. Take more samples and calculate a mean (average)

Page 77 Quick Test

1. Through photosynthesis in green plants
2. Evaporation
3. The predators run short of food and so some die

Page 79 Quick Test

1. **Any one of:** increasing population; rise in standard of living
2. Acid rain
3. **Any one of:** increased demand for wood; land for crops or farming animals; minerals
4. It increases it because there are fewer trees to take up carbon dioxide by photosynthesis

Pages 80–83 Review Questions

Page 80 Pathogens and Disease

1. a) An organism that carries a disease from one host to another without being infected itself **[1]**
   b) Mosquito **[1]**
   c) Insect repellent / killing the mosquitoes **[1]**; mosquito nets **[1]**

Page 80 Human Defences Against Disease

1. a) dead **[1]**; white **[1]**; antibodies **[1]**; live **[1]**
   b) By becoming infected with the disease **[1]**
2. a) 1250 **[1]**
   b) 8th to 10th May **[1]**
   c) People were not immune **[1]**; they had not been vaccinated / no vaccine was available **[1]**
   d) Viruses are inside cells for much of the time **[1]**; so drugs / antibodies find it more difficult to reach them **[1]**

Page 81 Treating Diseases

1. a) Because the drug may work differently on humans compared with other animals. **[1]**
   b) A placebo is a tablet / liquid containing no drug **[1]**; used for comparison (control) **[1]**; to make sure any positive responses are not just a psychological response to taking a pill / medicine **[1]**
   c) Need to balance the risk with the benefit gained **[1]**; a slight risk of side effects may be acceptable if the benefit is that it relieves great pain **[1]**

Page 82 Photosynthesis

1. a) Number of bubbles increase as carbon dioxide levels increase **[1]**; then the number of bubbles per minute levels off **[1]**

> Be careful not to say that photosynthesis stops. It is still taking place – it is just not increasing in rate.

   b) **Any two of:** she kept the lamp the same distance away **[1]**; used the same pondweed **[1]**; kept the temperature the same **[1]**
   c) There is some carbon dioxide dissolved in the water **[1]**

Page 82 Respiration and Exercise

1. a) Carbon dioxide **[1]**
   b) i) Anaerobic / fermentation **[1]**
      ii) Alcohol / ethanol **[1]**
   c) To provide food **[1]**; for the yeast **[1]**

Page 83 Homeostasis and the Nervous System

1. a) Sensory neurone **[1]**
   b) A synapse **[1]**

Page 83 Hormones and Homeostasis

1. a) Glycogen **[1]**
   b) Liver **[1]**
   c) Glucagon **[1]**

> Be very careful with the words glucagon, glycogen and glucose. Any spelling mistakes may mean that your answer is unclear and marks may be lost.

   d) The pancreas does not produce insulin. **[1]**

Page 83 Hormones and Reproduction

1. a) FSH / follicle stimulating hormone **[1]**
   b) Oestrogen **[1]**
   c) i) FSH can be given **[1]**; to increase production of eggs **[1]**
      ii) Combined pill containing oestrogen and progesterone **[1]**; inhibits FSH, so ovulation stops **[1]**

Pages 84–87 Practice Questions

Page 84 Sexual and Asexual Reproduction

1. a) A = nucleus **[1]**; B = cytoplasm **[1]**; C = chromosomes **[1]**; D = cell membrane **[1]**
   b) fusion **[1]**; DNA **[1]**; variation **[1]**

Page 84 Patterns of Inheritance

1. a) XX and XY **[1]**
   b) Correctly labelled diagram: female on left (XX) and male on right (XY) **[1]**
2. Four correctly drawn lines **[3]** (2 marks for two correct lines and 1 mark for one correct line)
   both alleles are the same – homozygous
   two different alleles – heterozygous

what the organism looks like – phenotype
an allele that is always expressed if present – dominant
3. a) NN **[1]**
   b) Correct parent: nn **[1]**;correct gametes: N, n, n, n **[1]**; correct offspring: Nn, nn, Nn, nn **[1]**
   c) $\frac{2}{4} \times 100$ **[1]**; = 50% **[1]**

Page 85 Variation and Evolution

1. The giraffes showed variation **[1]**; longer-necked individuals were more likely to survive as they could get more food **[1]**; they breed and pass on the gene / characteristic to their offspring **[1]**

Page 86 Manipulating Genes

1. a) Enzymes **[1]**
   b) **Any two of:** they easily take up genes in plasmids **[1]**; they use them to make proteins **[1]**; they replicate rapidly **[1]**

Page 86 Classification

1. a) **Any one of:** dead organisms are covered in mud and compressed **[1]**; from hard parts of animals that do not decay **[1]**; from parts of organisms that have not decayed because the conditions prevented it **[1]**; when parts of an organism are replaced by other materials as they decay **[1]**; preserved traces of organisms, e.g. footprints **[1]**
   b) May be deep in rock / not many to find as they have been destroyed **[1]**
   c) **Any three of:** geographical changes **[1]**; new predators **[1]**; new diseases **[1]**; more competitors **[1]**; catastrophic events **[1]**; human actions **[1]**
   d) Great auk / dodo **[1]** (Accept any other sensible answer)

Page 86 Ecosystems

1. a) **Any two of:** extreme temperatures **[1]**; lack of food **[1]**; lack of water **[1]**
   b) i) The thick stem enables it to store water / makes it tough for animals to bite and eat **[1]**
      ii) **Any one of:** a smaller surface area so less water loss **[1]**; stops animals eating it **[1]**

Page 87 Cycles and Feeding Relationships

1. a) i) A **[1]**
      ii) D **[1]**
      iii) B **[1]**
      iv) C **[1]**
   b) Combustion **[1]**

Page 87 Disrupting Ecosystems

1. tropical **[1]**; trees **[1]**; carbon dioxide **[1]**; biodiversity **[1]**; extinct **[1]**; habitats **[1]**

Answers

d) They might encourage companies and individuals to choose other options **[1]**; leading to less carbon dioxide being produced **[1]**

Page 157 Earth's Resources
1. **a)** Water with low levels of salts and microorganisms **[1]**
    b) Potable **[1]**
    c) To remove solids **[1]**
    d) To kill microorganisms **[1]**
2. **a)** A naturally occurring rock that contains metal, or metal compounds, in sufficient amounts to make it economically worthwhile extracting them **[1]**
    b) The ore is heated **[1]**; with carbon (which is more reactive than copper) **[1]**
    c) Bacteria **[1]**; are used to remove copper from a leachate solution **[1]**

Page 157 Using Resources
1. **a)** They allow people to make informed decisions **[1]**
    b) Some aspects of an LCA are hard to quantify (e.g. the impact of pollution) **[1]** (Accept any other sensible answer)
2. **Any three of:** it saves energy **[1]**; it saves money **[1]**; it ensures that resources are not wasted **[1]**; it reduces damage to the environment **[1]** (Accept any other sensible answer)

Page 159 Quick Test
1. A scalar quantity just has size, but a vector has size and direction.
2. Magnetism / gravity / electrostatic attraction is a non-contact force; a magnet will attract or repel another magnet without needing to touch it / one mass will gravitationally attract another / one charge will attract or repel another charge without physical contact.
3. 128N
4. 100N to the east

Page 161 Quick Test
1. $150 \times 2 = 300J$
2. **a)** weight = mass × gravitational field strength (g) / $W = mg$; $W = 50 \times 10 = 500N$; work done = force × distance / $W = Fs$; $W = 500 \times 4$ $W = 2000Nm$ or $2000J$
    b) Chemical energy to gravitational potential energy
3. The springs must compress as they go over a bump and then return to their original shape. If they did not, they would be bent out of shape by bumps on the road and no longer work.

Page 163 Quick Test
1. Distance is the total distance travelled in any direction; displacement is the total distance in a straight line from the start point and includes the direction from the start point.

2. $\frac{1000}{50} = 20m/s$
3. Speed

Page 165 Quick Test
1. $6^2 - 2^2 = 2 \times a \times 8$, $36 - 4 = 16a$, $32 = 16a$, $a = \frac{32}{16} = 2m/s^2$
2. The object represented by graph A is accelerating at twice the rate of the object represented by graph B.
3. To identify anomalous results, which can then be removed and allow a mean to be taken.

Page 167 Quick Test
1. $2000 \times 20 = 40\,000$kg m/s
2. As the fish swims it exerts a force backwards on the water. This creates an equal and opposite force from the water on the fish, that pushes the fish forwards.
3. **a)** $600 \times 8 = 4800$Nm
    b) $\frac{4800}{12} = 400$kg

Page 169 Quick Test
1. It might increase the stopping distance, as it increases the thinking distance by causing a distraction.
2. Answer should be between 2160N and 3375N using city speeds as 12–15m/s.
3. One person holds a ruler with the bottom of the ruler level with the other person's hand. They let go of the ruler and the other person catches it when they see it move. The reaction time is calculated from the distance the ruler falls before being caught.
    OR
    One person presses a switch that turns on a light and starts a timer. The other person presses a switch that stops the timer when they see the light.

Page 171 Quick Test
1. Diagram should show chemical to gravitational (+ wasted heat)
2. $E = 1.5 \times 4200 \times (100 - 20) = 6300 \times 80 = 504\,000J$
3. Reproducible means that another person could follow your method and get the same answer. It is important because it helps to show that the results are valid.

Page 173 Quick Test
1. It is released as heat and lost to the surroundings.
2. To build it you need to flood valleys and displace inhabitants / destroy habitats.

Page 174 Alkanes
1. CH_4 **[1]**

> Saturated hydrocarbons only contain single covalent bonds, so each carbon atom is bonded to four other atoms.

2. **a)** Crude oil is heated **[1]**; the vapour passes up the fractionating column **[1]**; each fraction cools and condenses at a different temperature and is collected **[1]**
    b) Fuel for some cars / lorries / trains **[1]**
3. **a)** $C_3H_8 + 5O_2 \rightarrow 3CO_2 + 4H_2O$ **[3]** (1 mark for each correct number shown in bold)
    b) It is poisonous / toxic **[1]**

Page 174 Cracking Hydrocarbons
1. Add bromine water **[1]**; shake **[1]**; propene / alkenes decolorise the bromine water **[1]**; propane / alkanes do not react **[1]**
2. Single covalent and double covalent **[1]**

Page 175 Chemical Analysis
1. **a) i)** Absorbent paper **[1]**
        ii) Solvent / water **[1]**
    b) It does not run / ink would dissolve and affect the results **[1]**
    c) $\frac{4.5}{10.0}$ **[1]**; = 0.45 **[1]**
    d) It produces two spots / it has two components **[1]**

Page 175 The Earth's Atmosphere
1. **a)** 2.7 billion **[1]**
    b) carbon dioxide **[1]**; oxygen **[1]**
2. **a)** By volcanic activity **[1]**
    b) Carbon dioxide **[1]**
    c) By bacteria **[1]**
    d) Carbon dioxide levels went down **[1]**; as carbon was locked-up in rocks **[1]**
3. **a)** Insoluble carbonates **[1]**; soluble hydrogen carbonates **[1]**
    b) Algae take carbon dioxide out of the atmosphere for photosynthesis **[1]**

Page 176 Greenhouse Gases
1. **a)** CO_2 **[1]**
    b) They absorb / trap **[1]**; infrared radiation **[1]**
2. Deforestation **[1]**
3. **a)** Carbon dioxide **[1]**
    b) Trees photosynthesise **[1]**; taking in carbon dioxide from the atmosphere, which reduces the level of carbon dioxide **[1]**

Page 177 Earth's Resources
1. Ultraviolet radiation **[1]**
2. **a)** It is a good electrical conductor **[1]**; it bends easily **[1]**
    b) It is a good thermal conductor **[1]**
    c) Copper is less reactive than carbon **[1]**
    d) Plants **[1]**; grow and take up the metal **[1]**; which is collected by burning the plants **[1]**
3. Digging up large amounts of rock **[1]**; moving large amounts of rock **[1]**; and disposing of large amounts of waste material all have a negative impact on the environment **[1]**

Page 177 Using Resources
1. a) Life cycle assessment **[1]**
   b) **Any three of**: cost of raw materials **[1]**; cost of production **[1]**; availability of materials **[1]**; durability **[1]**; potential to recycle / reuse **[1]** **(Accept any other sensible answer)**

Page 178 Forces – An Introduction
1. A scalar quantity has size only **[1]**; a vector quantity has size and direction **[1]**; mass is the measure of how much matter is in an object **[1]**; weight is the force of gravity and, as a force, it acts in a particular direction **[1]**
2. a) F_1 is twice the size of F_2 **[1]**; and acts in the opposite direction **[1]**
   b) Zero **[1]**
3. Correctly drawn vertical and horizontal arrows **[1]**

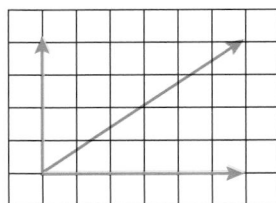

Page 178 Forces in Action
1. a) It increases **[1]**
   b) 150 000J **[1]**
   c) $\frac{150\,000}{50}$ **[1]**; = 3000N **[1]**
2. a) $\frac{2597}{10}$ = 259.7kg **[1]**; = 260kg (to the nearest kg) **[1]**
   b) mass = 259.7 + 100 = 359.7kg **[1]**; weight = 359.7 × 10 = 3597N **[1]**
   c) They will be the same / equal **[1]**

Page 179 Forces and Motion
1. a) 5 miles **[1]**
   b) 1 mile **[1]**; east **[1]**
2. a) i) 52 × 3.65 = 189.8 miles **[1]**
   ii) 0 miles **[1]**
   iii) $\frac{189.8}{1.5}$ **[1]**; = 126.5mph (to 1 decimal place) **[1]**
   b) This occurs when direction is changing **[1]**; but speed is constant **[1]**; so, could occur on bends / corners **[1]**
3. a) 3 seconds **[1]**
   b) C to D **[1]**
   c) $\frac{3}{5}$ **[1]**; = 0.6m/s **[1]**
   d) Change in speed / acceleration and deceleration **[1]**

Page 179 Forces and Acceleration
1. $v^2 = 2as + u^2$ **[1]**; $v^2 = (2 \times 4 \times 100) + 10^2$ **[1]**; $v^2 = 900$ **[1]**; $v = \sqrt{900} = 30$m/s **[1]**

2. a) Correct axes and labels **[1]**; accurately plotted points **[1]**; straight lines joining points **[1]**

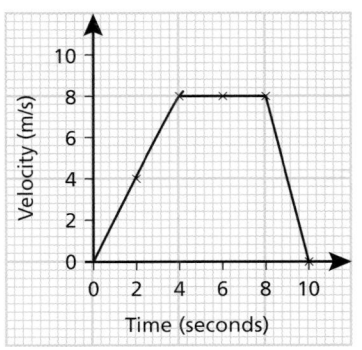

   b) A constant speed of 8m/s **[1]**
   c) $\frac{8}{4}$ **[1]**; = 2m/s² **[1]**
   d) Area under graph divided into three sections **[1]**;
   = $\frac{(8 \times 4)}{2} + (8 \times 4) + \frac{(8 \times 2)}{2}$ **[1]**;
   = 56m **[1]**

Page 180 Terminal Velocity and Momentum
1. a) acceleration = $\frac{\text{change in velocity}}{\text{time}}$ /
   $a = \frac{\Delta v}{t}$ **[1]**; $a = \frac{(4-0)}{0.4}$ **[1]**; = 10m/s² **[1]**
   b) As the object gets faster, the resistive forces increase **[1]**; eventually they balance the weight **[1]**; and the object reaches terminal velocity **[1]**
2. a) 25 000N **[1]**
   b) The helicopter pushes the air downwards **[1]**; which creates an equal and opposite force from the air on the helicopter, which pushes the helicopter upwards **[1]**; when the force exerted on the air downwards is equal to the weight of the helicopter, the helicopter hovers at a constant height **[1]**

Page 180 Stopping and Braking
1. **Any three of**: alcohol **[1]**; fatigue **[1]**; drugs **[1]**; distractions **[1]** (Accept specific distractions, e.g. mobile phone use)
2. The car travelling at a higher speed has more energy **[1]**; the brakes have to take this energy from the car to stop it **[1]**; and they convert it to heat **[1]**

Page 181 Energy Stores and Transfers
1. a) i) Chemical Energy **[1]**
   ii) Kinetic Energy **[1]**
   iii) Sound Energy **[1]**
   iv) Heat Energy **[1]**
   b) Each square represents 100kJ **[1]**; so, kinetic energy = 300kJ **[1]**
2. a) She begins with gravitational energy **[1]**; as she descends this is converted into kinetic energy **[1]**; once she reaches the bottom, she has the maximum amount of kinetic energy and the minimum amount of

gravitational energy **[1]**; as she travels up the other side, kinetic energy is converted back to gravitational energy **[1]**; during this some energy is converted to sound **[1]**; and heat, which are lost to the surroundings **[1]**
   b) kinetic energy = 0.5 × mass × (speed)² /
   $E_k = \frac{1}{2}mv^2$ **[1]**; = 0.5 × 60 × 4² **[1]**; = 480J **[1]**
   c) gravitational potential energy = mass × gravitational field strength × height / $E_p = mgh$ **[1]**; = 60 × 10 × 2 **[1]**; = 1200J **[1]**

Page 181 Energy Transfers and Resources
1. It will eventually run out and cannot be replaced **[1]**
2. It depends on the thermal conductivity of the materials **[1]**; the thickness of the materials **[1]**; and the difference in temperature between the inside and outside **[1]**

Page 183 Quick Test
1. At right-angles to the direction of energy transfer.
2. $\frac{1}{20}$ = 0.05s
3. If the waves are visible, count the number of waves passing a fixed point in a second. Measure the wavelength. Multiply these two measurements together.

Page 185 Quick Test
1. Radio waves, microwaves, infrared, visible light, ultraviolet (UV), X-rays, gamma rays
2. A line drawn at right-angles to the surface at the point of origin.
3. Away from the normal.

Page 187 Quick Test
1. Microwaves are absorbed by food and heat it up. (Accept any other sensible answer)
2. They are easily absorbed by water.
3. They have more energy and are ionising, so they can cause mutations in cells resulting in cancer

Page 189 Quick Test
1. 10A
2.

Cell
Variable resistor
Ammeter (A)
Diode
Voltmeter (V)

3. $\frac{10}{2}$ = 5Ω

Answers

Page 191 Quick Test
1. Diode
2. Thermistor

Page 193 Quick Test
1. The resistance of R_2 is one-fifth that of R_1; it has a 1V potential difference across it
2. $P = I^2R = 2^2 \times 6 = 24W$
3. $P = IV$, $P = 30 \times 230 = 6900W$

Page 195 Quick Test
1. An alternating current changes direction rapidly and goes backwards and forwards; a direct current travels in one direction only.
2. 230V
3. $P = IV = 2.5 \times 1.2 = 3W$; $E = pt = 3 \times 10 = 30J$
4. $E = pt = 2000 \times 50 = 100\,000J$

Page 197 Quick Test
1. 2.5V
2. A Sankey diagram with an electrical input showing a 50/50 split between useful energy (heat and kinetic) and waste energy (heat and sound).

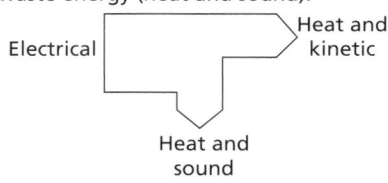

3. 2hrs = $2 \times 60 \times 60 = 7200s$, $E = Pt = 500 \times 7200 = 3\,600\,000J$
4. The power is generated in large stations; the electricity is transferred at high voltage / low current.

Pages 198–201 Review Questions

Page 198 Forces – An Introduction
1. A contact force only occurs where objects are touching [1]; a non-contact force can be felt at a distance [1]
2. a) i) $70 \times 1.6 = 112N$ [1]
   ii) $70 \times 10 = 700N$ [1]
   b) The force from the astronaut's legs, and the speed they would leave the ground at, is unchanged [1]; but the force acting downwards is much less, so the astronaut travels higher before coming back down [1]
3. Correctly drawn driving force [1]; frictional force [1]; and resultant force [1]; resultant force of 15N stated [1]

Frictional 5N ← ▬▬▬ → Driving 20N
Resultant 15N

Page 198 Forces in Action
1. Five floors = $5 \times 3.5 = 17.5m$ [1]; work done = 1200×17.5 [1]; = $21\,000J$ [1]
2. Weight = $40 \times 10 = 400N$ [1]; work done = 400×3 [1]; = $1200J$ [1]
3. There needs to be two forces pulling in opposite directions [1]; or the spring would just move [1]

Page 199 Forces and Motion (continued)
4. The limit of proportionality is the point that once exceeded [1]; the extension is no longer directly proportional to the applied force [1]
5. The rubber band obeys Hooke's law up to a force of 5N [1]; after this it has passed the limit of proportionality [1]; and the amount of force required to keep increasing the length increases [1]

Page 199 Forces and Motion
1. Velocity is a vector so has direction and magnitude [1]; speed has only magnitude [1]
2. a) $3 + 2 + 1 = 6$ miles [1]
   b) Correctly drawn vector diagram with three arrows showing the stages of the journey [1]; and one arrow showing the displacement [1]

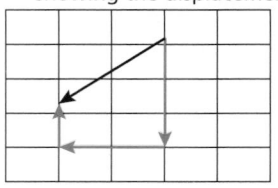

3. $\frac{4500}{3}$ [1]; = 1500m/s [1]
4. a) High speed / velocity [1]
   b) Stationary / at rest [1]
   c) Changing speed / acceleration or deceleration [1]
5. A planet in orbit / a roundabout / a car on a bend (any other sensible example) [1]; it is changing direction so the velocity changes [1]; but the magnitude of velocity i.e. the speed is constant [1]
6. Inertia [1]

Page 199 Forces and Acceleration
1. a) True [1]
   b) False [1]
   c) False [1]
2. $0 - 20^2 = (2 \times -2 \times s)$ [1]; $s = \frac{-400}{-4}$ [1]; $s = 100m$ [1]
3. a) Speeding up [1]
   b) Slowing down [1]
   c) Constant forwards / positive velocity [1]
   d) Constant backwards / negative velocity [1]
   e) Distance travelled [1]

Page 200 Terminal Velocity and Momentum
1. The blades on the propeller push the air backwards [1]; this creates an equal and opposite force of the air pushing on the hovercraft [1]; which moves it forwards [1]
2. momentum = mass × velocity / $p = mv$ [1]
3. Momentum of the cannon ball = $1 \times 120 = 120kgm/s$ [1]; since momentum is conserved, the momentum of the cannon is 120kgm/s in the opposite direction [1];
   $v = \frac{p}{m} = \frac{120}{240}$ [1]; = 0.5m/s [1]

Page 200 Stopping and Braking
1. **Any three of:** rain / snow [1]; old / worn tyres [1]; old / worn brakes [1]; poor road condition [1]
2. He is incorrect [1]; doubling the speed quadruples the kinetic energy [1]; so the brakes need to do four times the work and it takes four times the distance to stop [1]
3. The reaction time is the time it takes a person to react to something [1]; the thinking distance is the distance a vehicle travels during this reaction time before the driver applies the brakes [1]; the stopping distance is the sum of thinking and braking distance [1]
4. Thinking distance = $12 \times 0.5 = 6m$ [1]; braking distance = $6 + 14$ [1]; = 20m [1]

Page 201 Energy Stores and Transfers
1. a) Work done by motor = the gravitational potential energy [1]; $2200 \times 10 \times 16.2$ [1]; = $356\,400J$ [1]
   b) $356\,400J$ [1]
   c) $v^2 = \frac{2 \times 356\,400}{2200} = 324$ [1]; $v = \sqrt{324} = 18$m/s [1]
2. a) Extension = 0.6m [1]; energy stored = $0.5 \times 100 \times 0.6^2$ [1]; =18J [1]
   b) $E_k = 18J = 0.5 \times 0.0576 \times v^2$ [1]; $v^2 = \frac{18}{(0.5 \times 0.0576)} = 625$ [1]; $v = \sqrt{625} = 25$m/s [1]
   c) Energy is transferred to the surroundings [1]; as heat and sound [1]

Page 201 Energy Transfers and Resources
1. 1 mark for each correctly completed row [4]

Input Energy	Situation	Useful Output Energy
Chemical / Kinetic	Walking uphill	Gravitational
Gravitational	Sliding down a slide	Kinetic
Elastic	Firing a catapult	Kinetic
Chemical	Fireworks / sparklers	Light and sound

Pages 202–205 Practice Questions

Page 202 Waves and Wave Properties
1. When a stone is dropped in water the resulting waves move outwards [1]; but no hole is left in the middle [1] (Accept any other sensible example)
2. particles [1]; oscillate [1]; fixed [1]; energy [1]
3. a) Correct labels showing amplitude [1]; and wavelength [1] (Accept an arrow for amplitude from the midpoint to the maximum negative displacement

and any arrow for wavelength beginning and ending at an equivalent position of the repeating wave)

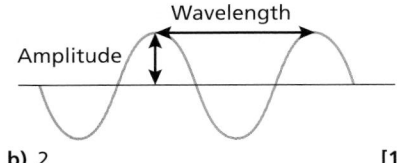

b) 2 **[1]**

c) 2 × 2 = 4 seconds **[1]**

4. a) frequency = $\frac{1}{period}$ **[1]**; = $\frac{1}{0.05}$

= 20Hz **[1]**

b) wavelength = $\frac{wave\ speed}{frequency}$ / $\lambda = \frac{v}{f}$ **[1]**;

= $\frac{330}{20}$ = 16.5m **[1]**

5. Stand a known distance from a large wall or cliff face and bang two pieces of wood together **[1]**; measure the time between banging the wood and hearing the echo **[1]**; divide this time by two **[1]**; and then divide into the distance to the wall to give the speed of sound **[1]**

Page 202 Electromagnetic Waves

1. Correct drawing showing normal where light enters the block **[1]**; ray line bending towards normal **[1]**; normal where light exits the block **[1]**; ray line bending away from normal **[1]**

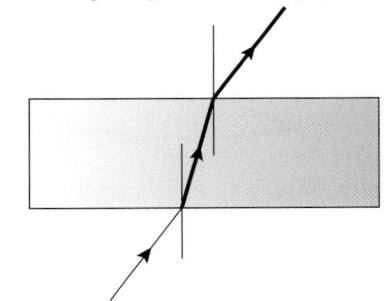

2. Its speed changes **[1]**

3. It is faster in a vacuum **[1]**

Page 203 The Electromagnetic Spectrum

1. a) Microwaves **[1]**

b) An opinion **[1]**

c) i) Short to medium term use by adults does not cause brain tumours. **[1]**

ii) Microwaves are not ionising radiation **[1]**; they are low frequency **[1]**; and do not carry enough energy to cause the DNA of a cell to mutate when they are absorbed **[1]**

Page 204 An Introduction to Electricity

1. a) The variable resistor allows a larger number of measurements **[1]**; with smaller intervals, more precise results can be recorded **[1]**

b) This prevents the circuit from becoming hot **[1]**; which could affect resistance **[1]**

c) Resistance (R) = 50Ω in darkness and 1Ω in maximum brightness **[1]**; maximum difference = 50 − 1 = 49Ω **[1]**

Page 204 Circuits and Resistance

1. a) It is constant **[1]**

b) A line drawn that passes through zero **[1]**; but is less steep than the first line **[1]**

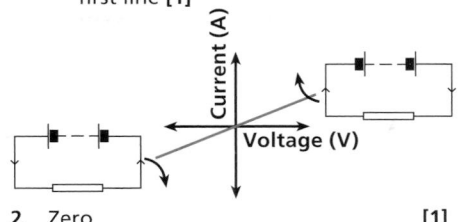

2. Zero **[1]**

Page 205 Circuits and Power

1. a) Diagram showing a correctly connected 6V battery **[1]**; and three bulbs in series **[1]**

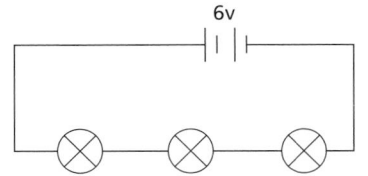

b) $\frac{6}{3}$ = 2V **[1]**

c) 2A **[1]**

d) resistance = $\frac{potential\ difference}{current}$ /

$R = \frac{V}{I}$ **[1]**; = $\frac{2}{2}$ **[1]**; = 1Ω **[1]**

e) 1 + 1 + 1 = 3Ω **[1]**

Page 205 Domestic Uses of Electricity

1. Live – brown **[1]**; neutral – blue **[1]**; earth – yellow and green **[1]**

2. The live wire is at a high potential **[1]**; and a person is at zero **[1]**; touching the wire creates a large potential difference **[1]**; so a large current flows to earth through the person **[1]**

3. To complete the circuit **[1]**

4. 0V **[1]**

Page 205 Electrical Energy in Devices

1. a) TV **[1]**

b) Comparing power × time for each appliance **[1]**; the tumble dryer has the greatest value **[1]**

c) Time (t) = 1 × 60 × 60 = 3600s **[1]**; energy transferred = power × time / $E = Pt$ **[1]**; $E = 5000 × 3600$ **[1]**; = 18 000 000J **[1]**

Pages 206–217 Revise Questions

Page 207 Quick Test

1. The magnet is very strong.

2.

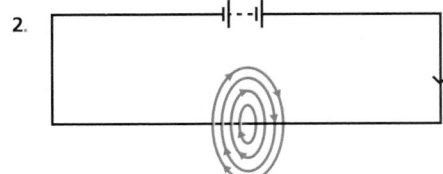

3. Increase current / more turns of wire in the coil.

Page 209 Quick Test

1. Increase current / use larger magnets

2. The coil would not be free to spin and the current would not stay in the same direction, so the coil would move to one position and stay there.

3. force = magnetic flux density × current × length of wire / $F = BIl$; $F = 2 × 10 × 8$ = 160N

Page 211 Quick Test

1. Because there are no gaps between the particles.

2. By putting the object in a liquid. The amount of displaced liquid is equal to the volume. Then find the object's mass and use the equation density = $\frac{mass}{volume}$ to calculate the density.

3. The particles would move faster and hit surfaces with a bigger force and more frequently; the pressure would increase because $P = \frac{F}{A}$.

Page 213 Quick Test

1. Positive / +1

2. An element is defined by the number of protons it has. Isotopes of an element have the same number of protons but different numbers of neutrons.

3. The mass number tells us the total number of protons plus neutrons in the nucleus.

4. a) That some alpha particles were reflected.

b) This showed that most of the mass of an atom is in one place and has a positive charge.

Page 215 Quick Test

1. Gamma rays

2. Because beta has a longer range in air and can penetrate the skin to damage internal organs; alpha has a very short range in air and is unlikely to damage internal organs unless the radioactive source is inside the body.

3. Contamination leaves the substance radioactive but irradiation does not.

Page 217 Quick Test

1. It is not possible to predict exactly when it will occur.

2. 2 months

3. $\frac{1}{32}$

Pages 218–221 Review Questions

Page 218 Waves and Wave Properties

1. a) i) Transverse **[1]**

ii) Move their hand up and down faster **[1]**

iii) Move the end of the spring up and down through a greater distance **[1]**

b) i) Longitudinal **[1]**

ii) Sound **[1]** (Accept any other sensible answer)

Answers

Page 218 Electromagnetic Waves

1. As waves move from one medium to another, they change speed [1]; if entering at an angle, the part that crosses the boundary first changes speed first [1]; if it slows down, the wave will turn towards the normal [1]; if it speeds up, the wave will turn away from the normal [1]
2. The refractive index of the block should be controlled [1]; by using the same block [1]; if the block is changed, the amount of refraction could change, even when the angle is the same [1]

Page 219 The Electromagnetic Spectrum

1. a) Visible light is absorbed by the plant [1]; providing it with the energy it needs to grow [1]
   b) Infrared waves are absorbed by the bread [1]; heating it up and toasting it [1]
   c) Radio waves are absorbed by an aerial [1]; creating an electrical signal in the aerial [1]
2. high [1]; most [1]; bacteria [1]
3. Ionising radiations [1]; can knock electrons from atoms ionising them [1]; when ionising radiation is absorbed by the nucleus of a cell [1]; the ionisation can cause DNA to mutate and cells to become cancerous [1]
4. Alternating current   [1]

Page 219 An Introduction to Electricity

1. a) 1 hour = 3600s [1]; $Q = 0.1 \times 3600$ [1]; = 360C [1]
   b) Twice the voltage would be needed [1]
2. a) 0.1A   [1]
   b) 20 ohms   [1]

Page 220 Circuits and Resistance

1. a) i) Graph 3   [1]
      ii) Graph 1   [1]
      iii) Graph 2   [1]
   b) Graph 1   [1]
   c) Graph 2   [1]

Page 220 Circuits and Power

1. a) Diagram showing correctly connected 2V battery [1]; and two bulbs in parallel [2] (1 mark for each correctly drawn pathway and bulb)

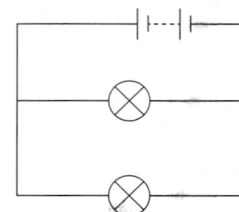

   b) 2V   [1]
   c) 1A   [1]
   d) resistance = $\dfrac{\text{potential difference}}{\text{current}}$ /
      $R = \dfrac{V}{I}$ [1]; = $\dfrac{2}{1}$ = 2Ω [1]

Page 221 Domestic Uses of Electricity

1. 230V [1]; 50Hz [1]
2. a) Correctly labelled earth wire   [1]
   b) Correctly labelled live wire   [1]
   c) Correctly labelled neutral wire   [1]
   d) Correctly labelled fuse   [1]
   e) Correctly labelled cable grip   [1]

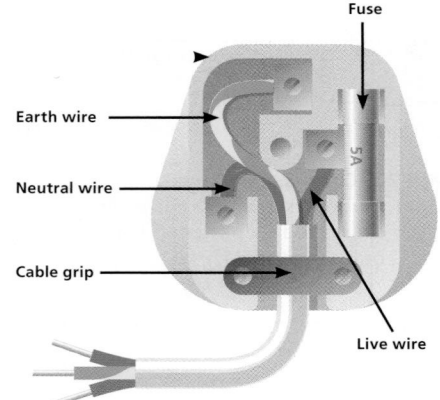

Page 221 Electrical Energy in Devices

1. power = potential difference × current / $P = VI$ [1]
2. a) i) Step-up transformer   [1]
      ii) Step-down transformer   [1]
   b) It reduces the current [1]; to reduce energy loss on transmission [1]
   c) It reduces the voltage [1]; to safe levels for consumers to use [1]

Pages 222–225 Practice Questions

Page 222 Magnetism and Electromagnetism

1. Correctly drawn diagram showing lines from north to south pole [1]; coming out of the north pole into the south pole [1]; more widely spaced the further they are from the magnet [1]

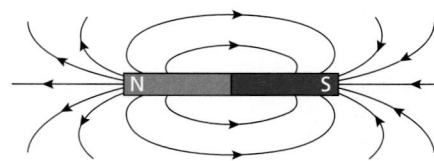

2. When a material becomes a magnet due to the presence of a permanent magnet or an electric current   [1]
3. a) A bigger current [1]; putting the compass closer to the wire [1]
   b) By reversing the current   [1]
   c) A downward pointing arrow drawn on the diagram   [1]

Page 222 The Motor Effect

1. a) Fleming's left hand rule   [1]
   b) Upwards   [1]
   c) Bigger current / stronger magnets [1]
   d) Reverse the current / reverse the magnets   [1]

Page 223 Particle Model of Matter

1. Cannot be squashed   [1]
2. An object less dense will float on the water and will not be submerged [1]; so the volume measured would be too low [1]
3. The input energy is being used to give the particles enough energy to escape from the liquid [1]; as the water becomes a gas, the particles move much faster [1]
4. When the steam condenses into water, it releases the energy it needed to be a gas [1]; so steam contains more energy than the same mass of water at the same temperature [1]
5. As the temperature increases, the particles move faster and hit the wall of the container more often/with more energy [1]; this increases the pressure [1]; and the force that the particles exert on the lid [1]
6. Object A – No [1]; Object B – Yes [1]; Object C – Yes [1]

Remember, density = $\dfrac{\text{mass}}{\text{volume}}$

Page 224 Atoms and Isotopes

1. a) Diagram correctly labelled with protons [1]; neutrons [1]; and electrons [1]

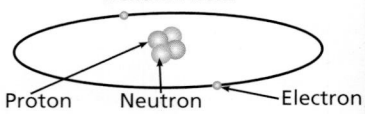

   b) Positive / +1   [1]
   c) 1×10^{-10}m   [1]
   d) It would gain electrons   [1]
   e) Mass number 4 [1]; atomic number 2 [1]
2. a) 17   [1]
   b) 17   [1]
   c) Chlorine-37 has 20 neutrons [1]; chlorine-35 has 18 neutrons [1]

Page 224 Nuclear Radiation

1. Three correctly drawn lines [2] (1 mark for one correct line)
   Alpha – Highest likelihood of being absorbed and causing damage when passing through living cells.
   Beta – Moderately ionising – can pass through the skin and damage organs inside the body.
   Gamma – Weakly ionising – likely to pass straight through living cells without being absorbed.
2. Arrows drawn showing alpha being stopped by paper [1]; beta being stopped by aluminium [1]; and gamma being stopped / partially stopped by lead [1]

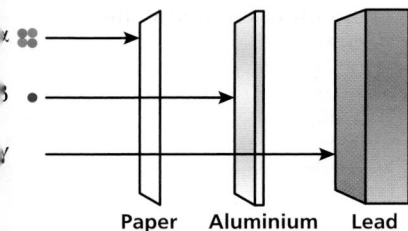

Paper   Aluminium   Lead

Page 225 Half-Life

1. a) The count has fallen to a quarter of the original, representing 2 half-lives [1]; one half-life = $\frac{4}{2}$ = 2 minutes [1]
   b) Background radiation [1]
   c) It has a very long half-life of hundreds of years or longer [1]

Pages 226–229 Review Questions

Page 226 Magnetism and Electromagnetism

1. Diagram labelled with north pole [1]; with field lines with arrows running from north to south [1]; lines further apart as they get further away from centre [1]

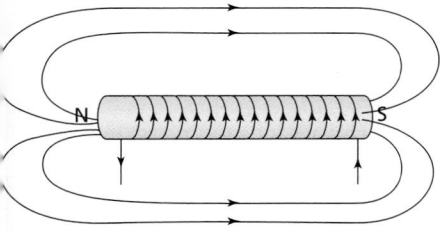

2. Magnet A is stronger than Magnet B [1]

Page 226 The Motor Effect

1. a) An upward pointing arrow [1]

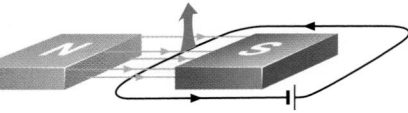

   b) It will reverse it [1]
2. a) How strong the magnetic field is [1]; shown by how close together the field lines are [1]
   b) Teslas [1]
3. magnetic flux density = $\frac{\text{force on conductor}}{\text{current} \times \text{length of wire}}$ / $B = \frac{F}{Il}$ [1];

   magnetic flux density = $\frac{0.2}{4 \times 0.04}$ [1];

   = 1.25T [1]

   Don't forget to convert length from centimetres to metres before performing the calculation.

Page 227 Particle Model of Matter

1. Diagrams showing particles in a rigid pattern with no spaces for a solid [1]; particles in a random pattern with small spaces, but with particles touching other particles, for a liquid [1]; and particles in a random pattern with big spaces between them for a gas [1]

SOLID   LIQUID   GAS

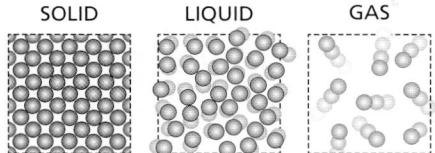

2. a) As a solid is heated, the particles gain energy and begin to vibrate more and more [1]; until they are able to move around in a random pattern [1]; but not move away from one another [1]
   b) As a liquid is heated, the particles move faster and faster [1]; and eventually have enough energy to move away from each other [1]; and escape as a gas [1]
3. Measure the mass of the object [1]; fill a measuring cylinder to a known mark [1]; drop the object into the cylinder and measure the increase in volume [1]; divide mass by volume to find density [1]
4. a) Initially, the temperature increases [1]; then the temperature stays at 0 degrees as energy is being used to melt the ice [1]; once all the ice has melted, the temperature increases once more [1]; at 100 degrees, the temperature levels off again [1]; energy is now being used to turn the water into steam [1]; once all of the water has turned into steam, the temperature begins to rise again as more energy is given to the system [1]
   b) The pressure will increase [1]
   c) 334000 × 0.5 [1]; = 167 000J [1]

Page 228 Atoms and Isotopes

1. a) Diagram marked with lines showing paths of alpha particles, with some passing straight through [1]; some changing direction as they strike the nucleus [1]; and some bouncing back [1]

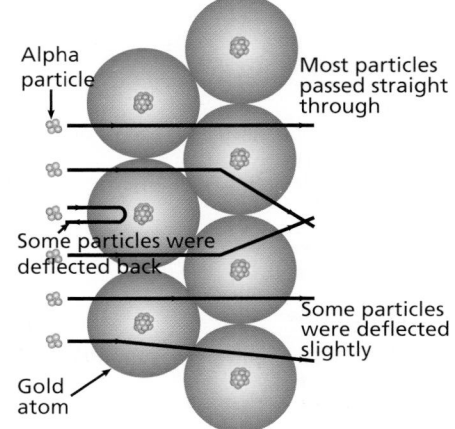

Alpha particle

Most particles passed straight through

Some particles were deflected back

Some particles were deflected slightly

Gold atom

   b) Most of the mass of the atoms is concentrated in one place [1]; and has a positive charge [1]
   c) Electrons orbit the nucleus [1]; in energy levels / shells [1]
2. a) 92 [1]
   b) U-233 [1]
   c) They both have the same number (92) of protons [1]; but U-238 has three more neutrons than U-235 (146 versus 143) [1]
   d) An uncharged atom has 92 electrons [1]; 92 – 3 = 89 electrons [1]

   An ion with a charge of +3 has lost three electrons.

Page 229 Nuclear Radiation

1. Three correctly drawn lines [2] (1 mark for one correct line)
   Alpha – A helium nucleus (2 protons and 2 neutrons).
   Beta – An electron emitted from the nucleus.
   Gamma – High-frequency electromagnetic radiation.
2. Because radioactive decay is a random process [1]; there is only a probability that a particular nucleus will decay in a certain time [1]

Page 229 Half-Life

1. a) The count rate has fallen by a factor of 4 [1]; representing 2 half-lives [1]; therefore, age = 2 × 6000 = 12000 years [1]
   b) 18000 years is equivalent to 3 half-lives [1]; so the count will be 8 times lower [1]; $\frac{200}{8}$ = 25 counts per minute [1]
   c) 60000 years is ten half-lives [1]; and after that time the carbon-14 has almost completely decayed / its count-rate is no more than background radiation [1]

Pages 230–237 Mixed Exam-Style Biology Questions

1. a) 0.02mm [1]
   b) Vacuoles [1]; chloroplasts [1]; cell walls [1]
   c) Top to bottom: vacuole [1]; cell membrane [1]; nucleus [1]
2. a) Large surface area [1]; rich blood supply [1]
   b) i) Further increases the surface area [1]; so diffusion is more rapid [1]
      ii) Some substances are taken up by active transport [1]; which requires energy [1]
   c) Less / slower absorption of digested food [1]
3. a) $\frac{(13 - 6)}{6} \times 100$ [1]; = 116.7% [1]
   b) 90 – 30 = 60 minutes [1]

Answers

c) It is sent to the liver **[1]**; where it is broken down using oxygen **[1]**

d) To make it a valid experiment / fair test **[1]**

e) The second athlete **[1]**

4. a) They are both in the same genus (*Sciurus*) **[1]**

b) Food **[1]**

c) They are not the same species **[1]**

d) Advantage: it should only kill the grey squirrels **[1]**; Disadvantage: it could kill other animals **[1]** (Accept any other sensible answers)

e) Some red squirrels might have the enzymes to digest acorns **[1]**; they are more likely to compete successfully and survive **[1]**; they will pass on the gene for making the enzymes **[1]**; and over generations the population would become able to digest the acorns **[1]**

5. Concentration of carbon dioxide or the temperature. **[1]**

6. a) i) Cc **[1]**

　　ii) cc **[1]**

b) Female sufferer **[1]**

c) Parents: Cc × cc **[1]**;

Gametes: C or c × c **[1]**;

Offspring: Cc and cc **[1]**;

Probability of a baby with cystic fibrosis is 50% **[1]**

7. a) 210 **[1]**

b) Cretaceous **[1]**

c) From their fossils **[1]**

d) Lizard **[1]**

e) Archosaurs **[1]**

8. a) Contains a weakened or dead pathogen **[1]**; stimulates the production of antibodies **[1]**; by white blood cells **[1]**; if the normal pathogen infects then antibodies are produced **[1]**

b) i) May contain a weakened pathogen **[1]**; so they get a mild form of the disease **[1]**

　　ii) They might be worried about the side effects **[1]**; they do not want the children to have three sets of side effects close together **[1]**

　　iii) Cheaper / less time-consuming for medical staff / worried that they may not come back after first or second vaccination **[1]**

c) To achieve herd immunity / so that the pathogen cannot be passed on **[1]**; to prevent an outbreak **[1]**

9. a) Help the frog to climb / cling onto trees **[1]**

b) Hides it from predators **[1]**

c) To catch insects **[1]**

10. a) Accurately plotted points **[2]**; best curve / straight lines drawn connecting points **[1]**

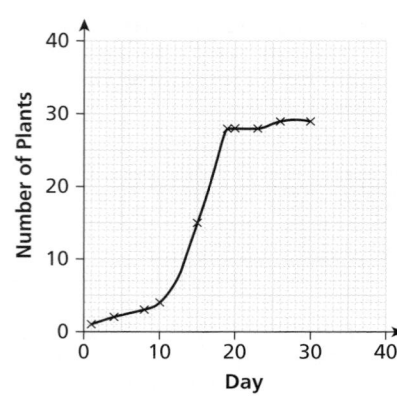

b) Increasing at a steady rate **[1]**

c) Any two of: lack of minerals **[1]**; lack of space **[1]**; competition for light **[1]**

11. a) Four correctly drawn lines **[3]** (2 marks for two correct lines and 1 mark for one correct line)
Increasing the amount of metal recycled – Fewer quarries are dug to provide raw materials
Reducing sulfur dioxide emissions – Less acid rain is produced
Using fewer pesticides and fertilisers – Less pollution of rivers flowing through farmland
Increasing the amount of paper recycled – Fewer forests are cut down

b) Peat decomposes in the gardens **[1]**; and carbon dioxide is given off **[1]** OR Peat bogs are destroyed to obtain the peat **[1]**; which is a threat to plants and animals that live in that habitat **[1]**

c) Producing food without damaging the environment **[1]**; or putting at risk future food supplies **[1]**

12. a) 46 **[1]**

b) Half the number of chromosomes in a human body cell. **[1]**

c) Which sex chromosomes they have **[1]**; XY is male and XX is female **[1]**

d) Parents: XX × XY **[1]**;

Gametes: X × X or Y **[1]**;

Offspring: XX or XY **[1]**

Pages 238–245 Mixed Exam-Style Chemistry Questions

1. a) 2,8,8,1 **[1]**

b) It has one electron **[1]**; in its outer shell **[1]**

c) A correctly drawn diagram **[1]**

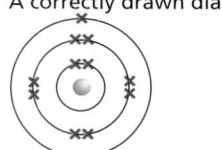

d) Both have one electron in their outer shell **[1]**

2. An atom that has gained electrons **[1]**; or lost electrons **[1]**

3. 3 protons **[1]**; 2 electrons **[1]**; 4 neutrons **[1]**

4. a) tin **[1]**; carbon dioxide **[1]**

b) Tin oxide is reduced **[1]**; carbon is oxidised **[1]**

5. Sodium sulfate **[1**

6. Properties ticked should be: very strong **[1]**; good electrical conductor **[1]**; good thermal conductor **[1]**; and nearly transparent **[1]**

7. a) It contains lots of strong bonds **[1]**; that require lots of energy to overcome them **[1]**

b) No, because the ions cannot move **[1**

c) Yes, because the ions can move **[1**

d) Yes, because the ions can move **[1**

8. a) 20 **[1**

b) 21 **[1**

c) 20 **[1**

9. a) 14 + (3 × 1) **[1]**; = 17 **[1]**

b) 17g **[1**

c) 0.5 × 17 = 8.5g **[1**

10. a) $2Ca + O_2 \rightarrow 2CaO$ **[2**
(1 mark for each correct number shown in bold)

b) $2Na + Br_2 \rightarrow 2NaBr$ **[2**
(1 mark for each correct number shown in bold)

c) $H_2 + Br_2 \rightarrow 2HBr$ **[1**
(1 mark for each correct number shown in bold)

d) $3H_2 + N_2 \rightarrow 2NH_3$ **[2**
(1 mark for each correct number shown in bold)

e) $2K + I_2 \rightarrow 2KI$ **[2**
(1 mark for each correct number shown in bold)

11. Distillation **[1**

12. a) 2,1 **[1**

b) It has one electron in its outer shell **[1**

13. a) Correctly drawn diagrams showing particles in a solid **[1]**; liquid **[1]**; and gas **[1]**

Solid	**Liquid**	**Gas**

b) Quickly **[1]**; in all directions **[1]**

c) liquid **[1]**; gaseous **[1]**

d) Water has covalent bonds **[1]**; and a simple molecular structure **[1]**; little energy is required to overcome the forces of attraction between molecules **[1]**; iron has metallic bonds **[1]**; and a metallic structure **[1]**; lots of energy is required to overcome the strong metallic bonds / bonds between the metal ions and the delocalised electrons **[1]**

14. a) Gained **[1]**; 1 electron **[1]**

Remember that electrons have a negative charge.

b) Lost **[1]**; 2 electrons **[1]**

15. a) Ag **[1**

b) Metallic [1]

c) It has a regular structure [1]; so layers can slide over each other [1]

d) A mixture that contains at least one metal [1]

16. a) $Mg + 2HCl \rightarrow MgCl_2 + H_2$ [1]

b) $2HgO \rightarrow 2Hg + O_2$ [2]
(1 mark for each correct number shown in bold)

17. Single covalent [1]

18. a) i) Group 2 [1]
   ii) Group 6 [1]
   iii) A diagram showing the correctly drawn outer electrons for a magnesium ion [1]; and oxide ion [1]

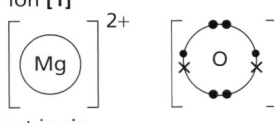

b) Giant ionic [1]

c) Ions cannot move in a solid [1]; but they can move when the substance is molten [1]

19. a) Covalent [1]

b) A correctly drawn diagram [1]

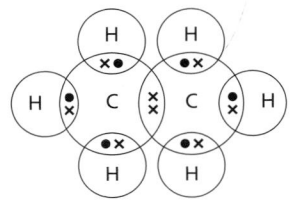

20. a) Fractional distillation [1]

b) $2C_2H_6 + 7O_2 \rightarrow 4CO_2 + 6H_2O$ [3]
(1 mark for each correct number shown in bold)

c) It is a poisonous gas [1]

21. a) $\frac{39}{39} = 1$ mole $= 6.02 \times 10^{23}$ atoms [1]

b) $\frac{16}{32} = 0.5$ mole $= 3.01 \times 10^{23}$ atoms [1]

22. C_3H_8 [1]

23. a) i) Bromine [1]
   ii) Lead [1]
   b) i) Br_2 [1]
   ii) $2e^-$ [1]

24. a) Particles must collide [1]; and have enough energy to react [1]

b) Particles are closer together [1]; so they collide more often [1]

25. a) Each lead ion gains two electrons [1]; to form lead atoms [1]

b) Each iodide ion loses electrons [1]; so it is oxidised [1]

   Remember: OIL RIG

c) So the ions can move [1]

26. a) A naturally occurring rock that contains metal or metal compounds in sufficient amounts to make it economically worthwhile extracting them [1]

b) The ore is heated [1]; with carbon [1]

c) Lead is less reactive than carbon [1]

d) When plants are used to take in metals as they grow [1]

e) Aluminium reacts with oxygen / forms a layer of aluminium oxide [1]; which stops any further reactions [1]

27. Increasing the surface area of solids [1]

28. a) So it does not run / because ink would dissolve and affect the results [1]

b) $\frac{8}{10}$ [1]; $= 0.8$ [1]

c) There are two dots / two components [1]

29. a) Na_2SO_4 [1]

b) Group 6 [1]; it has 6 electrons in its outer shell [1]

30. a) A strong acid produces H^+ ions [1]; and is fully ionised [1]

b) The gas carbon dioxide is made [1]; and escapes [1]

Pages 246–253 Mixed Exam-Style Physics Questions

1. a) Terminal velocity [1]

b) acceleration = $\frac{\text{change in velocity}}{\text{time taken}}$ /
$a = \frac{\Delta v}{t}$ [1]; $a = \frac{(10-50)}{5}$ [1]; $= -8\text{m/s}^2$ / a deceleration of 8m/s^2 [1]

c) i) As they freefall, they accelerate and the drag force increases [1]; eventually the resistive force equals the skydiver's weight [1]; so the resultant force is zero and they no longer accelerate [1]

   ii) It is harder to move through air when the parachute is open / less streamlined [1]; so drag balances / equals weight at a lower speed [1]

d) weight = mass × gravitational field strength / $w = mg$ [1];
$75 \times 10 = 750\text{N}$ [1]

2. a) Decelerating [1]

b) acceleration = $\frac{\text{change in velocity}}{\text{time taken}}$ /
$a = \frac{\Delta v}{t}$ [1]; $a = \frac{(50-10)}{(6-2)}$ [1];
$= \frac{40}{4}$ [1]; $= 10\text{m/s}^2$ [1]

c) Distance = area under graph for this region [1]; $= 50 \times 3$ [1]; $= 150\text{m}$ [1]

3. a) The faster the car, the greater the thinking distance [1]; they are directly proportional [1]

b) i) distance = speed × time = $9 \times 1 = 9\text{m}$ [1]
   ii) 6m [1]
   iii) It has increased [1]

4. a) work done = force × distance (along the line of action of the force) /
$W = Fs$ [1]

b) 400×5 [1]; $= 2000\text{J}$ [1]

c) Kinetic [1]

d) Chemical [1]

5. a) i) 15J [1]
   ii) 60J [1]
   iii) 25J [1]
   b) 25% [1]
   c) 75% [1]

6. a) work done = force × distance /
$W = Fs$ [1]; $W = 250 \times 5 = 1250\text{J}$ [1]

b) It is transferred into the kinetic energy of the bicycle [1]

7. a) $\frac{3}{12}$ [1]; $= 0.25\text{m/s}^2$ [1]

b) power = $\frac{\text{energy transferred}}{\text{time}}$ /
$P = \frac{W}{t}$ [1]; $P = \frac{2400}{12} = 200\text{W}$ [1]

c) resultant force = mass × acceleration / $F = ma$ [1]; force used on first car = $800 \times 0.25 = 200\text{N}$ [1]; acceleration of second car = $\frac{3}{18} = 0.17\text{m/s}^2$ [1]; mass of second car = $\frac{200}{0.17} = 1200\text{kg}$ [1]

8. a) Nuclear provided more energy than renewables (5.5% more of the national share) [1]

b) **Any two of:** wind turbines [1]; solar panels [1]; waves [1]; tides [1]; geothermal [1]

c) Advantage: reliable / high output / no greenhouse gases produced / large available fuel supply [1]; Disadvantage: radioactive waste / expensive to build and decommission / risk of nuclear accident [1]

d) gravitational energy [1]; → kinetic energy [1]

e) It reduces energy loss on transmission [1]; but higher voltages are more dangerous and must be transformed back to safer voltages at the other end [1]

9. a) 1900×4.5 [1]; $= 8550\text{J}$ [1]

b) $\frac{8500}{9}$ [1]; $= 950\text{W}$ [1]

c) 950×2 [1]; $= 1900\text{J}$ [1]

10. a) i) light [1]; electrical [1]
   ii) electrical [1]; chemical [1]
   iii) electrical [1]; kinetic [1]

b) i) $\frac{1500}{10}$ [1]; $= 150\text{W}$ [1]
   ii) Heat / sound [1]

11. a) It is reduced [1]

b) To reduce energy loss on transmission [1]

c) Step-down transformer [1]

12. a) Correctly labelled voltmeter [1]; diode [1]; and ammeter [1]

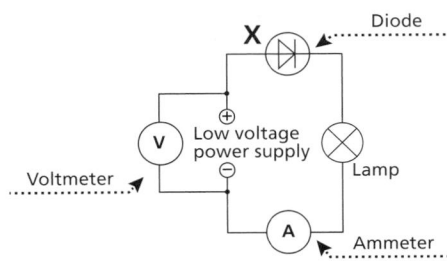

b) When the voltage is negative, the resistance is extremely high [1]; and no current flows [1]; with a positive voltage, the resistance is extremely high at first, so no current flows [1]; but after a certain point, the resistance falls and current flows [1]

Answers

13. a) Circuit with correctly connected battery **[1]**; ammeter **[1]**; and voltmeter **[1]**

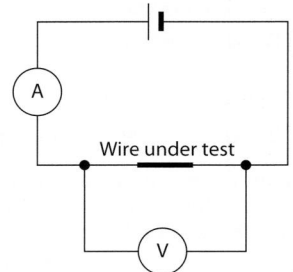

Wire under test

b) i) potential difference = current × resistance / $V = IR$ **[1]**

ii) resistance = $\dfrac{\text{potential difference}}{\text{current}}$ /

$R = \dfrac{V}{I}$ **[1]**; $R = \dfrac{9}{0.30} = 30\Omega$ **[1]**

iii) The resistance has increased **[1]**

14. a) potential difference = current × resistance / $V = IR$ **[1]**;
$0.5 \times 8 = 4V$ **[1]**

b) $9 - 4 = 5V$ **[1]**

c) 0.5A **[1]**

15. a) A1 = 0.5A **[1]**; A4 = 0.5A **[1]**

b) 20×0.3 **[1]**; = 6V **[1]**

c) 6V **[1]**

16. a) Microwaves are absorbed by food **[1]**; when this happens energy is transferred from the microwaves to the food **[1]**; causing its temperature to increase **[1]**

b) wave speed = frequency × wavelength / $v = f\lambda$ **[1]**;
$v = 2450\,000\,000 \times 0.122$ **[1]**;
= 298\,900\,000 m/s or 2.989×10^8 m/s **[1]**

17. a) Correctly drawn continuation of ray AB after entering the block **[1]**; and CD after entering the block **[1]**

b) Correctly drawn lines indicating wave fronts **[1]**; that are closer together than those outside the block **[1]**

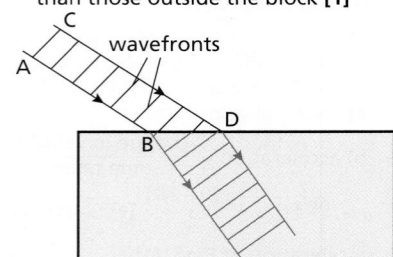

wavefronts

c) The wave slows down when it enters the block, reducing the wavelength **[1]**; the part of the wave that enters first, slows down first, causing the wave to bend towards the normal **[1]**

18. a) Weight **[1]**

b) It accelerates **[1]**

c) Its acceleration **[1]**; increases **[1]**

d) i) acceleration = $\dfrac{\text{change in velocity}}{\text{time taken}}$

$a = \dfrac{\Delta v}{t}$ **[1]**

ii) Time taken to reach 8100m/s from stationary = $9 \times 60 = 540$s **[1]**;

acceleration = $\dfrac{\text{change in velocity}}{\text{time taken}}$

$a = \dfrac{\Delta v}{t}$ **[1]**; $a = \dfrac{8100}{540} = 15$m/s² **[1]**

iii) The velocity is the speed in a specific direction **[1]**; with speed, direction does not matter **[1]**

The Periodic Table of the Elements

Key

relative atomic mass
atomic symbol
name
atomic (proton) number

| 1 | H hydrogen 1 (mass 1) |

1	2											3	4	5	6	7	0
7 **Li** lithium 3	9 **Be** beryllium 4											11 **B** boron 5	12 **C** carbon 6	14 **N** nitrogen 7	16 **O** oxygen 8	19 **F** fluorine 9	4 **He** helium 2
23 **Na** sodium 11	24 **Mg** magnesium 12											27 **Al** aluminium 13	28 **Si** silicon 14	31 **P** phosphorus 15	32 **S** sulfur 16	35.5 **Cl** chlorine 17	20 **Ne** neon 10
39 **K** potassium 19	40 **Ca** calcium 20	45 **Sc** scandium 21	48 **Ti** titanium 22	51 **V** vanadium 23	52 **Cr** chromium 24	55 **Mn** manganese 25	56 **Fe** iron 26	59 **Co** cobalt 27	59 **Ni** nickel 28	63.5 **Cu** copper 29	65 **Zn** zinc 30	70 **Ga** gallium 31	73 **Ge** germanium 32	75 **As** arsenic 33	79 **Se** selenium 34	80 **Br** bromine 35	40 **Ar** argon 18
85 **Rb** rubidium 37	88 **Sr** strontium 38	89 **Y** yttrium 39	91 **Zr** zirconium 40	93 **Nb** niobium 41	96 **Mo** molybdenum 42	[98] **Tc** technetium 43	101 **Ru** ruthenium 44	103 **Rh** rhodium 45	106 **Pd** palladium 46	108 **Ag** silver 47	112 **Cd** cadmium 48	115 **In** indium 49	119 **Sn** tin 50	122 **Sb** antimony 51	128 **Te** tellurium 52	127 **I** iodine 53	84 **Kr** krypton 36
133 **Cs** caesium 55	137 **Ba** barium 56	139 **La*** lanthanum 57	178 **Hf** hafnium 72	181 **Ta** tantalum 73	184 **W** tungsten 74	186 **Re** rhenium 75	190 **Os** osmium 76	192 **Ir** iridium 77	195 **Pt** platinum 78	197 **Au** gold 79	201 **Hg** mercury 80	204 **Tl** thallium 81	207 **Pb** lead 82	209 **Bi** bismuth 83	[209] **Po** polonium 84	[210] **At** astatine 85	131 **Xe** xenon 54
[223] **Fr** francium 87	[226] **Ra** radium 88	[227] **Ac*** actinium 89	[261] **Rf** rutherfordium 104	[262] **Db** dubnium 105	[266] **Sg** seaborgium 106	[264] **Bh** bohrium 107	[277] **Hs** hassium 108	[268] **Mt** meitnerium 109	[271] **Ds** darmstadtium 110	[272] **Rg** roentgenium 111	[285] **Cn** copernicium 112	[286] **Uut** ununtrium 113	[289] **Fl** flerovium 114	[289] **Uup** ununpentium 115	[293] **Lv** livermorium 116	[294] **Uus** ununseptium 117	[222] **Rn** radon 86
																	[294] **Uuo** ununoctium 118

* The Lanthanides (atomic numbers 58 – 71) and the Actinides (atomic numbers 90 – 103) have been omitted.
Relative atomic masses for Cu and Cl have not been rounded to the nearest whole number.

Physics Equations

You must be able to recall and apply the following equations using standard units:

Word Equation	Symbol Equation
weight = mass × gravitational field strength	$W = mg$
work done = force × distance (along the line of action of the force)	$W = Fs$
force applied to a spring = spring constant × extension	$F = ke$
distance travelled = speed × time	$s = vt$
acceleration = $\dfrac{\text{change in velocity}}{\text{time taken}}$	$a = \dfrac{\Delta v}{t}$
resultant force = mass × acceleration	$F = ma$
HT momentum = mass × velocity	$p = mv$
kinetic energy = 0.5 × mass × (speed)^2	$E_k = \frac{1}{2}mv^2$
gravitational potential energy = mass × gravitational field strength × height	$E_p = mgh$
power = $\dfrac{\text{energy transferred}}{\text{time}}$	$P = \dfrac{E}{t}$
power = $\dfrac{\text{work done}}{\text{time}}$	$P = \dfrac{W}{t}$
efficiency = $\dfrac{\text{useful output energy transfer}}{\text{useful input energy transfer}}$	
efficiency = $\dfrac{\text{useful power output}}{\text{total power output}}$	
wave speed = frequency × wavelength	$v = f\lambda$
charge flow = current × time	$Q = It$
potential difference = current × resistance	$V = IR$
power = potential difference × current	$P = VI$
power = (current)^2 × resistance	$P = I^2R$
energy transferred = power × time	$E = Pt$
energy transferred = charge flow × potential difference	$E = QV$
density = $\dfrac{\text{mass}}{\text{volume}}$	$\rho = \dfrac{m}{V}$

The following equations will appear on the equations sheet that you are given in the exam. You must be able to select and apply the appropriate equation to answer a question correctly.

Word Equation	Symbol Equation
(final velocity)² – (initial velocity)² = 2 × acceleration × distance	$v^2 - u^2 = 2as$
elastic potential energy = 0.5 × spring constant × (extension)²	$E_e = \frac{1}{2}ke^2$
change in thermal energy = mass × specific heat capacity × temperature change	$\Delta E = mc\Delta\theta$
period = $\dfrac{1}{\text{frequency}}$	
HT force on a conductor (at right-angles to a magnetic field) = magnetic flux density × current × length	$F = BIl$
thermal energy for a change of state = mass × specific latent heat	$E = mL$
HT potential difference across primary coil × current in primary coil = potential difference across secondary coil × current in secondary coil	$V_s I_s = V_p I_p$

Glossary and Index

Electron microscope   a device that fires electrons at a specimen to obtain a high resolution image   **18–19**

Electrostatic   a force of attraction between oppositely charged species   **96–97, 101**

Element   a substance made of only one type of atom   **88–89, 212–213**

Embryonic stem cells   cells found in an embryo that can differentiate into any type of cell   **21**

Endocrine system   a system of glands that release hormones directly into the bloodstream   **52**

Endothermic reaction   a reaction that takes in energy from the surroundings   **46, 120–121, 122**

Energy   a measure of the capacity of a body or system to do work   **170–171**

Energy level diagram   (reaction profile) a diagram showing the relative energies of the reactants and the products of a reaction, the activation energy and the overall energy change of the reaction   **121**

Equation   a scientific statement that uses chemical names or symbols to sum up what happens in a chemical reaction   **88**

Equilibrium   when the rate of forward reaction is equal to the rate of backwards reaction   **127**

Eukaryotic   describes cells that have a nucleus and sub-cellular organelles such as mitochondria   **17**

Evolution   a gradual change in a group of organisms over a long period of time   **68–69**

Exothermic reaction   a reaction that gives out energy to the surroundings   **48, 120–121, 123**

Extension   the distance over which an object (like a spring) has been extended / stretched   **160–161**

Extinct   describes a species that has died out   **72**

Extraction   the process of obtaining / taking out / removing, e.g. obtaining a metal from an ore or compound   **115**

Extremophile   an organism that can live in very extreme environments   **75**

F

Fermentation   the conversion of sugar to alcohol and carbon dioxide in yeast   **48**

HT Fertility drug   a drug that makes it more likely for sexual intercourse to result in pregnancy   **55**

Filtration   separation techniques used to separate insoluble solids from soluble solids   **89**

Flagella   a long tail-like structure that is used by some cells for movement   **17**

Flux density   a measure of the density of the field lines around a magnet   **206, 208**

Follicle stimulating hormone (FSH)   a hormone released by the pituitary gland that causes an egg to develop in the ovaries   **54**

Food chain   the feeding relationships between organisms   **77**

Force   an influence that occurs when two objects interact   **158–159**

Formulation   a mixture that has been carefully designed to have specific properties   **140**

Fossils   the remains of animals / plants preserved in rock   **69**

Fractional distillation   a separation technique used to separate mixtures which contain components with similar boiling points   **89, 136**

HT Free body diagram   a diagram used to show the relative magnitude and direction of all the forces acting on an object in a given situation   **159**

Frequency   the number of times that a wave / vibration repeats itself in a specified time period   **182**

Fresh water   water that contains low levels of dissolved salts   **146**

Fullerene   a form of carbon in which the carbon atoms are joined together to form hollow structures, e.g. tubes, balls and cages   **100**

G

Gamete   a specialised sex cell formed by meiosis   **64**

Gamma radiation / gamma ray   high frequency, short wavelength electromagnetic waves; a type of nuclear radiation, emitted from a nucleus   **186–187, 214–215**

Gene   part of a chromosome, made of DNA, which codes for a protein   **20, 65**

Genetic engineering   the process of moving a gene from one organism to another   **70–71**

Genetically modified (GM)   describes organisms that have had specific areas of their genetic material changed using genetic engineering techniques   **70–71**

Genome   all the genetic material found in an organism or a species   **65**

Genotype   the combination of alleles an individual has for a particular gene, e.g. BB, Bb or bb   **66**

Genus   a group of closely relate species   **72**

Global warming   the increase in the average temperature on Earth due to a rise in the levels of greenhouse gases in the atmosphere   **79**

Gradient   a measure of the steepness of a sloping line; the ratio of the change in vertical distance over the change in horizontal distance; gradient $= \dfrac{\text{difference in the } y\text{-axis value}}{\text{difference in the } x\text{-axis value}}$ **125, 164**

Graphene   a form of carbon; a single layer of graphite, just one atom thick   **100**

Graphite   a form of carbon, with a giant covalent structure, that conducts electricity   **99**

Gravitational potential energy (GPE)   the energy gained by raising an object above ground level (due to the force of gravity)   **170**

Gravity   the force of attraction exerted by all masses on other masses, only noticeable with a large body, e.g. the Earth or Moon   **158**

Greenhouse gases   gases (e.g. carbon dioxide, water vapour or methane) that, when present in the atmosphere, absorb the outgoing infra-red radiation, increasing the Earth's temperature   **144–145**

H

Haemoglobin   the red pigment in red blood cells, that carries oxygen to the organs   **28**

HT Half equation   used to show what happens to one of the reactants in a chemical reaction   **102**

Half-life   the average time it takes for half the nuclei in a sample of radioactive isotope to decay; the time it takes for the count rate / activity of a radioactive isotope to fall by 50% (halve)   **216**

Halogens   elements in Group 7 of the periodic table   **93**

Health   the absence of disease *and* a state of complete physical, mental and social well-being   **30**

Heterozygous   when an individual carries two different alleles for a gene, e.g. Bb   **66**

Homeostasis   the process of keeping the internal conditions of the body constant   **50**

Homozygous   when an individual carries two copies of the same allele for a gene, e.g. BB or bb   **66**

Hormone   a chemical messenger produced by a gland that travels in the blood to its target organ   **52–55**

Hydrocarbon   a molecule that contains only carbon and hydrogen atoms   **136–137**

I

Immune system   the body's defence system against infections and diseases (consists of white blood cells and antibodies)   **42–43**

Immunity   the ability to attack a pathogen before it causes disease due to a previous encounter with the pathogen   **43**

Incompressible cannot be compressed / squeezed / compacted **210**

Indicator a chemical that is one colour in an acid and another colour in an alkali **116–117**

Induce to produce (a potential difference) or transmit (magnetism) **206–207**

Induced magnet an object that becomes magnetic when placed in a magnetic field **206–207**

Inelastically deformed describes an object that cannot return to its original shape when the forces that caused it to change shape are removed (because the limit of proportionality has been exceeded) **160**

HT Inertia the tendency of a body to stay at rest or in uniform motion unless acted upon by an external force **163, 165**

Infrared the part of the electromagnetic spectrum with a longer wavelength than light but a shorter wavelength than radio waves **144, 186–187**

Interdependence when one organism relies on another for certain resources / factors **74**

Intermolecular between molecules **98–99**

Internal energy the sum of the energy of all the particles that make up a system, i.e. the total kinetic and potential energy of all the particles added together **170–171**

Inversely proportional a relationship between two variables, where one variable increases and the other decreases **165**

HT Inverse square law when a light source is moved to double the distance then the light intensity reduces by a quarter **47**

HT In vitro fertilisation (IVF) a process in which an egg is fertilised by sperm outside of the body **55**

Ion formed when an atom loses or gains one or more electrons to become charged **91, 96–97, 212–213**

Ionic bond the force of attraction between positive and negative ions **96–97**

HT Ionic equation a simplified version of a chemical equation, which just shows the species that are involved in the reaction **102**

HT Ionised split up into ions **117**

Ionising refers to radiation that can cause atoms to lose or gain atoms, becoming ions **187**

Irradiation to expose an object to nuclear radiation (the object does not become radioactive) **215**

Isolated refers to an object that has no conducting path to earth **64–65**

Isotope describes atoms of the same element that have the same atomic number but a different mass number **91, 212**

K

Kinetic energy the energy of motion of an object, equal to the work it would do if brought to rest **170**

L

Lactic acid a compound produced when cells respire without oxygen (i.e. anaerobically) **49**

Latent heat of fusion the amount of heat energy needed for a substance to change from solid to liquid **211**

Latent heat of vaporisation the amount of heat energy needed for a substance to change from liquid to gas **211**

HT Le Chatelier's Principle a rule that can predict the effect of changing conditions on a system that is in equilibrium **127**

Life cycle assessment (LCA) used to assess the environmental impact a product has over its whole lifetime **148**

HT Limiting factor a factor that prevents a reaction going any faster **46–47**

HT Limiting reactant the reactant that is completely used up in a reaction; it stops the reaction from going any further and any further products from being produced **105**

Limit of proportionality the point up to which the extension of an elastic object is directly proportional to the applied force (once exceeded, the relationship is no longer proportional) **160–161**

Lipase an enzyme that breaks down fat into fatty acids and glycerol **26–27**

Lock and key theory a model used to explain how enzymes work, where the active site is the lock and the substrate is the key **26**

Longitudinal a wave in which the oscillations are parallel to the direction of energy transfer, e.g. sound waves **182**

Luteinising hormone (LH) a hormone that stimulates the release of an egg in the menstrual cycle **54**

M

Magnification how many times larger an image is than the real object **19**

Malignant describes a tumour that can spread to other areas of the body **31**

Malleable can be hammered into shape **101**

Mass a measure of how much matter an object contains, measured in kilograms (kg) **158**

Mass number the total number of protons and neutrons in the nucleus of an atom **91**

Matter describes the material that everything is made up of **94–95**

Medium a material or substance **185**

Meiosis cell division that forms daughter cells with half the number of chromosomes of the parent cell **64**

Melting point the specific temperature at which a pure substance changes state, from solid to liquid and from liquid to solid **94–95**

Mendeleev the Russian chemist who devised the periodic table **92**

Meniscus the curved upper surface of a liquid standing in a tube, from which measurements of volume are taken **210**

Menstrual cycle the monthly cycle of an egg being released in females; controlled by hormones **54–55**

Meristems areas of cells in plants that can divide to form new cells **21**

Metabolism the sum of all the chemical reactions occurring in the body **49**

Metallic bond the strong attraction between metal ions and delocalised electrons **101**

Microwaves electromagnetic radiation in the wavelength range 0.3 to 0.001 metres, used in satellite communication and cooking **186–187**

Mitochondria the structures in the cytoplasm where energy is produced from chemical reactions **16**

Mitosis cell division that forms two daughter cells, each with the same number of chromosomes as the parent cell **20–21**

Mixture a combination of two or more elements or compounds, which are not chemically combined together **89**

Mobile phase in chromatography, the phase that does move (e.g. the solvent) **140**

HT Mole (mol) the unit for measuring the amount of substance; one mole of any substance contains the same number of particles; the mass of one mole of a substance is equal to the relative formula mass in grams **104–105**

Molten liquefied by heat **97**

HT Momentum the product of an object's mass and velocity **166–167**

Monohybrid inheritance the pattern of inheritance shown when a characteristic is controlled by a single gene **66**

HT Motor effect the force experienced by a current-carrying conductor when it is placed in a magnetic field, which is used to create movement in electrical motor **208–209**

MRSA (Methicillin-Resistant Staphylococcus Aureus) – an antibiotic-resistant bacterium; a 'superbug' **44**

N

National Grid the network of high voltage power lines and transformers that connects major power stations, businesses and homes **197**

Natural selection the survival of individual organisms that are best adapted to their environment **68–69**

HT Negative feedback a set of events that detects a variable and then corrects any change in the variable away from a set value **50**

Neutralisation occurs when an acid reacts with exactly the right amount of base **116–117**

Neutron a subatomic particle with a relative mass of 1 and no charge; a type of nuclear radiation, which can be emitted during radioactive decay **90–91, 212–213**

Noble gases unreactive non-metal elements that have a full outer shell of electrons and are found in Group 0 of the periodic table **92**

Non-communicable refers to a disease that cannot be passed on from one individual to another **30–31**

Non-contact force a force that occurs between two objects that are not in contact (not touching), e.g. gravitational and electrostatic forces **158**

Non-specific defences the first line of defence against pathogens in general, includes skin, hair, mucus, etc. **42**

Normal at right-angles to / perpendicular to **184–185**

Nucleus (cell) the control centre of the cell **16**

Nucleus (atom) the positively charged, dense region at the centre of an atom, made up of protons and neutrons, orbited by electrons **212–213**

O

Oestrogen a hormone secreted by the ovaries that inhibits the production of FSH and triggers the production of LH **54**

Ohmic conductor a resistor in which the current is directly proportional to the potential difference at a constant temperature **191**

Optimum describes the conditions at which an enzyme works best **26**

Organ a group of tissues gathered together to perform a particular function **25**

Organ system a group of organs that all perform related functions **25**

Oscillate to vibrate / swing from side to side with a regular frequency **182–183**

Osmosis the movement of water, through a partially permeable membrane, into a solution with a lower water concentration **22–23**

Ovulation the release of an egg (ovum) from the ovary into the fallopian tube **54–55**

Oxidation when a species loses electrons **114–115**

Oxygen debt oxygen deficiency caused by anaerobic respiration during intense / vigorous exercise **49**

P

Pacemaker a natural or artificial device that controls heart rate **29**

Parallel (circuit) a circuit in which the components are connected side by side on a separate branch / path, so that the current from the cell / battery splits with a portion going through each component **192**

Particle an extremely small body with finite mass and negligible (insignificant) size, e.g. protons, neutrons and electrons **94–95, 210–211**

Pathogen a disease-causing microorganism **40–41**

Penicillin an antibiotic extracted from the *Penicillium* fungus **44**

Period the time taken for a wave to complete one oscillation; the time it takes for a particle in the wave to move backwards and forwards once around its undisturbed position **182–183**

Permanent magnet an object that produces its own magnetic field **206**

Phagocytosis the process by which one cell, such as a white blood cell, surrounds and engulfs another cell **42**

Phenotype the physical expression of the genotype, i.e. the characteristic shown **66**

Photosynthesis a chemical reaction carried out by green plants in which carbon dioxide and water react to produce glucose and oxygen **143**

HT Phytomining the use of plants to extract metals from low-grade ores **147**

Pituitary gland a small gland at the base of the brain that produces hormones; known as the 'master gland' **52**

Placebo a dummy drug given to patients during drug trials **45**

Plasma the clear fluid part of blood that contains various dissolved substances, such as proteins and mineral ions **28**

Plasmid a small circle of bacterial DNA that is independent of the main bacterial chromosome **17**

Pole the two opposite regions in a magnet, where the magnetic field is concentrated; can be north or south **206–207**

Pollution the contamination of an environment, e.g. by chemicals or water **78**

Polydactyly a genetic disorder caused by a dominant allele, where affected people have extra fingers or toes **67**

Population a group of organisms of the same species living together in a habitat **75**

Potable water that is safe to drink **146–147**

Potential difference the difference in electric potential between two points in an electric field; the work that has to be done in transferring a unit of positive charge from one point to another, measured in volts (V) **189**

Power a measure of the rate at which energy is transferred or work is done **193, 195**

Precision refers to the degree of accuracy (of a measuring instrument), i.e. the minimum possible change that can be measured **189**

Predator an animal that hunts, kills and eats other animals (prey) **77**

Prey an organism that is hunted and killed by a predator for food **77**

Producer an organism that can make its own food **77**

Products the substances produced in a chemical reaction **88**

Progesterone a hormone that repairs the lining of the uterus after menstruation and prevents it breaking down **54**

Prokaryotic organisms, such as bacteria, that do not have a nucleus or organelles such as mitochondria **17**

Proportional describes two variables that are related by a constant ratio **164**

Protease an enzyme used to break down proteins into amino acids **26–27**

Proton a subatomic particle found in the nucleus of an atom, with an electrical charge of +1 and a relative mass of 1 **90–91, 212–213**

Pulmonary artery the blood vessel carrying deoxygenated blood from the heart to the lungs **29**

Pulmonary vein the blood vessel carrying oxygenated blood from the lungs to the heart **29**

Punnett square a type of diagram used to work out the outcome of genetic crosses **66**

Pure a substance that contains only one type of element or compound **101, 140**

Q

Quadrat a square frame (usually between 0.25m² and 1m²) used for sampling organisms in their natural environment **75**

R

Radioactive describes a substance that gives out radiation **214–215**

Rate of reaction a measure of how quickly a reactant is used up or a product is made **124–125**

Reactants the substances that react together in a chemical reaction **88**

Reaction profile (energy level diagram) a diagram showing the relative energies of the reactants and the products of a reaction, the activation energy and the overall energy change of the reaction **121**

Reactivity series a list in which elements are placed in order of how reactive they are **114–115**

Receptors cells found in sense organs, e.g. eyes, ears, nose **50–51**

Recessive describes an allele that will only be expressed if there are two present; represented by a lower case letter **66–67**

Reduction reaction in which a chemical species gains electrons **114–115**

Reflected when a wave meets a boundary between two different materials and is bounced back **184**

Refracted when a wave meets a boundary between two different materials and changes direction **184–185**

Refractive index a measure of the extent to which light is refracted by a material **184**

Relative atomic mass (A_r) the ratio of the average mass per atom of an element to one-twelfth the mass of an atom of carbon-12 **103**

Relative formula mass (M_r) the sum of the relative atomic masses of all the atoms shown in a chemical formula **103**

Renewable describes something that can be replaced **173**

Reproducible results are reproducible if the investigation / experiment can be repeated by another person, or by using different equipment / techniques, and the same results are obtained, demonstrating that the results are reliable **171**

Resistance a measure of how a component resists (opposes) the flow of electrical charge, measured in ohms (Ω) **189–191**

Resolution the smallest distance apart two objects can be and still be seen as separate objects **18–19**

Resultant a single force that represents the overall effect of all the forces acting on an object **159**

Reversible reaction a reaction that can go forwards or backwards **120–121, 126–127**

Ribosomes small structures found in the cytoplasm of living cells where protein synthesis takes place **16**

Risk factor a factor that will increase the chance of developing a disease **30–31**

Runners long shoots from plants, such as strawberries, that are used for asexual reproduction **64**

S

Saturated a hydrocarbon molecule that only contains single carbon–carbon (C–C) bonds **136–137**

Scalar a quantity, such as time or temperature, that has magnitude but no direction **158**

Selective breeding the breeding process used by scientists and farmers to produce organisms that show the characteristics that are considered useful **70**

Series (circuit) describes a circuit in which the components are connected one after then other, so the same current flows through each component **192**

Sex chromosomes the pair of chromosomes that determine the sex of organisms **67**

Silicon dioxide (silica) (SiO_2) a compound found in sand **99**

Simple distillation technique used to separate a substance from a mixture due to a difference in the boiling points of the components in the mixture **89**

Solenoid a coil of wire partially surrounding an iron core, which is made to move inside the coil by the magnetic field set up by a current; used to convert electrical to mechanical energy **207**

Specialised adapted for a particular purpose **24**

Species (biology) a group of organisms that can reproduce with each other to produce fertile offspring **72**

HT) Species (chemistry) the different atoms, molecules or ions that are involved in a reaction **102–103**

Specific heat capacity the amount of energy required to raise the temperature of one kilogram of substance by one degree Celsius **171**

Speed a scalar measure of the distance travelled by an object in a unit of time, measured in metres per second (m/s) **162–163**

Spring constant a measure of how easy it is to stretch or compress a spring; calculated as: $\dfrac{\text{force}}{\text{extension}}$ **161**

State of matter the structure and form of a substance, i.e. gas, liquid or solid **210–211**

Statins a drug used to help lower cholesterol levels in the blood **31**

Stationary phase in chromatography, the phase that does not move (e.g. the absorbent paper) **140**

Stem cell a human embryo cell or adult bone marrow cell that has yet to differentiate **21**

Stent a tube that is inserted into a blood vessel to keep it open **31**

Stomata openings / pores in the leaves of plants **33**

HT) Strong acid an acid, such as hydrochloric acid, nitric acid or sulfuric acid, that is completely ionised in water **117**

Sub-cellular structures structures found in cells that include the nucleus, mitochondria, chloroplasts and ribosomes **16**

Sublimation the process of changing state directly from a solid to a gas without first melting **211**

Surface area to volume ratio a way of comparing the surface area of an organism to its volume – the smaller this ratio, the harder it is to exchange substances with the environment at a fast enough rate **22**

Sustainable development development that meets the needs of the current generation without compromising the ability of future generations to meet their own needs **146–147**

Synapse the gap between two neurones **51**

System an object or group of objects **170–171**

T

HT) Tangent a straight line that touches a curve at one point **125**

Terminal velocity the constant maximum velocity reached by a body falling under gravity **166**

Testosterone a hormone produced by the testes that controls the male sexual characteristics **54**

Therapeutic cloning process in which clones are produced to treat diseases, but will not be allowed to develop into new offspring **21**

Three-domain system a new classification system that divides organisms into three domains rather than five kingdoms **72**

HT) Thyroxine a hormone released from the thyroid gland that controls the metabolic rate of the body **52**

Tissue a group of cells that have a similar structure and function **26**

Toxin a poisonous chemical, produced by certain pathogens **41**

Trachea the main tube or windpipe taking air from the mouth down to the lungs **29**

Transect line a fixed line along which sampling of populations, such as species abundance, is measured **75**

Transferred refers to how energy is changed, e.g. chemical energy can be transferred to electric energy **172–173**

Transformer a device that transfers an alternating current from one circuit to another, with an increase (step-up transformer) or decrease (step-down transformer) of voltage **197**

Translocation the method by which dissolved food is transported through the phloem in plants **33**

Transmitted when waves are sent out from a source or pass through a material **184**

Transparent see-through **100**

Transpiration the movement of water through a plant from root to leaf **33**

Transverse a wave in which the oscillations are at right-angles to the direction of energy transfer, e.g. water waves **182**

Collins

AQA GCSE Revision

Combined Science

Trilogy

Higher

AQA GCSE

Workbook

Ian Honeysett, Emma Poole,
Gemma Young and Nathan Goodman

Revision Tips

Rethink Revision

Have you ever taken part in a quiz and thought *'I know this!'* but, despite frantically racking your brain, you just couldn't come up with the answer?

It's very frustrating when this happens but, in a fun situation, it doesn't really matter. However, in your GCSE exams, it will be essential that you can recall the relevant information quickly when you need to.

Most students think that revision is about making sure you **know** stuff. Of course, this is important, but it is also about becoming confident that you can **retain** that *stuff* over time and **recall** it quickly when needed.

Revision That Really Works

Experts have discovered that there are two techniques that help with all of these things and consistently produce better results in exams compared to other revision techniques.

Applying these techniques to your GCSE revision will ensure you get better results in your exams and will have all the relevant knowledge at your fingertips when you start studying for further qualifications, like AS and A Levels, or begin work.

It really isn't rocket science either – you simply need to:

- **test yourself** on each topic as many times as possible
- **leave a gap** between the test sessions.

Three Essential Revision Tips

1. **Use Your Time Wisely**

   - Allow yourself plenty of time.
   - Try to start revising at least six months before your exams – it's more effective and less stressful.
   - Your revision time is precious so use it wisely – using the techniques described on this page will ensure you revise effectively and efficiently and get the best results.
   - Don't waste time re-reading the same information over and over again – it's time-consuming and not effective!

2. **Make a Plan**

   - Identify all the topics you need to revise (this All-in-One Revision & Practice book will help you).
   - Plan at least five sessions for each topic.
   - One hour should be ample time to test yourself on the key ideas for a topic.
   - Spread out the practice sessions for each topic – the optimum time to leave between each session is about one month but, if this isn't possible, just make the gaps as big as realistically possible.

3. **Test Yourself**

   - Methods for testing yourself include: quizzes, practice questions, flashcards, past papers, explaining a topic to someone else, etc.
   - This All-in-One Revision & Practice book provides seven practice opportunities per topic.
   - Don't worry if you get an answer wrong – provided you check what the correct answer is, you are more likely to get the same or similar questions right in future!

Visit our website to download your free flashcards, for more information about the benefits of these techniques, and for further guidance on how to plan ahead and make them work for you.

www.collins.co.uk/collinsGCSErevision

Contents

Contents

Contents

Cell Structure

1   **Figure 1** shows an animal cell.

Which part of an animal cell do the following descriptions refer to?
Write the letter **A**, **B** or **C** from the diagram next to each statement.

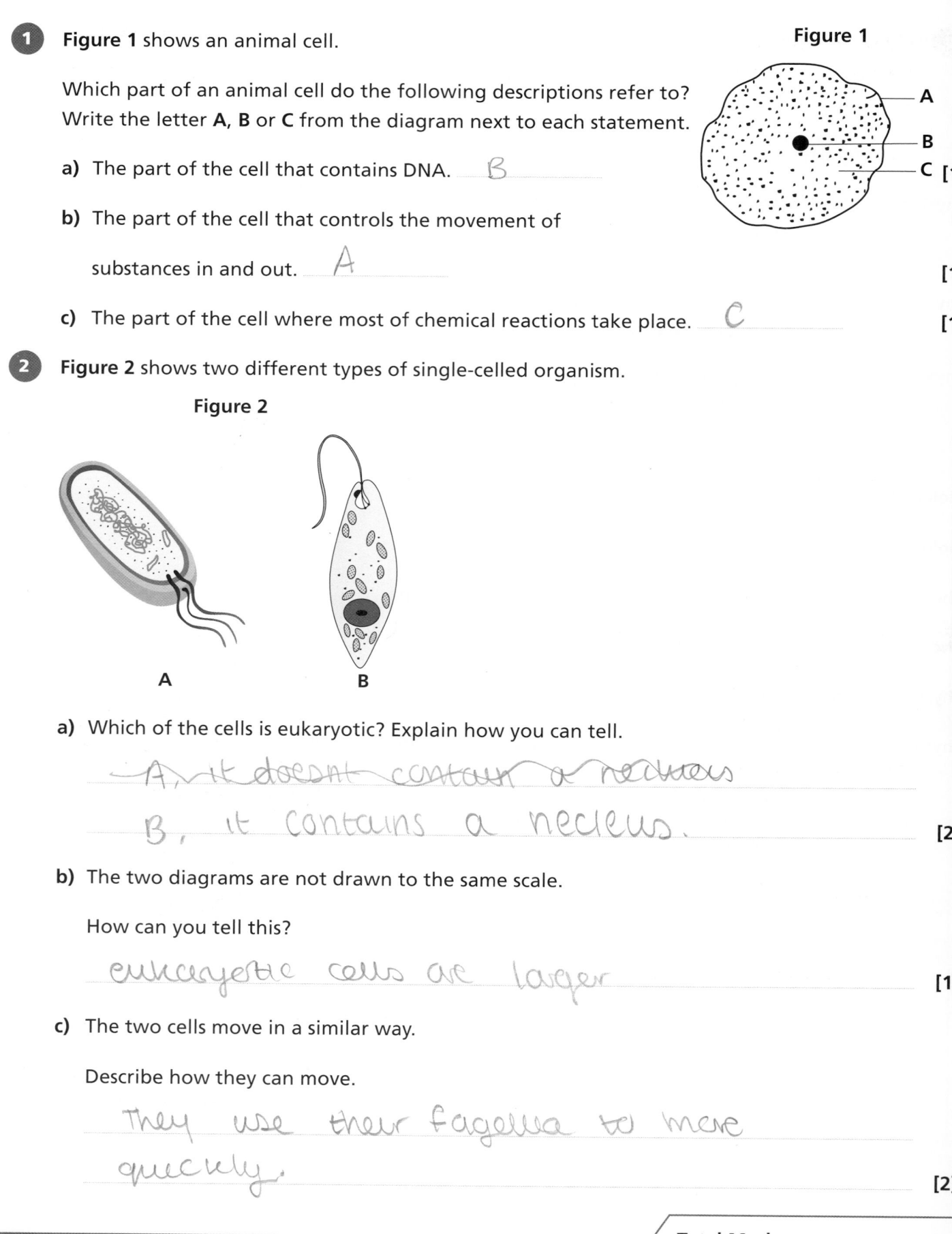

Figure 1

a)   The part of the cell that contains DNA.   *B*

          A

          B

          C [1

b)   The part of the cell that controls the movement of

    substances in and out.   *A*   [1'

c)   The part of the cell where most of chemical reactions take place.   *C*   [1]

2   **Figure 2** shows two different types of single-celled organism.

Figure 2

A   B

a)   Which of the cells is eukaryotic? Explain how you can tell.

    A, it doesn't contain a nucleus

    B, it contains a nucleus.   [2]

b)   The two diagrams are not drawn to the same scale.

    How can you tell this?

    eukaryotic cells are larger   [1]

c)   The two cells move in a similar way.

    Describe how they can move.

    They use their fagella to move

    quickly.   [2]

Total Marks   / 8

Investigating Cells

1 Rajesh uses his light microscope to view and draw a cheek cell.
He puts some cheek cells on to a slide and adds a few drops of a blue solution.

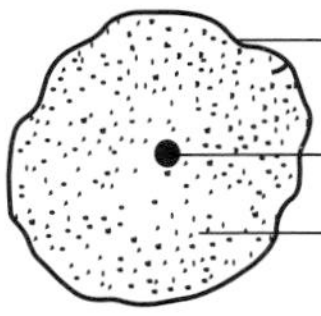

Figure 1 shows what Rajesh draws.

Figure 1

- Plasma membrane
- Nucleus
- Cytoplasm

a) Why does Rajesh add a few drops of blue solution to the cells?

to dye the cells to make it more visible.   [2]

b) A cheek cell is actually 0.03mm in diameter.

0.03mm   *0.03 × 1000*

i) What is the diameter in micrometres?

Answer: *30 µm*   [1]

ii) Calculate the magnification of Rajesh's drawing.

$$\text{magnification} = \frac{\text{size of image}}{\text{size of real object}}$$

30 µm / *0.03*

Answer: *1 000*   [2]

2 Roger is using a light microscope to look at some onion cells on a slide.
On the slide there is also a scale.
Each division on the scale represents **0.01mm**.

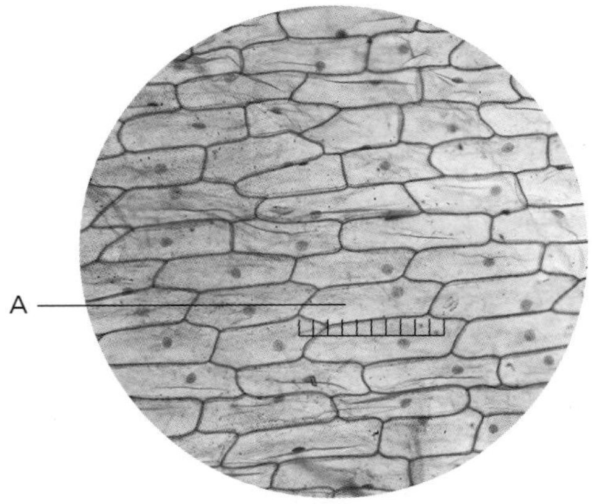

Figure 2 shows what Roger sees.

a) Estimate the length of cell **A** in **Figure 2** in millimetres.

Answer: _____   [1]

Figure 2

A

b) How wide is the nucleus of an onion cell?
Tick **one** box next to the best estimate.

100 micrometres	☐	1 micrometre	☑
10 micrometres	☐	0.1 micrometres	☐ [1]

Total Marks _____ / 7

Cell Division

1 A student investigates the growth of onions.
He puts an onion bulb in a jar of water.
The bulb starts to grow roots.

a) Cells in the root tip are dividing.

Which part of the cell cycle involves cells dividing?

_____ [1]

b) The student makes a slide of the onion root and looks at it with a light microscope.
He sees chromosomes inside some of the cells.

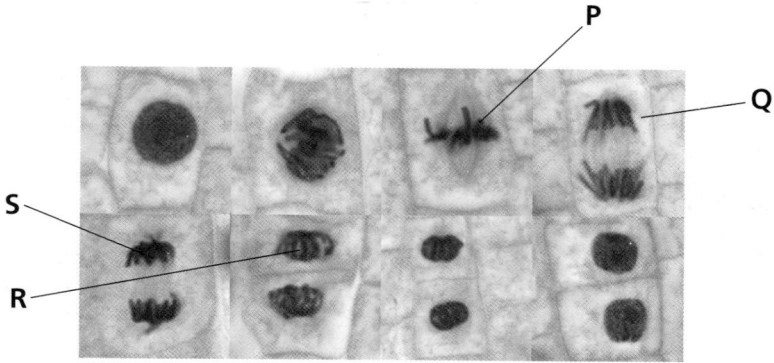

Cells **P, Q, R** and **S** are in different stages of cell division.

Put the stages in order.
One has been done for you.

		S	

[2]

2 Read the following quote and answer the questions below.

"People are always against new ideas such as using stem cells, but within a few years they will be used all the time to cure diseases."

a) How can stem cells be used to cure diseases?

_____

_____ [2]

b) Give **one** reason why people may be against using stem cells to cure diseases.

_____ [1]

Total Marks _____ / 6

Transport In and Out of Cells

1 The three blocks in **Figure 1** were cut from a block of agar jelly that had been dyed green.

Figure 1

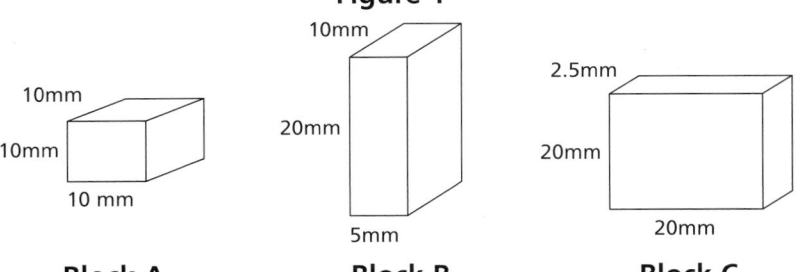

| Block A | Block B | Block C |

a) Calculate the surface area of block **A**. Put the results in **Table 1**.

Table 1

Block	Surface Area (mm²)	Volume (mm³)
A		1000
B	700	1000
C	1000	1000

[1]

b) The green dye turns yellow when acid moves through into the agar jelly.

What process would cause the acid to move through the agar jelly?

Answer: _____ [1]

c) The three blocks are submerged in an acid solution.

Which block would be the first to change colour completely? Explain your answer.

_____

_____

_____ [3]

2 Give **two** differences between:

a) Osmosis and diffusion.

_____

_____ [2]

b) Diffusion and active transport.

_____

_____ [2]

Total Marks _____ / 9

Levels of Organisation

1. Systems, **cells**, **organs** and **tissues** are all levels of organisation in living organisms.

   a) Write down these four levels in order of complexity, with the least complex first.

   _____ [1]

   b) Give the level of organisation of each of these structures.
      The first one has been done for you.

A lymphocyte	cell
The heart	
A neurone	
A leaf	
Epithelia on the skin	

   [4]

   c) **Table 1** shows the number of mitochondria in different types of cell.

   Which statement could best explain the data in the table?
   Tick **one** box.

   Table 1

Type of Cell	Number of Mitochondria
liver	1500
heart muscle	5000
skin	100

   Heart muscle cells are specialised to make protein. ☐

   Skin cells need large amounts of energy. ☐

   Liver cells do not require much energy. ☐

   Muscle cells are specialised for contraction. ☐

   [1]

2. **Figure 1** shows phloem and xylem cells.

   Describe how phloem and xylem cells are specialised for their functions in plants.

   Figure 1

   Phloem

   Xylem

   _____

   _____

   _____

   _____

   [4]

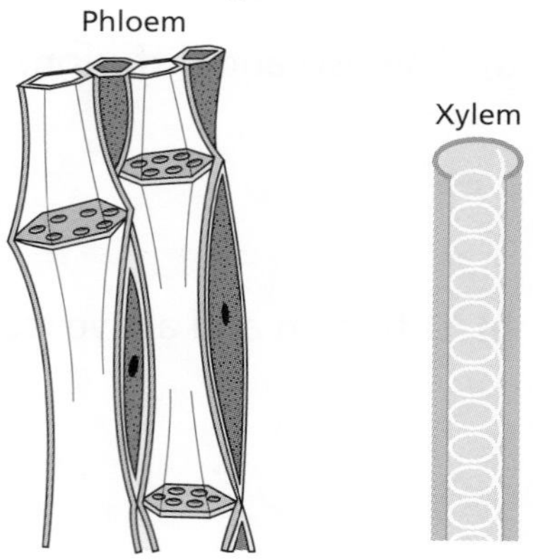

   Total Marks _____ / 10

Digestion

1 The enzyme amylase breaks down starch.

The graph in **Figure 1** shows the effect of temperature on the enzyme amylase.

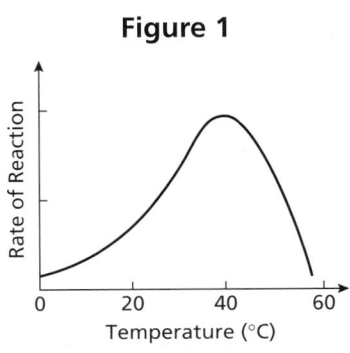

Figure 1

a) Describe the pattern shown in the graph.

....

.... **[2]**

b) Give the optimum temperature of this enzyme.

.... **[1]**

2 Lipids (fats) are digested in the body by the enzyme lipase.

a) Give one place in the body where lipase is made.

Answer: .... **[1]**

b) Some people want to be able to eat foods containing fat without gaining weight.
One way to do this is to replace fats in food with a substance called olestra.
Figure 2 shows an olestra molecule.

Figure 2

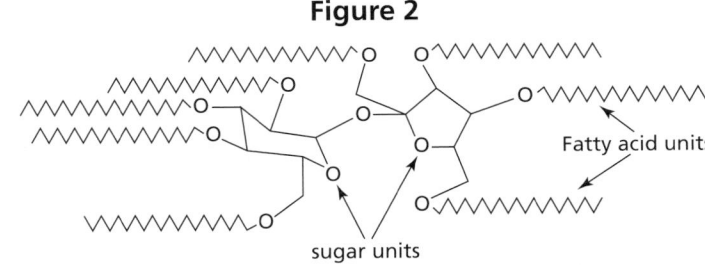

i) Write down **one** difference between an olestra molecule and a normal fat molecule.

.... **[1]**

ii) Suggest why lipase cannot digest olestra.

....

.... **[2]**

Total Marks .... / 7

Blood and the Circulation

1. Match the numbers on **Figure 1** with the blood vessels listed below.
Write the appropriate numbers in the boxes provided.

Figure 1

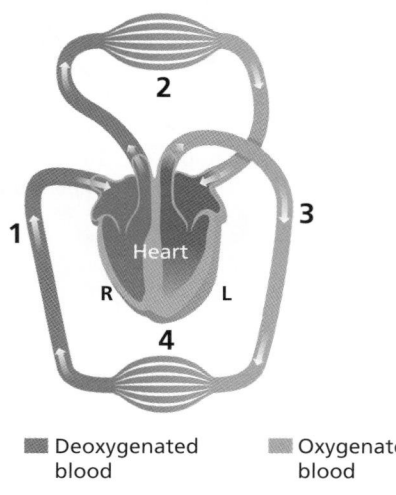

■ Deoxygenated blood   ■ Oxygenated blood

Aorta □

Vena cava □

Capillaries in the lungs □

Capillaries in the body □   **[3]**

2. Haemoglobin is found in red blood cells.

   a) Explain how haemoglobin supplies the tissues of the body with oxygen.

   _____

   _____

   _____   **[3]**

   b) Some people have a condition called sickle-cell anaemia.
   This affects their haemoglobin and can make their red
   blood cells change shape.

   Figure 2

   Explain why the red blood cells do **not** work so
   well after they change shape.

   Normal red blood cell   Sickled red blood cell

   _____

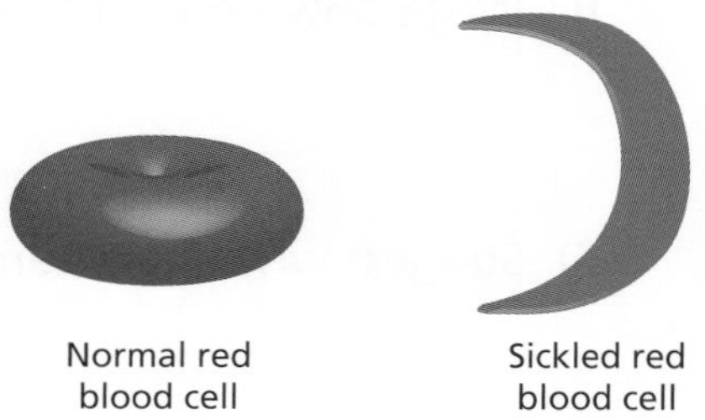

   _____   **[2]**

   Total Marks _____ / 8

Non-Communicable Diseases

1 A build-up of fatty material in the wall of the coronary arteries of the heart is called atheroma or atherosclerosis.

It has been known for many years that atheromas can cause a heart attack.

a) i) What is a heart attack?

.... **[1]**

ii) How can an atheroma cause a heart attack?

....

.... **[2]**

b) People used to think that the only cause of heart attacks was too much saturated fat, causing atheromas.

A scientist looked at groups of men who did different jobs involving different amounts of physical activity.

He recorded how many of these men suffered from heart attacks and how many had large atheromas. His results are shown on the graphs in **Figure 1**.

Figure 1

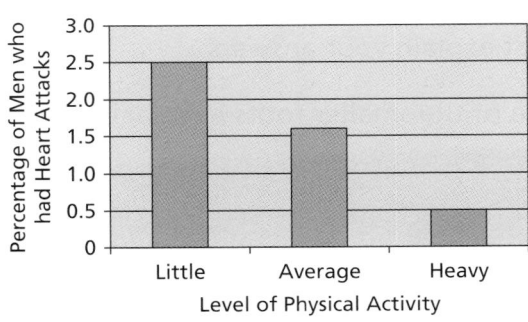

What conclusions can you make from these results?

....

....

.... **[3]**

Total Marks .... / 6

Transport in Plants

1. A student sets up an investigation into water transport in a plant as shown in **Figure 1**.
She measures the mass of the test tube and contents at the start and again the next day.
She repeats the experiment with the plant under different conditions.

The student's results are shown in **Table 1**.

Figure 1

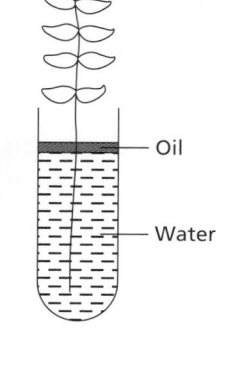

Table 1

Conditions	Mass of Tube and Contents (g)	
	At Start	After 1 Day
normal	29.7	26.5
windy	30.4	25.5
humid	29.2	27.2

a) Complete the following sentences about the student's results.

The plant loses mass because it loses _____ by a process called

_____ .

The greatest mass is lost from the plant in _____ conditions. **[3]**

b) Predict what the student's results would be under the following conditions compared with the normal conditions.
You must explain your answers.

i) Some of the smaller roots have been removed from the plant.

_____

_____ **[2]**

ii) Some of the leaves are painted with nail varnish.

_____

_____ **[2]**

2. Give **two** differences between the movement of water and the movement of sugars in a plant.

_____

_____ **[2]**

Total Marks _____ / 9

Pathogens and Disease

1 **a)** Draw **one** line from each disease to the type of microorganism that causes it.

Disease	Type of Microorganism
rose black spot	protozoan
salmonella	fungus
measles	virus
malaria	bacterium

[3]

b) Complete the sentences using the words from the list.

host   *Anopheles*   *Plasmodium*   **protozoan**   **red**   **vectors**

Mosquitoes are _____ because they carry malaria.

The mosquito's blood carries the malaria pathogen, which is called _____ .

The mosquito is a parasite because it feeds on a living organism,

called the _____ .

[3]

2 Human Immunodeficiency Virus (HIV) is a pathogen that can cause AIDS.

a) Give **two** ways that HIV can be passed on from one person to another.

_____

_____

[2]

b) AIDS stands for Acquired Immunodeficiency Syndrome.

Why is it called 'immunodeficiency'?

_____

_____

[2]

3 A gardener discovers that one of his rose bushes has black spot disease.
A friend advises him to burn any infected leaves rather than put them in the compost heap.

Why does she suggest this?

_____

_____

[2]

Total Marks _____ / 12

Human Defences Against Disease

1 **Figure 1** shows some of the main ways that pathogens can enter the body.

Figure 1

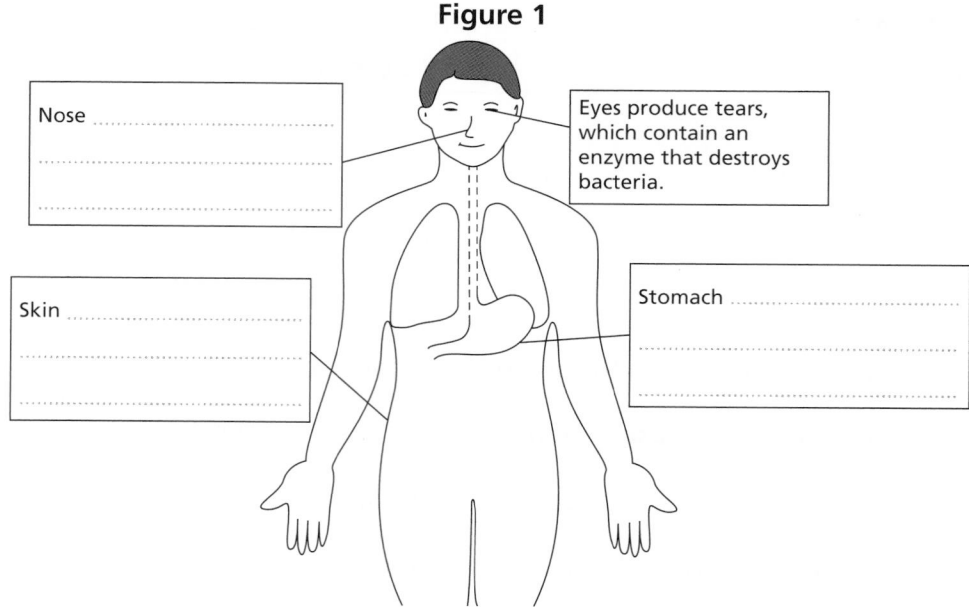

Nose ....
....
....

Eyes produce tears, which contain an enzyme that destroys bacteria.

Skin ....
....
....

Stomach ....
....
....

The diagram shows how the eyes work to stop pathogens entering the body.

Complete the diagram to show how the other **three** areas labelled prevent infection.   **[3]**

2 The four diagrams in **Figure 2** show stages in the immune response to a pathogen.

Figure 2

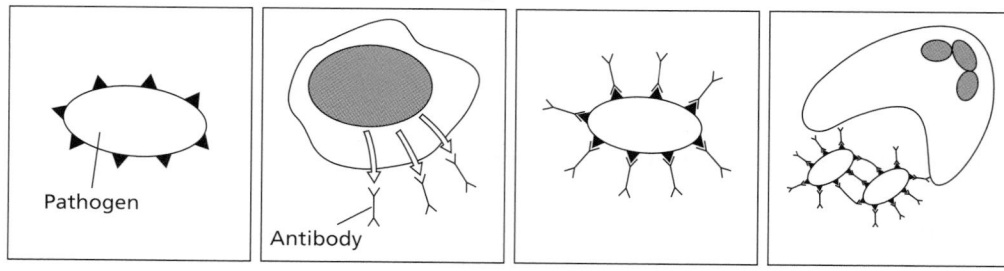

Pathogen

Antibody

Use the diagrams in **Figure 2** to explain how the immune system deals with a pathogen.

....

....

....

....

....   **[4]**

Total Marks .... / 7

Treating Diseases

1 This article is about the treatment of tuberculosis.

Antibiotic Could Beat TB

Scientists believe that they have found an antibiotic that could beat tuberculosis (TB).

TB is a disease of the respiratory system and is caused by a bacterium.

TB almost disappeared but is now killing more people. This is because the antibiotics that were used to treat TB are not effective now.

The new antibiotic is called linezolid and is being tested in a double-blind trial.

---

a) People who have TB can be treated with antibiotics, but HIV cannot be treated in this way.

Give the reason why.

_____ [2]

b) Explain why TB is now on the increase.

_____

_____ [2]

c) Describe **two** precautions that can be taken when using antibiotics to help prevent problems like this occurring.

_____

_____ [2]

d) The article says that the new drug is being tested in a double-blind trial.

Explain what is meant by a double-blind trial.

_____

_____

_____ [3]

> **Total Marks** _____ / 9

Photosynthesis

1 Plants make glucose by photosynthesis.

    a) Write down the word equation for this process.

        .... **[2]**

    b) i) Glucose is transported from the leaves to other parts of a plant.

        Suggest **two** parts of a plant that the glucose might be taken to.

        ....

        .... **[2]**

    ii) The glucose can also be built-up into different substances.
    These substances can then be used in many different ways.

        Name **two** of these substances and explain how they are used in a plant.

        ....

        ....

        ....

        .... **[4]**

2 The passage is about early investigations into plant growth.

> In the seventeenth century, the biologist, Jan Baptiste van Helmont, grew a tree in a bucket of soil. He fed the tree with rainwater only.
>
> In five years, the tree had grown but the amount of soil in the bucket did not decrease. Baptiste concluded that extra material in the tree must have come from the rainwater.
>
> In the eighteenth century, Jean Senebier showed that plants that live in water photosynthesise faster when they are supplied with extra carbon dioxide dissolved in the water.

    a) To what extent was Van Helmont's conclusion correct?

        ....

        .... **[2]**

    b) How did the work of Senebier show that Van Helmont's conclusion was not completely correct?

        .... **[1]**

> **Total Marks** .... / 11

Respiration and Exercise

1 When Joanne runs a race her muscles work hard.
Her muscles use aerobic respiration to release energy from glucose.

a) Complete the word equation for aerobic respiration. **[3]**

glucose + ... $\longrightarrow$ .... + ...

b) Joanne is training to run a marathon.

When she runs, her muscles start to make lactic acid and this passes into her blood.

The graph in **Figure 1** shows the lactic acid concentration in Joanne's blood during her first training run and during the first part of the race.

Figure 1

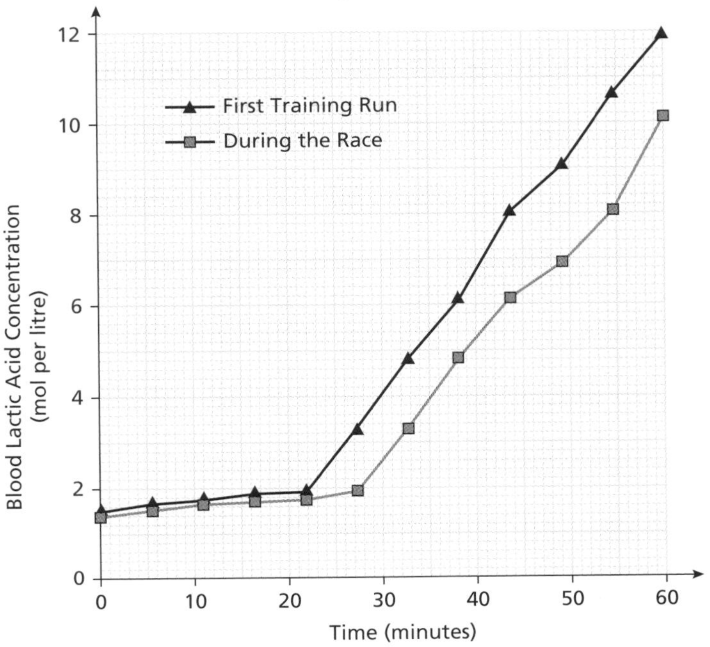

i) Use the graph to explain why Joanne can run more efficiently when she has trained for a race.

....

....

....

.... **[3]**

ii) Describe what happens to the lactic acid in the blood after Joanne stops running.

....

....

.... **[3]**

Total Marks .... / 9

Homeostasis and the Nervous System

1 **Figure 1** shows part of the nervous system.

The enlarged section shows a synapse. Transmitter molecules carry signals across the synapse.

Figure 1

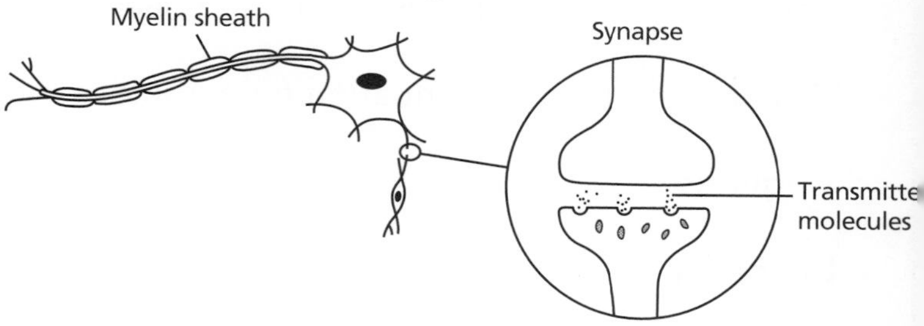

Myelin sheath

Synapse

Transmitter molecules

a) What are the functions of the myelin sheath (protective covering)?

....

....

[2]

b) Describe the function of the transmitter molecules at a synapse.

....

....

....

[3]

c) Many drugs or poisons work by affecting the action of synapses.

For parts **i)** and **ii)**, suggest how the different drugs or poisons described might work.

i) Some drugs stimulate the nervous system.
These drugs have a similar shape to neurotransmitter molecules.

....

....

[2]

ii) Neurotransmitter molecules are usually broken down by enzymes after they have stimulated the next neurone.
Some insect poisons block these enzymes.

....

....

[2]

Total Marks .... / 9

Hormones and Homeostasis

1 This is an extract from an article published in a magazine.

> **Ricky's Story**
>
> Ricky is 16 years old and found out that he had diabetes when he was 10.
>
> 'At first it was a big change because I had to learn to do my injections and not to eat all the sweet things I used to eat. Even with my diabetes, I still play sport regularly and I'm hoping to become a PE teacher.'

a) What type of diabetes does Ricky have?   Answer: ....   [1]

b) What would Ricky be injecting into his body?   Answer: ....   [1]

c) Suggest what problems there would be if Ricky did not inject himself.

....

....   [2]

d) Ricky's uncle Gary also has diabetes.
Gary is 50 years old and has only just developed diabetes.

    i) What type of diabetes does Gary have?   Answer: ....   [1]

    ii) How would the treatment for Gary's diabetes differ from Ricky's treatment?

....

....   [2]

2 Alcohol causes the pituitary gland to release too little ADH.

Explain what effects this might have on the body.

....

....

....   [3]

Total Marks .... / 10

Hormones and Reproduction

1 The graph in **Figure 1** shows the levels of two hormones in a woman's body at different stages of the menstrual cycle.

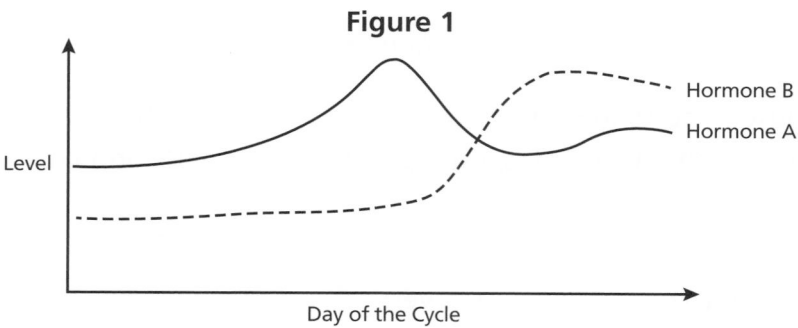

a) Both hormones are produced by the ovaries. Complete the following table.

	Name of Hormone	One Function of Hormone
Hormone A		
Hormone B		

[4

b) Mark on the graph the approximate time of the following events:

i) Mark the time of ovulation with an **X**. [1

ii) Mark the time of menstruation with a **Y**. [1

c) Two other hormones are involved in the menstrual cycle.

i) The levels of FSH are high in the first half of the cycle. Explain why.

_____

_____ [2]

ii) The levels of LH are highest in the middle of the cycle. Explain why.

_____ [1]

2 Explain how the hormones in the contraceptive pill can prevent pregnancy occurring.

_____

_____ [3]

Total Marks _____ / 12

Sexual and Asexual Reproduction

1 **Figure 1** shows a strawberry plant reproducing both sexually and asexually.

Figure 1

Plantlet

a) Describe how the plant reproduces asexually.

....

.... **[2]**

b) Describe how the plant reproduces sexually.

....

.... **[2]**

2 **Table 1** shows the number of chromosomes in different cells in a cat.
It also shows the type of cell division that produces these cells.

Complete the table.

Table 1

Cell Type	Number of Chromosomes	Type of Cell Division Used to Produce the Cell
Egg cell	19	
Eye cell		
Sperm cell		meiosis
Leg cell		mitosis

[4]

Total Marks .... / 8

Patterns of Inheritance

1 Gaucher disease (GD) is a genetic condition caused by a recessive allele.

People with Gaucher disease cannot make a particular enzyme.

The enzyme is needed to stop fat from building up in many organs, so the disease can be fatal.

Figure 1 shows part of a family tree showing some individuals that have Gaucher disease.

Figure 1

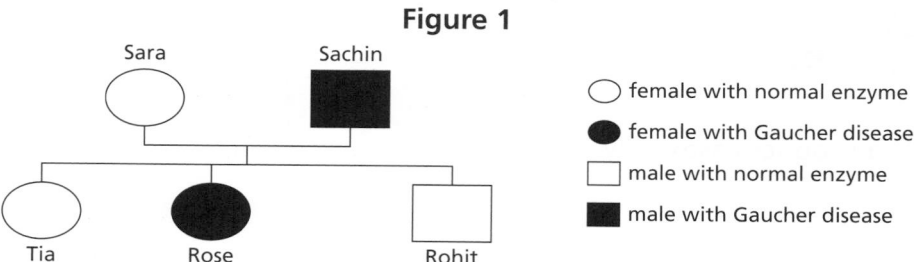

○ female with normal enzyme

● female with Gaucher disease

□ male with normal enzyme

■ male with Gaucher disease

a) Write down the names of all the people who are:

i) **Definitely** homozygous for this gene. Answer: _____ [1]

ii) **Definitely** heterozygous for this gene. Answer: _____ [1]

b) **Table 1** gives information about the different family members. Complete the table.

Table 1

Family Member	Possible Genotype	Phenotype
Sachin		
Tia	Gg	

[3]

c) Sara and Sachin are expecting another child.

Work out the probability that the child will have Gaucher disease. Use a table like **Table 1** to explain your thinking.

Probability: _____ [4]

d) A pregnant woman can have her unborn baby tested for Gaucher disease.

Suggest why she might decide **not** to have her baby tested even if it may have Gaucher disease.

_____

_____ [2]

Total Marks _____ / 11

Variation and Evolution

1 Bill and Ben are identical twins.
This means that they have inherited the same genes from their parents.

Write each of the characteristics from **Figure 1** in the correct column of the table.

Figure 1

Bill is 160cm tall

Bill and Ben have brown eyes

Ben has a scar

Ben's body mass is 60kg

Controlled by Their Genes	Caused by the Environment	Controlled by their Genes and the Environment

[4]

2 **Figure 2** shows two forms of a moth that rests on trees.
These moths are eaten by birds.

Figure 2

a) When large amounts of coal were burned the air was heavily polluted with soot.

Suggest what could happen to the colour of tree trunks in a heavily polluted area.

_____ [1]

b) A survey of dark coloured and light coloured moths was carried out in a polluted area.
An equal number of light and dark moths were collected, marked and released.
Several days later, moths were recaptured.

Suggest why more marked dark moths were recaptured than marked light moths.

_____

_____

_____ [2]

Total Marks _____ / 7

Manipulating Genes

1   The following sentences are about processes that involve manipulating genes

Complete the sentences with words from the list.

cloning   genetic engineering   selective breeding   tissue culture

Farmers manipulate the genes of animals, by carefully selecting which animals mate.

This is called _____ .

It is now possible to make bacteria produce human insulin by the process of _____ .   [2]

2   **Figure 1** shows some of the stages in the production of human insulin by bacteria. The stages are not in the correct order.

Figure 1

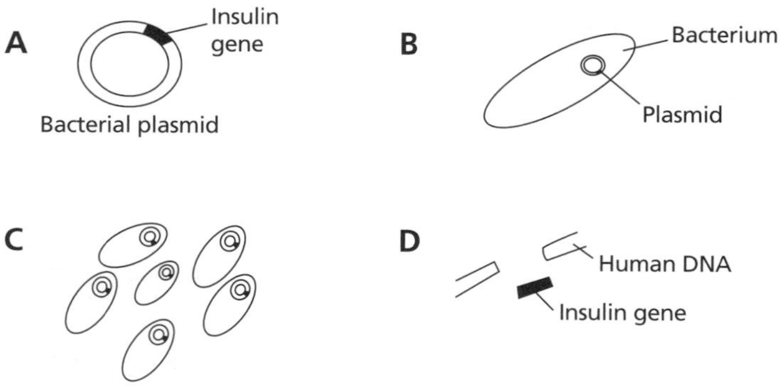

A   Insulin gene / Bacterial plasmid

B   Bacterium / Plasmid

C

D   Human DNA / Insulin gene

a)   Write down the order in which the steps should take place.
     The first one has been done for you.

   ____ **D** ____   _____   _____   _____   [2]

b)   Which of the statements about this method of making insulin are **incorrect**?
     Put a cross next to **two** statements.

   The process uses hormones to cut DNA.   ☐

   Vectors take pieces of DNA and insert them into other cells.   ☐

   The insulin molecules produced are a mixture of human and
   bacterial insulin.   ☐

   The bacteria can be grown on cheap waste products.   ☐   [2]

Total Marks _____ / 6

Classification

1 **a)** What is meant by the term 'species'?

_____

_____ **[2]**

b) **Table 1** shows the Latin names of some different cats.

Table 1

| **i)** Two of the cats are more closely related than the others.

Write down the common names of these two cats.

Common Name	Latin Name
bobcat	_Felis rufus_
cheetah	_Acinonyx jubatus_
lion	_Panthera leo_
ocelot	_Felis pardalis_

_____ **[2]**

ii) Explain your answer to part **i)** above.

_____ **[2]**

2 **a)** **Figure 1** shows an evolutionary tree, which includes an extinct animal called Archaeopteryx.
When Archaeopteryx became extinct humans did not exist.

Figure 1

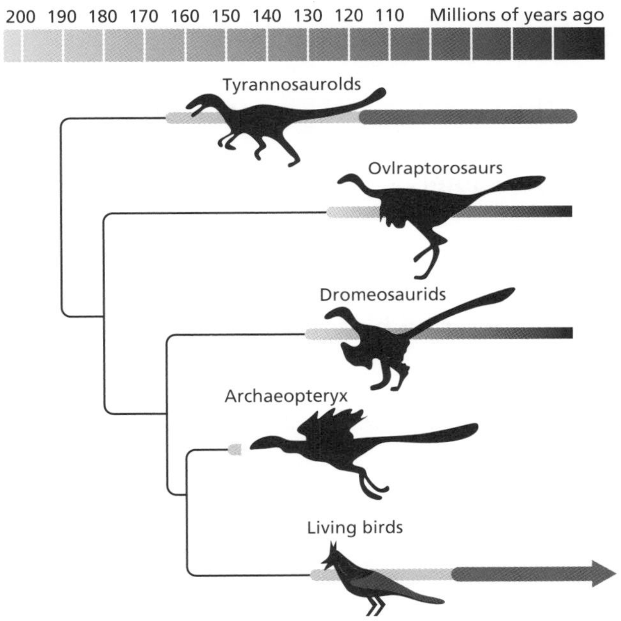

i) How many years ago did Archaeopteryx and birds share a common ancestor?

Answer: _____ **[1]**

ii) Which organism shown on the tree existed for the shortest period of time?

Answer: _____ **[1]**

b) Suggest **two** factors that may have led to its extinction.

_____

_____ **[2]**

Total Marks _____ / 10

Ecosystems

1 Camels live in desert areas of Africa.

a) Explain how the following special adaptations help the camel survive in the desert.

Figure 1

i) Webs of skin between their toes.

_____ **[1]**

ii) A store of fat in a hump on the top of their body.

_____ **[1]**

b) About one hundred years ago, camels were introduced into Australia.

Complete the sentences using words from the list.

abiotic   biotic   community   population   habitat   competed   reproduced

The camels that were introduced formed a large interbreeding _____.

The area that they were introduced into was a new _____ for them and they were well adapted to live there.

The _____ made up of other plants and animals living there started to suffer.

The camels _____ with cattle for _____ factors such as food. **[5]**

c) Scientists want to estimate how many organisms there are in an area of Australia.

i) Describe how they could estimate the number of cacti growing in that area.

_____

_____

_____

_____ **[6]**

ii) Why is it harder to estimate the number of camels in the area?

_____

_____ **[2]**

Total Marks _____ / 15

Cycles and Feeding Relationships

1 **Figure 1** shows part of the carbon cycle.

Complete the diagram by writing the names of the correct processes in the empty boxes.

Figure 1

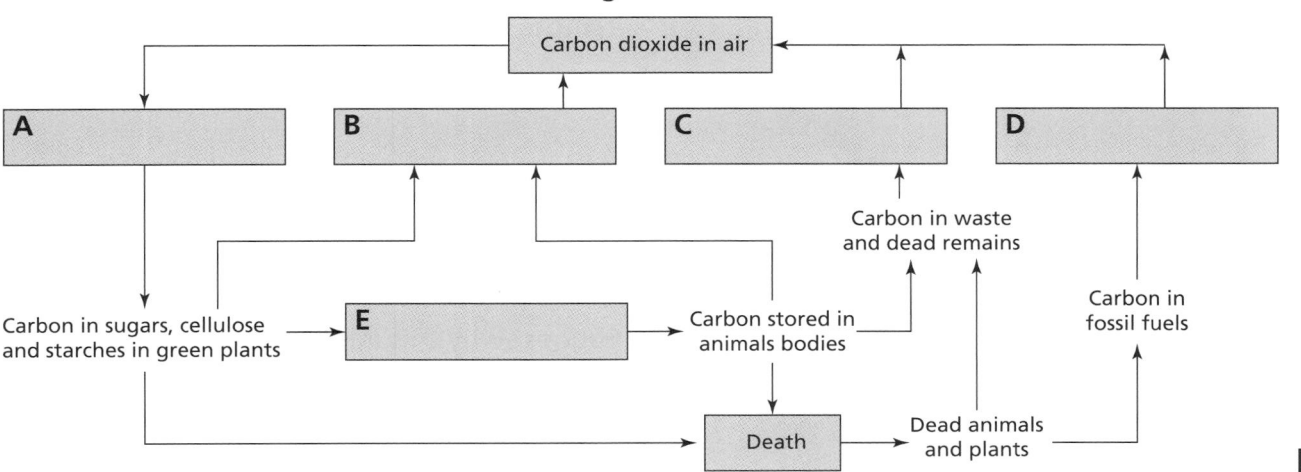

[5]

2 The graph in **Figure 2** shows how the number of rabbits and foxes in one square kilometre of woodland varies over a 40-month period.

Figure 2

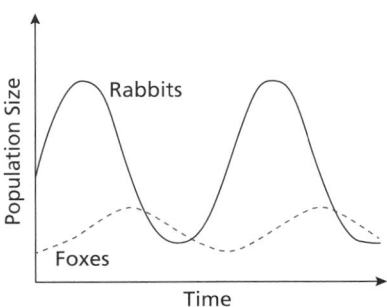

a) What name is given to this type of graph?

Answer: .... [1]

b) Explain why the fox numbers change throughout the 40 months.

....

....

.... [3]

Total Marks .... / 9

Disrupting Ecosystems

1 **Figure 1** shows the levels of carbon dioxide in the air above a small island in the Pacific Ocean. Scientists measured the carbon dioxide here to see if levels in the Earth's atmosphere are increasing.

Figure 1

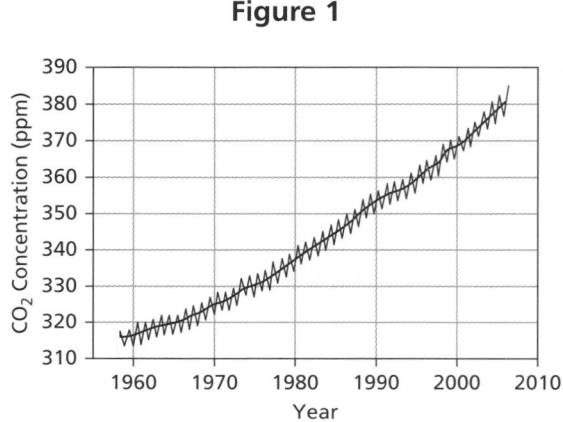

a) Suggest why scientists measured the levels at this remote island.

....

.... **[1]**

b) Levels of carbon dioxide go up slightly each winter and down each summer.

Suggest why this might be.

....

.... **[2]**

c) Explain how increasing carbon dioxide levels could lead to global warming.

....

.... **[3]**

2 When acid rain falls into a lake, the water becomes more acidic.
This may change the types of animals and plants that can live there.
Table 1 shows a number of aquatic animals and the range of pH in which they are found.

Table 1

Name of Animal	pH 4.0	pH 4.5	pH 5.0	pH 5.5	pH 6.0	pH 6.5
perch		✓	✓	✓	✓	✓
frog	✓	✓	✓	✓	✓	✓
clam					✓	✓
crayfish				✓	✓	✓

a) Which animal can survive the most acidic conditions?

Answer: .... **[1]**

b) Explain how acid rain is caused.

....

....

.... **[3]**

Total Marks .... / 10

Atoms, Elements, Compounds and Mixtures

1 A student heated a piece of copper in air.
The word equation for the reaction is:

copper + oxygen ➜ copper(II) oxide

a) Which equation correctly represents the reaction?
Tick **one** box.

$Cu + O → CuO$ ☐

$2Cu + O_2 → 2CuO$ ☐

$2Cu + 2O → 2CuO$ ☐

$Cu_2 + O_2 → 2CuO$ ☐ **[1]**

b) The mass of the copper before heating was 15.6g.
The mass of copper oxide produced was 18.9g.

Calculate the mass of oxygen that was used in the reaction.

Answer: _____ g **[1]**

c) What was the resolution of the balance the student used to measure the masses?

Answer: _____ g **[1]**

2 A student was asked to produce pure water from salt solution.

a) Describe how they would do this using the equipment in **Figure 1**.

_____

_____

_____

_____ **[3]**

Figure 1

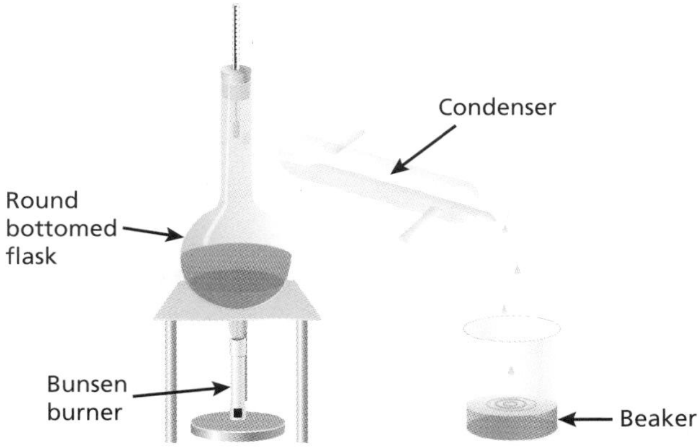

b) State **one** hazard when carrying out this procedure.

_____ **[1]**

Total Marks _____ / 7

Atoms and the Periodic Table

1 Early models of atoms showed them as tiny spheres that could not be divided into simpler substances.

In 1897, Thomson discovered that atoms contained small negatively charged particles. He proposed a new model shown in **Figure 1**.

Figure 1

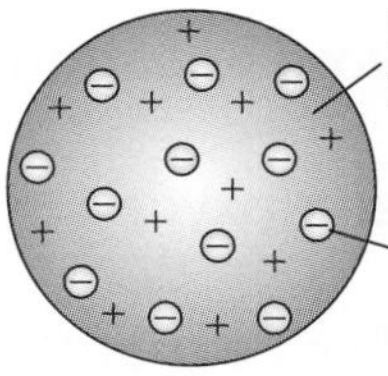

Sea of positive charge

X

a) Why did the model of the atom have to change?

....

....

.... **[2]**

b) Name the particle labelled **X**.

Answer: .... **[1]**

2 In 1909, Geiger and Marsden bombarded a thin sheet of gold with positively charged alpha particles. They found that most passed through, but some were deflected back.

a) Explain why this result was unexpected.

....

.... **[1]**

b) Explain why they would have repeated their experiment several times.

....

....

.... **[2]**

c) Describe the new model of the atom that resulted from this experiment.

....

....

....

.... **[3]**

Total Marks .... / 9

The Periodic Table

1 An unknown element has the electronic configuration 2,8,8,7.

Where would it be found in the periodic table shown in **Figure 1**?

Tick **one** box.

A ☐ B ☐

C ☐ D ☐

Figure 1

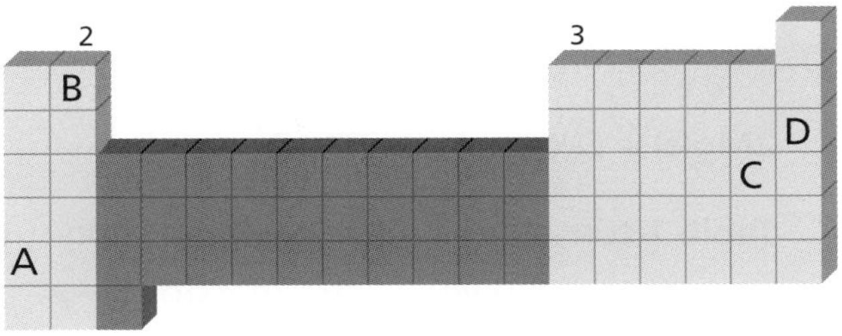

[1]

2 **Table 1** shows some data on the physical properties of elements.

Table 1

Physical Property	Element W	Element X	Element Y	Element Z
Melting Point (°C)	−38.82	−189.34	180.5	1538
Density (g/cm³)	13.53	1.40	0.53	7.87
Conductor of Electricity?	Yes	No	Yes	Yes

a) Which element, **W**, **X**, **Y** or **Z**, is a non-metal?
   Give a reason for your answer.

   ....

   .... [2]

b) Which element, **W**, **X**, **Y** or **Z**, is mercury?
   Give a reason for your answer.

   ....

   ....

   .... [3]

c) One of the elements is the Group 1 metal lithium.
   Complete the word equation to show the reaction of lithium with water.

   lithium + water → lithium .... + ... [2]

   Total Marks .... / 8

States of Matter

1 A student burns magnesium.

It reacts with oxygen in the air to form a white powder called magnesium oxide.

a) Add the missing state symbols to the equation for the reaction.

$2Mg(s) + O_2(\underline{\hspace{1cm}}) \rightarrow 2MgO(\underline{\hspace{1cm}})$ **[2]**

Table 1 contains some information about the melting and boiling points of the substances involved in the reaction.

Table 1

	Magnesium	Oxygen	Magnesium Oxide
Melting Point (°C)	650	−219	2830
Boiling Point (°C)	1091	−183	3600

b) State the temperature at which magnesium would change from a solid into a liquid.

Answer: .... °C **[1]**

c) State the temperature that oxygen gas would have to be cooled to in order for it to condense.

Answer: .... °C **[1]**

d) Explain, in terms of bonding, the difference in boiling points between oxygen and magnesium oxide.

....

....

....

....

....

....

....

....

....

....

[6]

Total Marks .... / 10

Ionic Compounds

1 A student investigated the properties of some different compounds.

a) Which of the following compounds contain ionic bonds?
Tick **two** boxes.

water (H_2O) ☐ glucose ($C_6H_{12}O_6$) ☐

calcium chloride ($CaCl_2$) ☐ sodium carbonate (Na_2CO_3) ☐

hydrogen chloride (HCl) ☐ **[2]**

b) The student discovered that the ionic compounds would not conduct electricity when solid but they would when dissolved in water.

Explain why.

....

....

.... **[3]**

2 **Figure 1** shows the outer electrons in an atom of the Group 2 element magnesium and in an atom of the Group 7 element bromine.
Magnesium forms an ionic compound with bromine.

Describe what happens when **one** atom of magnesium reacts with **two** atoms of bromine.
Give your answer in terms of electron transfer.
Give the formulae of the ions formed.

Figure 1

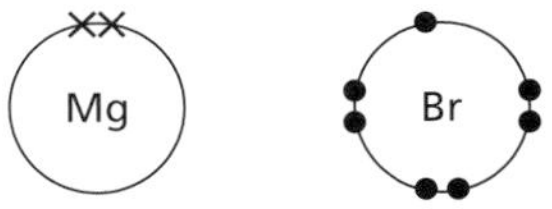

....

....

....

....

....

.... **[5]**

Total Marks .... / 10

Covalent Compounds

1 Fluorine and bromine are elements found in Group 7 of the periodic table.
At room temperature fluorine is a gas and bromine is a liquid.

Why, at room temperature, is fluorine a gas and bromine a liquid?
Tick **one** box.

The covalent bonds between bromine are stronger. ☐

Bromine has a giant covalent structure and fluorine is a simple molecule. ☐

The forces between bromine molecules are stronger. ☐

Fluorine contains fewer molecules than bromine. ☐ **[1]**

2 Bromine reacts with hydrogen to form hydrogen bromide.

Figure 1

a) Complete the dot and cross diagram in **Figure 1** to show
the covalent bonding in a molecule of hydrogen bromide.

Show the outer shell electrons only. **[2]**

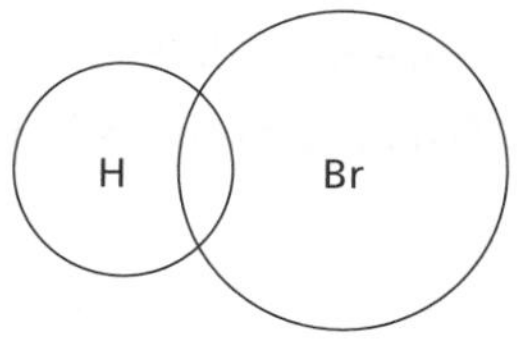

b) State the formula for a molecule of hydrogen bromide.

Answer: _____ **[1]**

3 Diamond has a giant covalent structure made up of carbon atoms.
It has a high melting point and does not conduct electricity when molten.

Draw **one** line from each property to the explanation of that property.

Property	Explanation of Property
	Strong covalent bonds between many carbon atoms
High melting point	Atoms are free to move
	Weak bonds between carbon atoms
Does not conduct electricity when molten	There are no charged particles that are free to move
	Carbon atoms are in a regular arrangement

[2]

Total Marks _____ / 6

Metals and Special Materials

1 **Figure 1** shows the bonding in the metal copper.

Figure 1

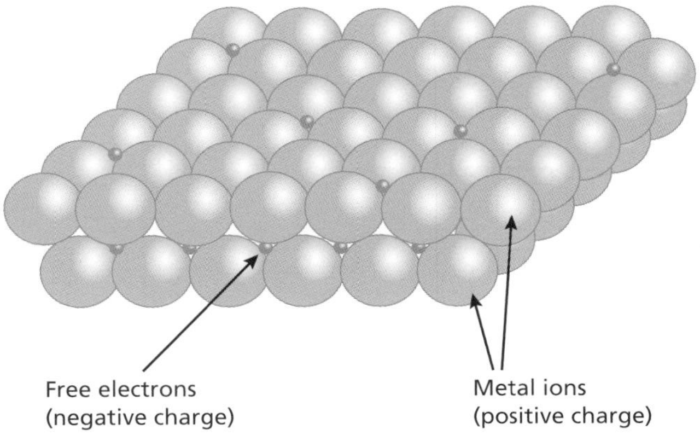

Free electrons
(negative charge)

Metal ions
(positive charge)

Copper is useful as a material for making saucepans.
This is because it has the following properties:

- high melting point
- good thermal conductor
- malleable (can be easily shaped).

Explain, in terms of its metallic bonding, why copper has these properties.

[6]

Total Marks .... / 6

Conservation of Mass

1 A student was asked to carry out the reaction shown in the word equation:

zinc + hydrochloric acid → zinc chloride + hydrogen

a) What is the ionic equation for the reaction?
Tick **one** box.

$Zn(s) + 2HCl(aq) → ZnCl_2(aq) + H_2(g)$ ☐

$Zn^{2+}(aq) + 2Cl^-(aq) → ZnCl_2(aq)$ ☐

$Zn(s) + 2H^+(aq) → Zn^{2+}(aq) + H_2(g)$ ☐

$Zn(s) + 2Cl^-(aq) → ZnCl_2(aq)$ ☐ **[1]**

b) The student plans to use the equipment in **Figure 1**.
The teacher tells the student not to use the bung.

Explain why.

_____

_____

_____

_____ **[2]**

Figure 1

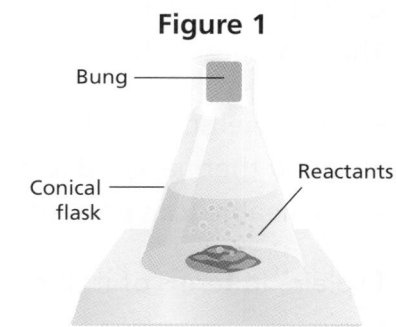

c) The student carries out the reaction on a balance, as shown in **Figure 2**.

Figure 2

Explain what happens to the mass reading during the reaction.

_____

_____

_____

_____

_____ **[3]**

Total Marks _____ / 6

Amount of Substance

1 **Figure 1** shows a cube of pure silver with a side length of 2cm.

a) Calculate the volume of the cube.

Figure 1

2cm

Answer: _____ cm^3 **[1]**

b) The density of silver is $10.49 g/cm^3$.

Use the formula **density** $= \dfrac{\textbf{mass}}{\textbf{volume}}$ to calculate the mass of the silver cube to 2 decimal places.

Answer: _____ g **[2]**

c) Calculate how many moles of silver are in the cube to 2 decimal places.
(Relative atomic mass, A_r, of silver = 108)

Answer: _____ moles **[2]**

d) There are 6.02×10^{23} atoms in one mole of silver.

Which of the following also contains 6.02×10^{23} atoms?
(Relative atomic masses, A_r: C = 12, O = 16)
Tick **two** boxes.

1 mole of oxygen (O_2) ☐

0.5 mole of oxygen (O_2) ☐

12g of carbon (C) ☐

6g of carbon (C) ☐   **[2]**

Total Marks _____ / 7

Reactivity of Metals

Figure 1

1 This question is about how metals are extracted from their ores.
Figure 1 shows the reactivity series of metals.

Figure 1

Most Reactive
Sodium
Calcium
Magnesium
Aluminium
Zinc
Iron
Lead
Copper
Gold
Platinum
Least Reactive

a) Zinc can be found naturally as the compound zinc(II) oxide.

Use the information in **Figure 1** to decide which of these metals can be used to displace zinc from zinc(II) oxide.
Tick **two** boxes.

magnesium ☐   iron ☐

copper ☐   sodium ☐   [2]

b) In reality, carbon is used to extract zinc from zinc(II) oxide.
The equation for the reaction is:

zinc(II) oxide + carbon → zinc + carbon monoxide

i) State the substance that is oxidised during the reaction.

Answer: _____ [1]

ii) State the substance that is reduced during the reaction.

Answer: _____ [1]

2 A student reacted zinc with a solution of copper chloride:

$Zn + CuCl_2 \rightarrow ZnCl_2 + Cu$

Draw **one** line from each change that takes place in this reaction to the correct half equation.

Change	Half Equation
	$Cu \rightarrow Cu^{2+} + 2e^-$
Oxidation	$Zn \rightarrow Zn^{2+} + 2e^-$
	$Zn^{2+} + 2e^- \rightarrow Zn$
Reduction	$Zn^{2-} \rightarrow 2e^- + Zn$
	$Cu^{2+} + 2e^- \rightarrow Cu$

[2]

Total Marks _____ / 6

The pH Scale and Salts

1 Universal indicator was added to a sample of an unknown solution.
The universal indicator turned orange.

What is the pH value of the solution?
Tick **one** box.

1 ☐   7 ☐

4 ☐   10 ☐   **[1]**

2 A student was asked to prepare pure, dry crystals of a soluble salt using insoluble copper(II) oxide and hydrochloric acid.
First, the student measured out $25cm^3$ of acid into a conical flask.

a) Describe how the student should complete the preparation of the salt.

....

....

....

....

....

....

....

....

....   **[4]**

b) State **one** safety precaution that the student should take during the preparation.

....

....   **[1]**

c) Name the salt produced.

Answer: ....   **[1]**

> **Total Marks** .... / 7

Electrolysis

1 Aluminium is extracted from its ore, aluminium oxide, using electrolysis, as shown in **Figure 1**.

Figure 1

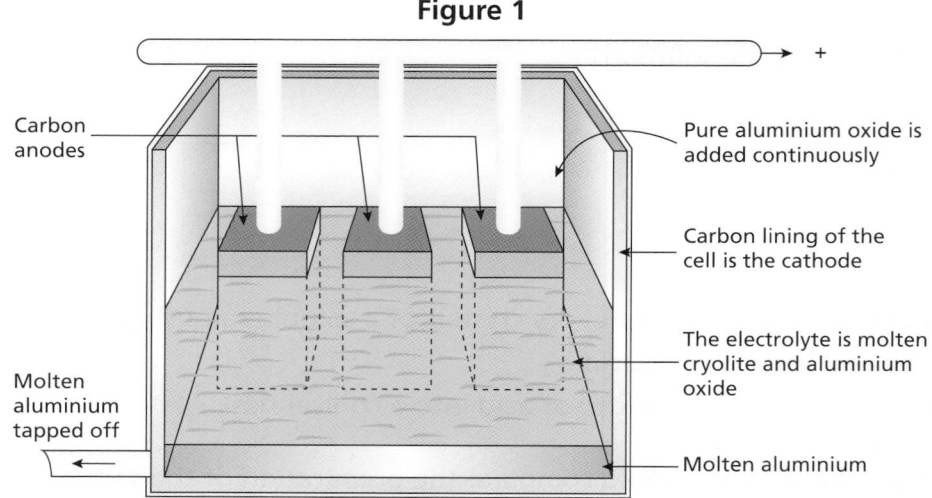

a) Explain why aluminium cannot be extracted from aluminium oxide using reduction with carbon.

_____

_____ [1]

b) Balance the half equation for the reaction that occurs at the carbon anodes.

_____ O^{2-} → _____ e^- + O_2 [2]

c) Explain why the extraction of aluminium from aluminium oxide is an expensive process and describe the methods used by manufacturers to reduce the cost.

_____

_____

_____

_____

_____

_____

_____

_____ [3]

Total Marks _____ / 6

Exothermic and Endothermic Reactions

1 A student carried out an investigation into how the reactivity of metals affects how exothermic or endothermic their reaction with dilute hydrochloric acid is.

The student used the apparatus shown in **Figure 1**.

The student's results are shown in **Table 1**.

Figure 1

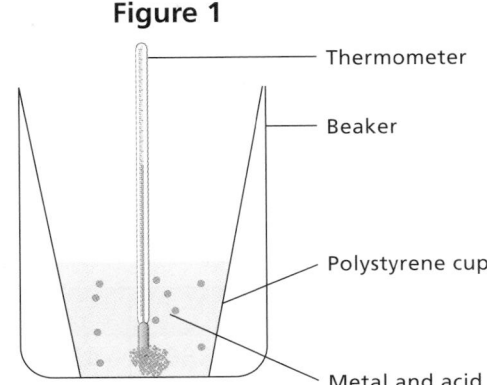

Thermometer

Beaker

Polystyrene cup

Metal and acid

Table 1

Metal	Temperature at Start (°C)	Highest Temperature Reached (°C)
Zinc	21.0	30.1
Copper	21.2	21.3
Magnesium	21.4	82.6
Iron	21.4	26.0

a) Describe the function of the polystyrene cup.

....

....

.... **[2]**

b) State **two** control variables the student should have used in this investigation.

....

.... **[2]**

c) The order of reactivity of these metals from highest to lowest is: magnesium, zinc, iron, copper.

Write a conclusion that answers the original question that the student was investigating.

....

....

....

.... **[2]**

Total Marks .... / 6

Measuring Energy Changes

1 **Figure 1** shows the displayed formulae for the decomposition of hydrogen bromide.

Figure 1

H–Br

H–Br ⟶ **H–H + Br–Br**

The bond enthalpies for the reaction are shown in **Table 1**.

Table 1

	H–Br	H–H	Br–Br
Energy (kJ/mol)	366	436	193

a) Calculate the energy needed to break the bonds in the reactants.

Answer: .... **kJ/mol** **[2**

b) Calculate the energy released when the bonds in the products are formed.

Answer: .... **kJ/mol** **[2**

c) Is this reaction **endothermic** or **exothermic**?

Explain your answer.

....

....

.... **[2]**

Total Marks .... / 6

Rate of Reaction

1 When hydrochloric acid reacts with sodium thiosulfate one of the products is sulfur, which is insoluble.

A student carried out an investigation into the following hypothesis:

As the concentration of acid increases, the rate of the reaction will increase.

The student added sodium thiosulfate to dilute hydrochloric acid of concentration 0.25mol/dm³ in a conical flask.
The conical flask was placed on a cross, as shown in **Figure 1**.
The student timed how long it took before they could no longer see the cross.

Figure 1

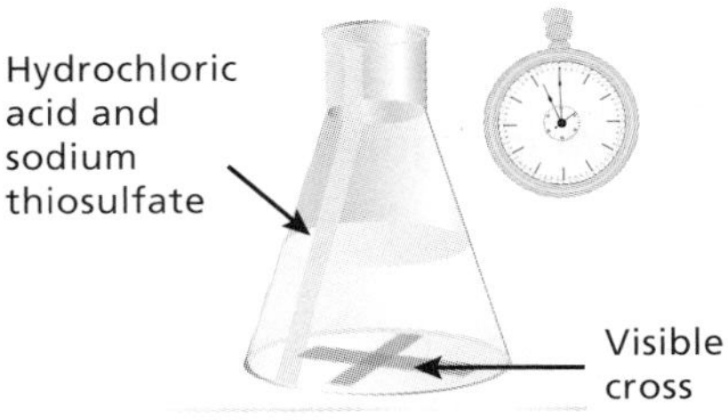

Hydrochloric acid and sodium thiosulfate

Visible cross

a) Describe what the student should do next in order to investigate the hypothesis.
   You should include a suitable range for the independent variable in your answer.

   _____

   _____

   _____

   _____

                                                                      [2]

b) State **one** safety precaution the student should take.

   _____

                                                                      [1]

c) Use collision theory to write a prediction for what will happen if the hypothesis
   is correct.

   _____

   _____

   _____

   _____

   _____

   _____

                                                                      [4]

Total Marks _____ / 7

Reversible Reactions

1 This question is about the reaction between ammonia and hydrogen chloride.
The equation for the reaction is:

$$NH_3(g) + HCl(g) \rightleftharpoons NH_4Cl(s)$$

The reaction can be carried out using the apparatus shown in **Figure 1**.

Figure 1

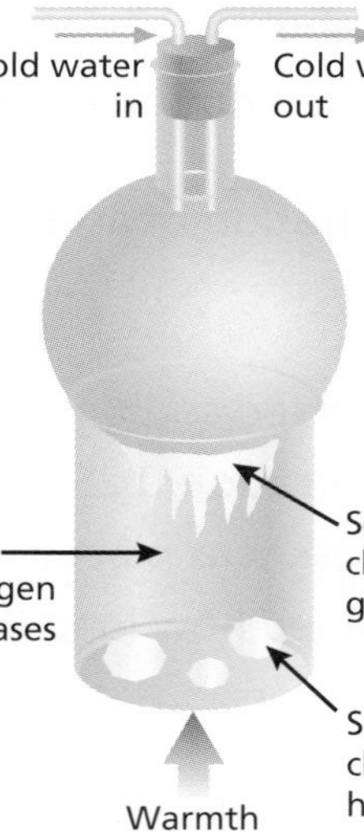

Cold water in
Cold water out

Ammonia and hydrogen chloride gases

Solid ammonium chloride formed by gases recombining

Solid ammonium chloride being heated

Warmth

a) What is meant by the symbol $\rightleftharpoons$ in the equation?

Answer: .... [1]

b) Which reaction, the **forward reaction** or the **reverse reaction**, is exothermic?
Give a reason for your answer.

....

....

.... [3]

c) The reaction taking place in **Figure 1** is at equilibrium.

What does this mean?
Tick **two** boxes.

Only ammonium chloride is being produced. ☐

The forward and reverse reactions are taking place at the same rate. ☐

The amounts of reactants and products are constant. ☐

The reaction has completed. ☐ [2]

d) Predict what will happen if more ammonia is added to the mixture.

....

....

.... [2]

Total Marks .... / 8

Alkanes

1 Fractional distillation is used to separate crude oil into useful mixtures called fractions.
It takes place in a column, as shown in **Figure 1**.

Figure 1

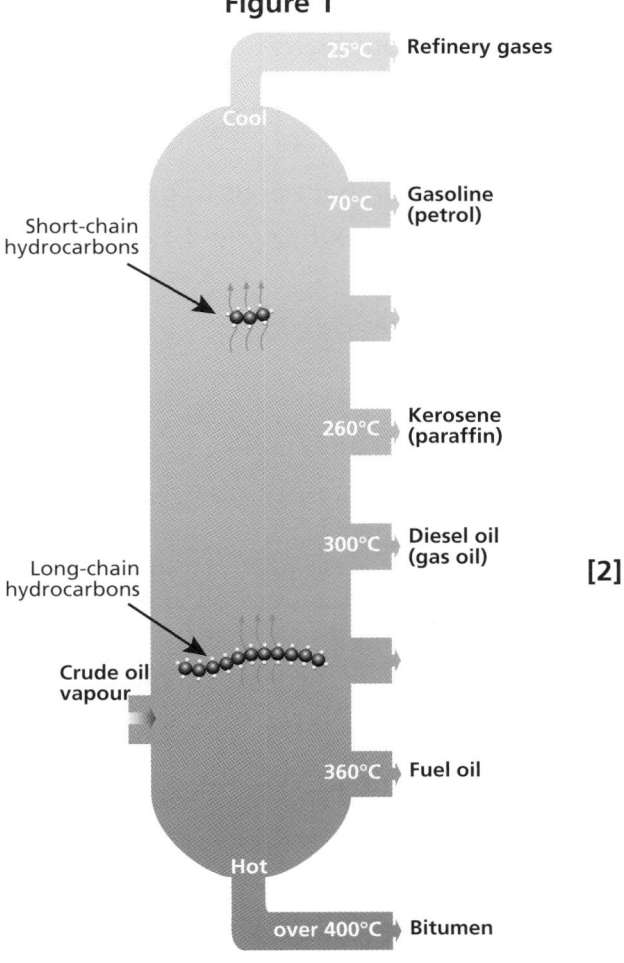

a) The molecules found in crude oil are mostly hydrocarbons.

What is a hydrocarbon?

....

....

....

....

[2]

b) This is what takes place inside the column:

1. Crude oil is heated until it forms a vapour.
2. The vapour rises up the tower.

Describe what happens next in order for the fractions to form.

....

....

....

....

....

....

[2]

2 Methane (CH_4) is the name of a hydrocarbon found in refinery gas.

Complete the balanced symbol equation for the complete combustion of methane.

$CH_4 + 2$ .... $\rightarrow CO_2 +$ .... H_2O

[2]

Total Marks .... / 6

Cracking Hydrocarbons

1 Hydrocarbon molecules can be cracked to form molecules with shorter chains.
The equation shows the cracking of $C_{20}H_{42}$.

$$C_{20}H_{42} \rightarrow C_8H_{18} + \boxed{} C_5H_{10} + C_2H_4$$

a) What number needs to go in the box to balance the equation?
Tick **one** box.

1 ☐

2 ☐

3 ☐

4 ☐   [1]

b) Which products are alkanes?
Tick **one** box.

C_8H_{18} and C_2H_4 ☐

C_5H_{10} and C_2H_4 ☐

C_8H_{18} only   ☐

C_8H_{18}, C_5H_{10} and C_2H_4 ☐   [1]

2 A student is given a test tube of colourless liquid hydrocarbon.

Describe a test the student could carry out to find out if the liquid is an alkane or
an alkene.
You should state the chemical used for the test and the results of the test for both an
alkane and an alkene.

_____

_____

_____

_____

_____

_____

[3]

Total Marks _____ / 5

Chemical Analysis

1 A student was asked to investigate five different inks using chromatography.

Figure 1 shows the apparatus they needed to use.

The student set up the apparatus correctly and left it for several minutes.

Figure 2 shows their results.

Figure 1

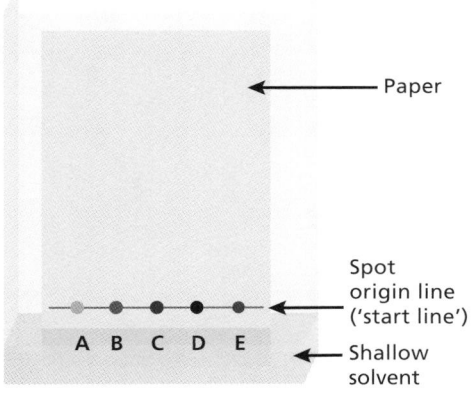

Figure 2

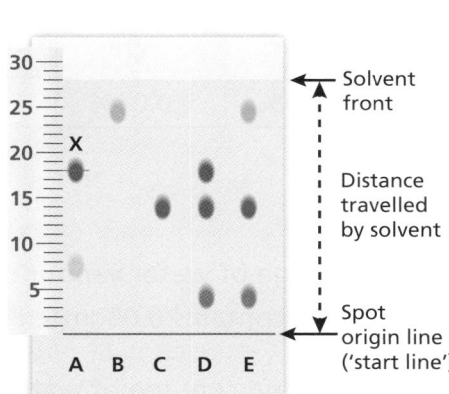

a) Which part of the apparatus is the stationary phase?

Answer: .... **[1]**

b) The student was not sure whether to use a pen or pencil to draw the origin line.

State which they should choose. Give a reason for your answer.

....

.... **[2]**

c) Which **two** inks are pure? Give a reason for your answer.

....

.... **[2]**

d) Calculate the R_f value of **X**.

Give your answer to 2 decimal places.

Answer: .... **[3]**

Total Marks .... / 8

The Earth's Atmosphere

1. **Table 1** shows some of the gases that are in the Earth's atmosphere today.

Table 1

	Oxygen	Nitrogen	Carbon Dioxide
A	80%	20%	0.04%
B	0.04%	80%	20%
C	20%	80%	0.04%
D	20%	0.04%	80%

Which letter, **A**, **B**, **C** or **D**, represents the correct proportions?
Tick **one** box.

A ☐

B ☐

C ☐

D ☐

[1

2. The percentage of water vapour in the atmosphere can vary.
4.2dm³ of air contains 0.05dm³ of water vapour.

Calculate the percentage of water vapour in the air.
Give your answer to 2 significant figures.

Answer: .... % [2

3. One theory for how the Earth's early atmosphere was formed is that gases were released by volcanoes.
Scientists have recently proposed a new theory: that the early atmosphere was formed by comets hitting the Earth.

a) What would the scientists need for this new theory to be accepted?

....

....

[2

b) Explain why we cannot be sure if either theory is correct.

....

....

....

....

[3]

Total Marks .... / 8

Greenhouse Gases

1. Many scientists believe that human activities are causing the mean global temperature of the Earth to increase.

   They think that this is because of an increase in the amounts of greenhouse gases in the atmosphere.

   Water vapour is an example of a greenhouse gas.

   a) Name **one** other greenhouse gas and state how human activity has increased its amount in the atmosphere.

   [2]

   In 2000, computer models were used to predict how the mean global temperature might change in the future.

   Figure 1 shows the results. Each line shows the predictions made using a different model.

   Figure 1

   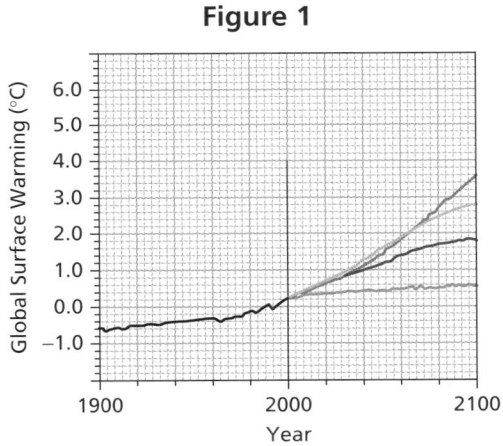

   b) Calculate the range of global surface warming as predicted for 2100 by the different models.

   From: .... To: .... [2]

   c) Explain why it is difficult to produce models to predict future climate change.

   [2]

Earth's Resources

1 In the UK, potable water is produced from an unpolluted source of fresh water. Potable water contains low levels of dissolved substances.

a) What is potable water?

.... [1]

b) Explain why potable water cannot be called pure in the chemical sense.

....

....

.... [2]

The stages of producing potable water from fresh water are:

1. Pass the water through filter beds to remove any solids.
2. Sterilise the water to kill microorganisms.

c) Give **one** way of sterilising the water.

.... [1]

In other parts of the world, sea water is used as a source of potable water.
To produce potable water from sea water a process called distillation is used.

d) Explain why distillation is used.

....

.... [1]

e) Explain why producing potable water from sea water is more expensive than producing it from fresh water sources.

....

....

.... [2]

Total Marks .... / 7

Using Resources

1. Shopping bags can be made out of paper or plastic.
   Table 1 is part of a LCA (life cycle assessment) comparing these two materials.

Table 1

	Paper Bag	Plastic Bag
Raw Materials	Wood pulp	Crude oil
Manufacture	• Wood pulp is added to water (up to 100 times the mass of the pulp) and mixed. • Clay, chalk or titanium oxide is added. • The mixture is squeezed and heated to remove the water.	• Crude oil is heated to 360°C during fractional distillation. • A fraction is cracked at 850°C to produce alkenes. • Polymerisation of alkenes at 150°C.
Transport	• Average mass = 55g • Seven trucks needed to transport two million bags.	• Average mass = 7g • One truck needed to transport two million bags.
Use During Lifetime	• Not normally reused.	• Can be reused many times.
Disposal at the End of Life	• Taken to landfill (biodegradable). • Can be incinerated (burned). • Can be recycled.	• Taken to landfill (not biodegradable). • Can be incinerated (burned). • Difficult to recycle.

Use the information in **Table 1** to compare the **advantages** and **disadvantages** of both types of bag.

[6]

Total Marks _____ / 6

Forces – An Introduction

1 When a plane is in flight, the engines provide a thrust force that pushes the aircraft forwards. The wings provide a 'lift' force that acts upwards.

a) Name **two** other forces that act on the plane.

In each case state whether it is contact force or non-contact force.

....

.... [4]

b) The plane has a mass of 120 000 kg. Calculate the weight of the plane (g = 10N/kg).

Weight = ... N [2]

c) The plane accelerates and ascends to a higher altitude.
During this time, the resultant forwards force is twice the size of the resultant upwards force.

Use this information to draw a scale
vector diagram.
Your diagram should show:

- The resultant forwards force.
- The resultant upwards force.
- A final resultant force that shows the
  combined effect of all the forces acting
  on the plane. [3]

2 A student carries out an investigation into forces.
They use an air track, which suspends a glider vehicle on a cushion of air so that it can move smoothly without touching the ground.

a) What force is the air track designed to reduce?

Answer: .... [1]

b) Use the idea of contact and non-contact forces to explain why the air track is effective at doing this.

....

.... [2]

Total Marks .... / 12

Forces in Action

1 A student carries out an investigation involving springs.
The student suspends a spring from a rod.
A force is applied to the spring by hanging a mass from it.

a) Describe how the student could test if the spring is behaving elastically.

....

.... **[2]**

b) In a second investigation, the student takes a set of measurements for force and extension.
The results are shown in **Table 1** below.

Table 1

Force (N)	0.0	1.0	2.0	3.0	4.0	5.0	6.0
Extension (cm)	0.0	4.0		12.0	16.0	22.0	31.0

  i) Add the missing value to the table. **[1]**

  ii) Explain why you chose this value.

  ....

  .... **[2]**

c) Complete the following sentences about the experiment in part **b)**.

The independent variable investigated was the ....

It had a range from .... to .... **[3]**

d) The student repeats the experiment in part **b)**.

Give **two** reasons why repeating an experiment can improve the accuracy of the results.

....

.... **[2]**

Total Marks .... / 10

Forces and Motion

1 A person takes their dog for a walk.
The graph in **Figure 1** shows how the distance from their home changes with time.

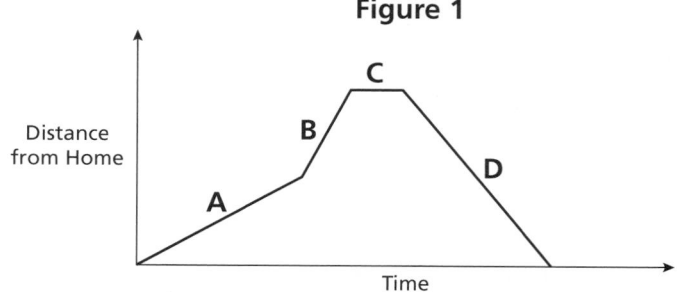

Figure 1

a) Work out the final displacement at the end of their journey.

Answer: _____ [1

b) Which part of the graph, **A**, **B**, **C** or **D**, shows them walking at the fastest rate?

Answer: _____ [1]

c) Describe their motion during section **C**.

Answer: _____ [1]

d) Describe how the velocity of section **A** compares to the velocity of section **D**.

_____

_____

_____ [3]

2 Drivers on a racetrack enter a hairpin bend travelling east.
The tight bend forces them to slow down.
When they exit the bend, they are travelling west and speed back up again.
The bend is 180m long.

a) If it takes 6 seconds to travel around the bend, what is the average speed of the car around the bend?

Speed = _____ m/s [2]

b) A driver enters the bend at 50m/s and exits the bend at 40m/s.

i) Work out the change in speed.

Change in speed = _____ m/s [1]

ii) Work out the change in velocity.

Change in velocity = _____ m/s [1]

Total Marks _____ / 10

Forces and Acceleration

1 An experiment is carried out to investigate how changing the mass affects the acceleration of a system.
A trolley is placed on a bench and is made to accelerate by applying a constant force using hanging masses. Different masses were then added to the trolley.

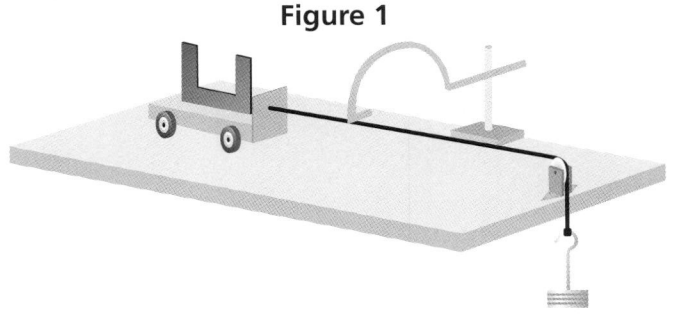
Figure 1

a) What is the independent variable?

Answer: .... **[1]**

b) What is the control variable?

Answer: .... **[1]**

c) What is the dependent variable?

Answer: .... **[1]**

2 A boat accelerates at a constant rate in a straight line.
This causes the velocity to increase from 4.0m/s to 16.0m/s in 8.0s.

a) Calculate the acceleration.
Give the unit.

Answer: .... **[3]**

b) A water skier being pulled by the boat has a mass of 68kg.

Use your answer to part **a)** to calculate the resultant force acting on the water skier whilst accelerating.

Answer: .... **[2]**

3 The manufacturer of a car gives the following information in a brochure:
The mass of the car is 950kg.
The car will accelerate from 0 to 33m/s in 11s.

a) Calculate the acceleration of the car during the 11s.

Answer: .... **[2]**

b) Calculate the force needed to produce this acceleration.

Answer: .... **[2]**

Total Marks .... / 12

Terminal Velocity and Momentum

1 The terminal velocity is the maximum velocity a falling object can reach.

Describe how a skydiver could increase their terminal velocity.

[2

2 A Saturn 5 rocket, as used on the Apollo space missions, has F1 rocket engines.
Each engine burns 5000kg of fuel every second.
The exhaust gases leave the thrusters at 1340m/s.

a) Calculate the momentum of the exhaust gases.

Answer: _____ kg m/s [3]

b) During the Apollo mission, five engines were used simultaneously.

Use your answer to part **a)** to calculate the momentum gained by the rocket in the first
10 seconds of travel (assuming there were no other resistive forces).

Answer: _____ kg m/s [3]

c) The entire launch vehicle has a mass of 3000000kg.

Use your answer to part **b)** to calculate the velocity after the first 10 seconds (assuming all
other factors are unchanged).

Answer: _____ m/s [3]

d) The rocket engines provide constant thrust.
However, after 100 seconds, the acceleration of the rocket is greater than the initial acceleration.

Give **two** reasons why this might be.
You must explain your answers.

[4]

Total Marks _____ / 15

Stopping and Braking

1 The graph in **Figure 1** shows how the velocity of a car changes from the moment the driver sees an obstacle blocking the road.

Figure 1

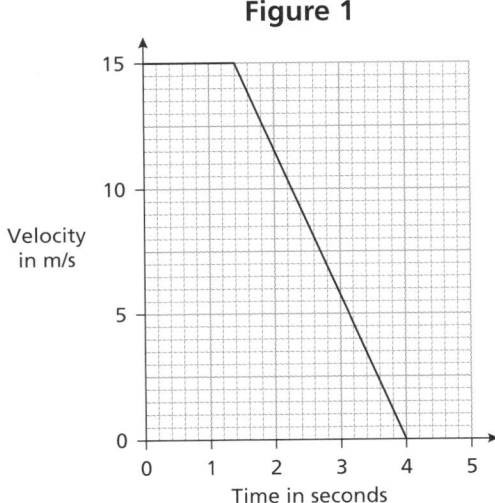

a) Work out the reaction time of the driver.

Answer: _____ **[1]**

b) Use the graph and your answer to part **a)** to calculate the thinking distance.

Answer: _____ **[2]**

c) Use the graph to work out the time it took for the car to stop from the moment the brakes were applied.

Answer: _____ **[1]**

d) The car and driver have a combined mass of 1300kg.

Work out the momentum of the car just before the brakes were applied.

Answer: _____ **[2]**

e) How would the time it took to stop the car be different it the road was wet or icy?

_____ **[1]**

f) The driver of the car was tired and had been drinking alcohol.

On the graph in **Figure 1**, sketch a second line to show how the graph would be different if the driver had been wide awake and fully alert. **[3]**

g) How would the graph look different if the vehicle had old / worn brakes and tyres?

_____

_____ **[2]**

Total Marks _____ / 12

Energy Stores and Transfers

1 An electric kettle is used to heat 2kg of water from 20°C to 100°C.

> **change in thermal energy = mass × specific heat capacity × temperature change**
> Specific heat capacity of water = 4200J/kg°C

a) All of the energy supplied to the kettle goes into the water.

Calculate the amount of electrical energy supplied to the kettle.

Answer: _____ [3]

b) On a different occasion, the kettle is filled with 2.5kg of water but is switched on for the same amount of time.

Use your answer to part **a)** to calculate what temperature the kettle heats the water to in this period.

Answer: _____ [3]

2 **Figure 1** shows a pendulum swinging backwards and forwards.
The pendulum is made from a 100g mass suspended by a light string.

Figure 1

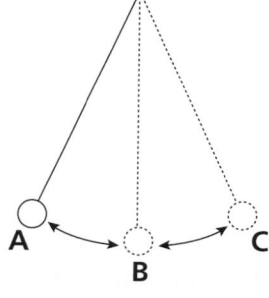

a) At position **C**, the mass is 5cm higher than at position **B**.

Calculate the difference in gravitational energy between positions **B** and **C**.
The gravitational field strength is = 10N/kg.

Answer: _____ [3]

b) The difference in potential energy is the same as the amount of kinetic energy gained by the mass as it swings down from **C** to **B**.

Calculate the velocity of the mass at position **B**.

Answer: _____ [3]

> **Total Marks** _____ / 12

Energy Transfers and Resources

1 Complete the sentences below to explain the energy transfers involved in a solar panel.

In a solar panel, _____ energy is converted into _____ energy.

Some energy is converted into _____ energy and lost to the surroundings.   **[3]**

2 A student tested four different types of fleece, **J**, **K**, **L** and **M**, to find out which would make the warmest jacket.
Each type of fleece was wrapped around a can.
The can was then filled with hot water.
The temperature of the water was recorded every 2 minutes for a 20-minute period.

The graph in **Figure 2** shows the student's results.

Figure 1

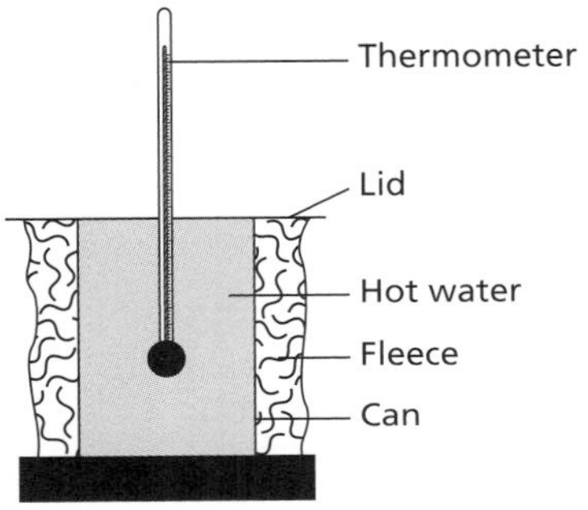

Thermometer

Lid

Hot water

Fleece

Can

Figure 2

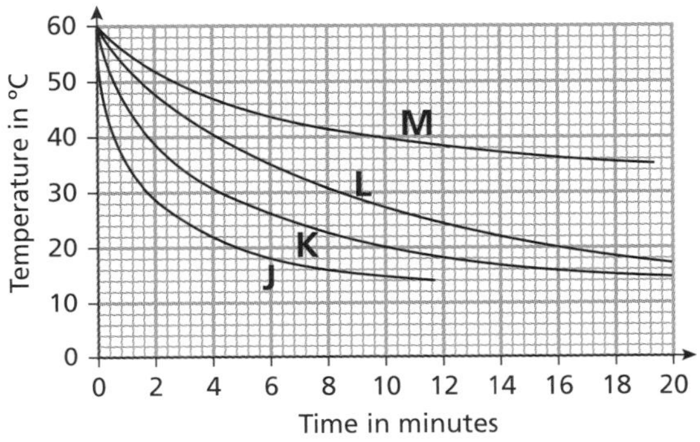

a) To be able to compare the results, it was important to use the same volume of water in each test.

Give **two** other variables that should have been kept the same in each test.

_____   **[2]**

b) Which type of fleece, **J**, **K**, **L** or **M**, should the student recommend for making a ski jacket?
You must explain your answer by making reference to the thermal conductivity of the different fleeces.

_____

_____

_____   **[3]**

Total Marks _____ / 8

Waves and Wave Properties

1 A wave machine in a swimming pool generates waves with a frequency of 0.5Hz.

    a) What does a frequency of 0.5Hz mean?

        _____ **[1**

    b) Give the equation that links the frequency, speed and wavelength of a wave.

        _____ **[1**

    c) The swimming pool is 50m long.
        It takes each wave 10 seconds to travel the length of the pool.

        Calculate the wave speed.

                                            Answer: _____ **[2]**

    d) Use your answers to parts **b)** and **c)** to calculate the wavelength of the waves.

                                            Answer: _____ **[2]**

    e) One section of the swimming pool is designed for young children and has much shallower water.
        A parent notices that the waves get closer together when they enter this section.

        What effect will this have on the wave speed?

        _____ **[1]**

2 Waves may be longitudinal or transverse.

  Describe the differences between longitudinal and transverse waves.

  _____

  _____

  _____

  _____

  _____ **[3]**

> **Total Marks** _____ / 10

Electromagnetic Waves

1 **Figure 1** shows two beakers. Each beaker has a drawing pin inside.

Figure 1

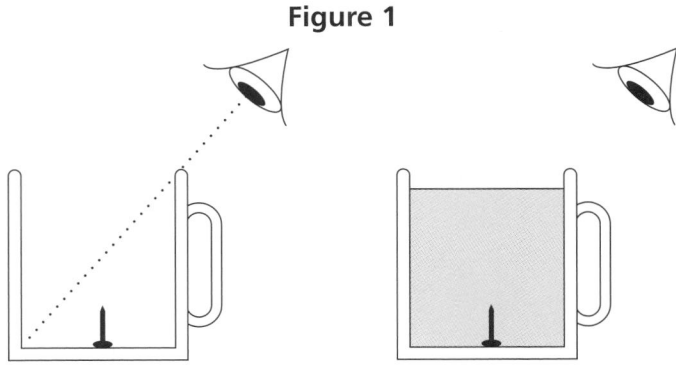

The first beaker is empty and the eye cannot see the drawing pin.
The second beaker is full of water and the drawing pin can be seen.

Explain how this is possible.
You may add a ray line to the diagram to help with your answer.

[3]

2 **Figure 2** shows the electromagnetic spectrum.

Figure 2

Radio waves	Microwaves	Infrared	Visible light	Ultraviolet	X-rays	Gamma rays

Complete the sentence below using words from the box.

amplitude	frequency	speed	wavelength

The arrow in the diagram points in the direction of increasing ....

and decreasing ....

[2]

Total Marks .... / 5

The Electromagnetic Spectrum

1 Radio waves and visible light are electromagnetic waves that are used for communication.

a) Name another type of electromagnetic wave that is used for communication.

Answer: _____ [1]

b) Name an electromagnetic wave that is **not** used for communication and give one of its uses.

_____

_____ [2]

2 After a person is injured, a doctor will sometimes request a photograph of the patient's bones.

a) Which type of electromagnetic radiation would be used to produce the photograph?

Answer: _____ [1]

b) What properties of this radiation enable it to be used to photograph bones?

_____

_____

_____ [2]

c) Another type of electromagnetic wave is ultraviolet light, which is used in sunbeds.

i) State **one** hazard of sunbed use.

_____ [1]

ii) Explain why ultraviolet light is more hazardous than visible light.

_____

_____ [2]

iii) Despite the risks many people still regularly use sunbeds.

Suggest **two** reasons why.

_____

_____ [2]

Total Marks _____ / 11

An Introduction to Electricity

1 Circle the correct words to complete the sentences.

Electric **current** / **charge** is the flow of electrical **charge** / **potential**.

The **greater** / **smaller** the flow, the higher the **current** / **voltage**. [4]

2 The element in a set of hair straighteners has a 5A current running through it and a 230V potential difference across it.

a) Write down the equation that links potential difference, current and resistance.

Answer: _____ [1]

b) Calculate the resistance of the element.

Answer: _____ [2]

c) Write down the equation that links charge, current and time.

Answer: _____ [1]

d) The straighteners are used for 2 minutes.

Calculate the charge that flows in this time.

Answer: _____ [2]

3 Circle the correct words to complete the sentences.

Potential difference determines the amount of **energy** / **power** transferred by the charge as it passes through a component.

The **greater** / **smaller** the potential difference, the higher the **current** / **voltage** that will flow. [3]

4 Write the name of the component that each circuit symbol represents.

a) ——o⁀o—— Answer: _____ [1]

b) —|⊦---|⊦— Answer: _____ [1]

c) —▭— Answer: _____ [1]

d) ^{↘↗}⊖— Answer: _____ [1]

> **Total Marks** _____ / 17

Circuits and Resistance

1 The circuit in **Figure 1** is used to measure the current and potential difference of various components.

Figure 1

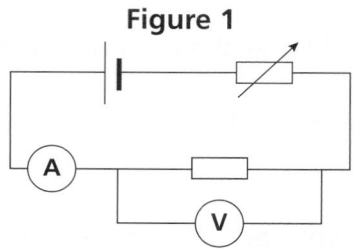

a) What is the purpose of the variable resistor?

_____

_____ **[2]**

b) Is the ammeter in **Figure 1** connected in series or parallel with the variable resistor?

Answer: _____ **[1]**

c) Is the voltmeter in **Figure 1** connected in series or parallel with the resistor?

Answer: _____ **[1]**

2 Draw **one** line from each component to the correct description.

Light dependent resistor (LDR)	Resistance decreases as temperature increases.
Thermistor	Resistance increases as the temperature increases.
Diode	Resistance decreases as light intensity increases.
Filament light	Has a very high resistance in one direction.

[3]

Total Marks _____ / 7

Circuits and Power

1 This question is about a hairdryer that heats air and blows it out the front through a nozzle.

a) The hairdryer has an input power of 1600W.

If a person takes two minutes to dry their hair, how much energy has been transferred?

Answer: _____ [3]

b) The hairdryer is 90% efficient. The remaining energy is output as sound.

How much sound energy is produced in two minutes of use?

Answer: _____ [2]

2 **Figure 1** shows a series circuit with two resistors, **X** and **Y**.

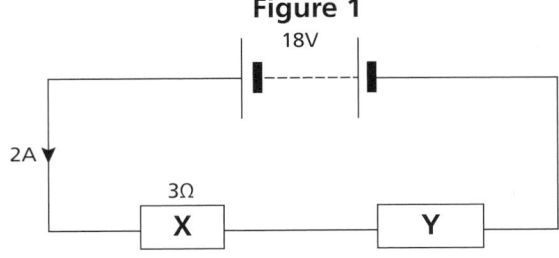

Figure 1

a) Calculate the potential difference across resistor **X**.

Answer: _____ [2]

b) Use your answer to part **a)** to work out the potential difference across component **Y**.

Answer: _____ [2]

c) Calculate the total resistance of the circuit.

Answer: _____ [2]

Total Marks _____ / 11

Domestic Uses of Electricity

1 A battery is connected to an oscilloscope and the trace in **Figure 1** is produced.

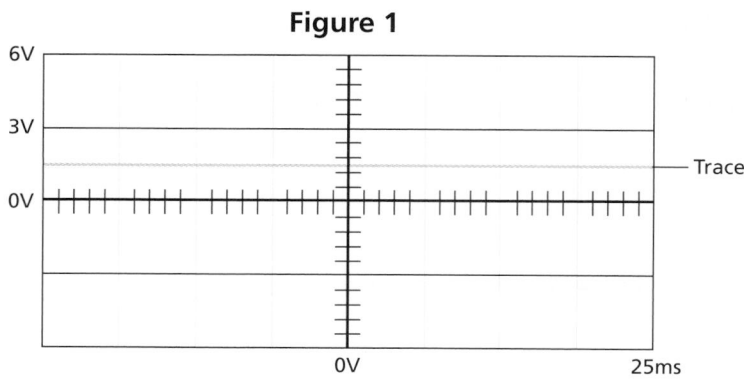

Figure 1

a) Use the trace to determine the potential difference of the battery and the type of current.

.... **[2]**

b) The battery is replaced by the mains supply and the trace in **Figure 2** is recorded by the oscilloscope.

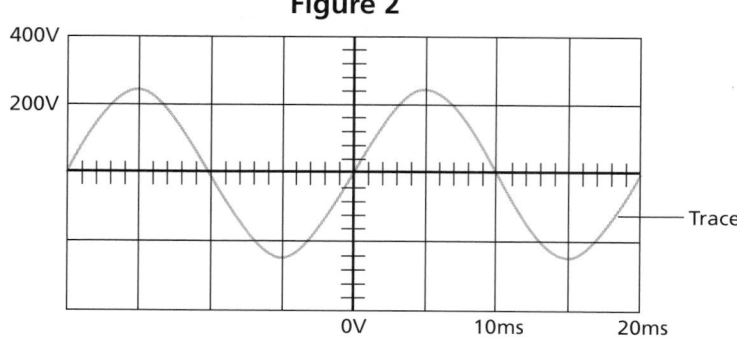

Figure 2

Use the trace to determine the potential difference and the type of current.

.... **[2]**

2 An appliance is switched off but the power cable is connected to the mains.

Explain how a live wire can still be dangerous.

....

....

....

....

.... **[4]**

Total Marks .... / 8

Electrical Energy in Devices

1 An electric blender transfers electrical energy into kinetic, heat and sound energy.

a) What is the useful energy output? Answer: .... **[1]**

b) What happens to the waste energy produced?

....

.... **[2]**

2 **Table 1** gives some information about an electric drill.

Table 1

Energy Input	
Useful Energy Output	
Wasted Energy	Heat and Sound
Power Rating	500W

a) Complete **Table 1** by adding the missing types of energy. **[2]**

b) Which of the following statements about the energy from the drill is **incorrect**?
Tick **one** box.

It spreads out and becomes more difficult to use. ☐

It disappears. ☐

It makes the surroundings warmer. ☐ **[1]**

c) How much energy does the drill use per second? Answer: .... **[1]**

3 The National Grid distributes electricity from power stations to consumers.
The voltage across the overhead cables of the National Grid is much higher than the output voltage from the power station generators.

Explain how this achieved and why it is important.

....

....

....

....

.... **[4]**

Total Marks .... / 11

Magnetism and Electromagnetism

1 The full name for the north pole of a magnet is the 'north-seeking pole'.

Explain what is meant by this.

_____

_____ **[2]**

2 The north pole of a permanent magnet is moved close to the north pole of another permanent magnet.

a) What would you expect to happen?

_____ **[1]**

b) A piece of iron is moved close to the north pole of a permanent magnet.

What would you expect to happen?

_____ **[1]**

3 **Figure 1** shows an electric bell.

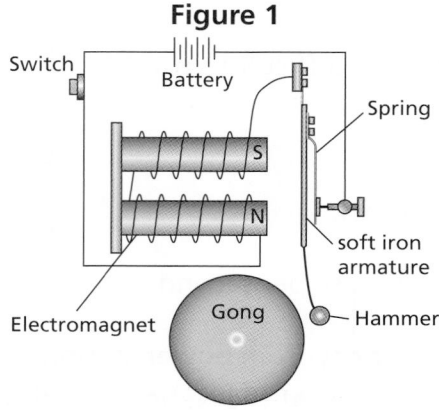

Figure 1

Describe what happens when the switch is pressed.

_____

_____

_____

_____ **[5]**

Total Marks _____ / 9

The Motor Effect

1 A current-carrying wire passes through a magnetic field at right-angles to the field and experiences a force.

The length of wire in the field is 5cm, the current is 2A and the magnetic field is 0.3mT.

a) Calculate the force on the wire.
Use the correct equation from the Physics Equation Sheet on page 449.

Answer: _____ **[2]**

b) The magnets are rearranged so that the current flowing in the wire is parallel to the field lines.

How will this affect the force on the wire?

_____ **[1]**

2 **a)** The left hand rule can be used to identify the direction of the force acting on a conductor carrying a current in a magnetic field.

State what is indicated by each of the fingers in the left hand rule.

_____

_____

_____ **[3]**

b) State **two** ways of increasing the speed of rotation of an electric motor.

_____

_____ **[2]**

c) State **two** ways of reversing the direction of an electric motor.

_____

_____ **[2]**

Total Marks _____ / 10

Particle Model of Matter

1 Heating a substance can cause it to change state from a solid to a liquid or from a liquid to a gas.

a) What is meant by 'specific latent heat of fusion'?

....

.... [2]

b) While a kettle boils, 0.012kg of water changes to steam.

Calculate the amount of energy required for this change.
Use the correct equation from the Physics Equation Sheet on page 449.
Specific latent heat of vaporisation of water = 2.3×10^6 J/kg

Answer: .... [2]

2 The graph in **Figure 1** shows how temperature varies with time as a substance cools. The graph is **not** drawn to scale.

Figure 1

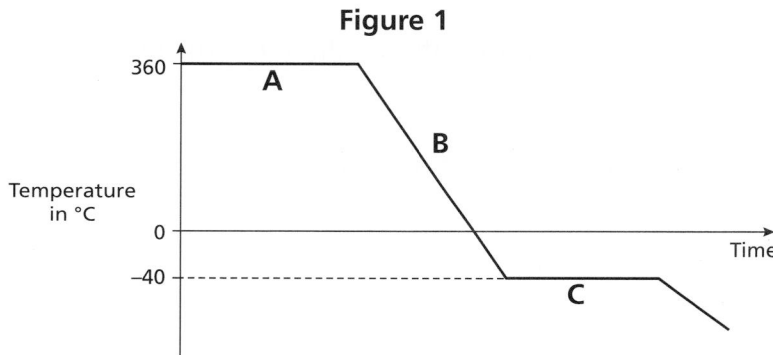

a) Explain what is happening to the substance in section **A** of the graph.

....

.... [2]

b) Explain what is happening to the substance in section **B** of the graph.

....

....

.... [2]

Total Marks .... / 8

Atoms and Isotopes

1 Atoms contain three types of particle.

a) Complete the table to show the relative charges of the subatomic particles.

Particle	Relative Charge
Electron	
Neutron	
Proton	

[3]

b) A neutral atom has no overall charge.

Explain why in terms of its particles.

....

....

....

[2]

c) Complete the sentences below.

An atom that loses or gains an electron becomes an ....

If it loses an electron, it has an overall .... charge. [2]

2 In the early part of the 20th century, Rutherford, Geiger and Marsden investigated the paths taken by positively charged alpha particles into and out of a very thin piece of gold foil. **Figure 1** shows the paths of three alpha particles.

Explain the different paths, **A**, **B** and **C**, of the alpha particles.

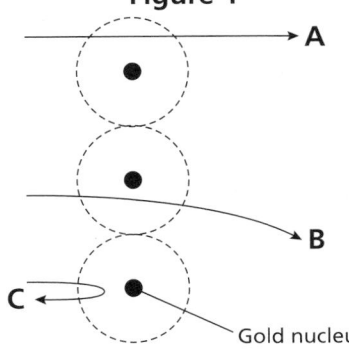

Figure 1

Gold nucleus

....

....

....

[3]

Total Marks .... / 10

Nuclear Radiation

1 Here is some information about potassium.

> Potassium is a metallic element in Group 1 of the Periodic Table.
> It has an atomic number of 19.
> Its most common isotope is potassium-39, $^{39}_{19}K$.
> Another isotope, potassium-40, $^{40}_{19}K$, is a radioactive isotope.

a) What is meant by 'radioactive isotope'?

_____ [1

b) During radioactive decay, atoms of potassium-40 change into atoms of calcium-40. Calcium-40 has an atomic number of 20 and a mass number of 40.

What type of radioactive decay has taken place?

Answer: _____ [1]

c) Potassium-39 does not undergo radioactive decay.

What does this tell us about potassium-39?

_____ [1]

d) Sodium-24 is another radioactive isotope.
It decays by gamma emission.

Give the name of the element formed when this decay takes place.

Answer: _____ [1]

2 Give the unit that is used to measure the activity of a radioactive isotope.

Answer: _____ [1]

3 List the decay mechanisms, **alpha**, **beta** and **gamma**, in order of penetrating power. Start with the most penetrative.

_____ [1]

> Total Marks _____ / 6

Half-Life

1 Iodine is found naturally in the world and is essential to life.
It is used by the thyroid gland for the production of essential hormones.
Iodine-127 is not radioactive but iodine-131 is.
Iodine-131 has as a half-life of 8 days.

a) During the Chernobyl nuclear disaster in 1986, an explosion caused a large quantity of the isotope iodine-131 to be released into the atmosphere.

   Is iodine-131 from the disaster still a threat to us today?
   Explain your answer.

   [3]

b) A sample of iodine-131 has a count-rate of 256 counts per minute.

   Work out the count-rate of the sample after 24 days.

   Answer: _____ [2]

c) The Isotope, Caesium-137, was also released during the disaster.
   Ceasium-137 has a half-life of 30 years.

   In what year will the activity of the caesium released in the disaster have fallen to $\frac{1}{8}$ of the initial activity?

   Answer: _____ [2]

d) Potassium has an atomic number of 19. It decays into calcium-40 by emitting an electron from the nucleus.

   Produce a balanced decay equation for the decay of potassium-40 into caesium-40.

   [3]

   Total Marks _____ / 10

Notes

Collins

GCSE
COMBINED SCIENCE   H
Biology: Paper 1 Higher Tier

Materials

Time allowed: 1 hour 15 minutes

For this paper you must have:

- a ruler
- a calculator.

Instructions

Answer **all** questions in the spaces provided.

Do all rough work on the page. Cross through any work you do not want to be marked.

Information

- There are **70** marks available on this paper.
- The marks for each question are shown in brackets [ ].
- You are expected to use a calculator where appropriate.
- You are reminded of the need for good English and clear presentation in your answers.
  When answering questions 03.2 and 05.3 you need to make sure that your answer:
  – is clear, logical, sensibly structured
  – fully meets the requirements of the question
  – shows that each separate point or step supports the overall answer.

Advice

In all calculations, show clearly how you work out your answer.

Biology Practice Exam Paper 1

01   Gonorrhoea is a disease caused by a microorganism.

Figure 1 shows the microorganism that causes gonorrhoea.

Figure 1

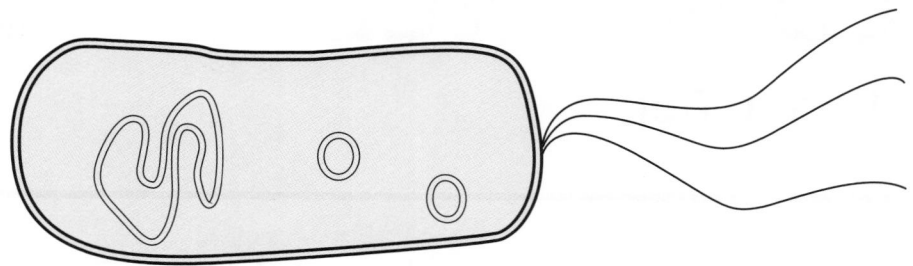

01.1 What type of microorganism is this?

Tick **one** box.

Bacterium ☐

Fungus ☐

Protist ☐

Virus ☐

[1 mark]

01.2 The magnification of **Figure 1** is ×14 000.

The length of the microorganism is shown by line **XY**.

$$\text{magnification} = \frac{\text{size of image}}{\text{size of real object}}$$

What is the real length of the microorganism in millimetres?

Real length = _____ mm   **[2 marks]**

01.3 The microorganism can reproduce once every 40 minutes in ideal conditions.

Calculate how many microorganisms could be produced from one microorganism in 4 hours.

Answer = _____ microorganisms   **[2 marks]**

Turn over for the next question

02 **Figure 2** shows the number of people who tested positive for salmonella food poisoning each month in one year.

Figure 2

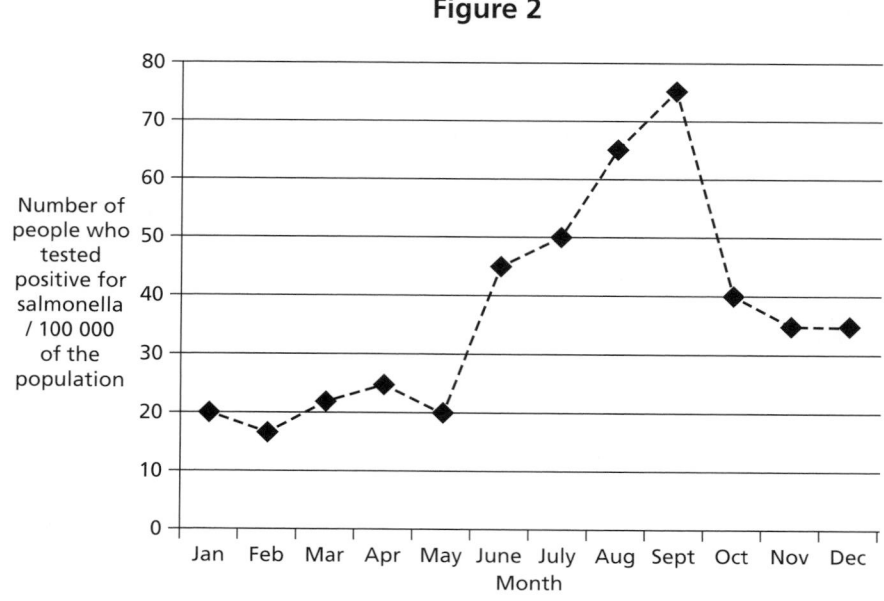

02.1 Describe the symptoms that the people who tested positive for salmonella were likely to show.

....

....

[2 marks]

02.2 Using your knowledge of how salmonella is passed on, suggest reasons for the pattern shown in **Figure 2**.

....

....

....

[2 marks]

03   Doctors are hoping that in the future they will be able to treat heart disease by injecting stem cells into a damaged heart.

Figure 3 shows where the stem cells are injected.

Figure 3

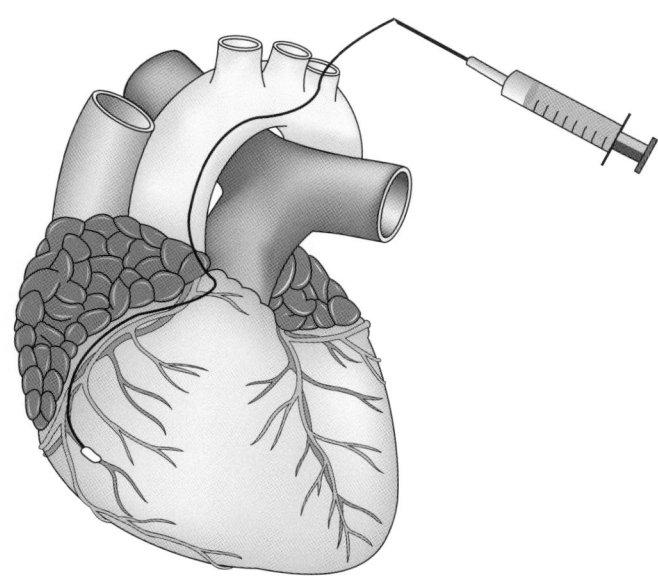

03.1 Which chamber of the heart are the cells being injected into?

Tick **one** box.

Left ventricle ☐

Right ventricle ☐

Left atrium ☐

Right atrium ☐   **[1 mark]**

Question 3 continues on the next page

03.2 Describe the cause of heart disease.

Include some of the risk factors associated with the disease.

_____

_____

_____

_____

_____

_____

_____

_____

_____

_____

_____

[6 marks

The stem cells for this treatment can be taken from an embryo that has been produced by cloning the patient's own cells.

03.3 Explain the benefit of using stem cells from an embryo cloned from the patient's own cells.

_____

_____

_____

[2 marks]

03.4 Suggest why some people may object to this treatment.

_____

_____

_____

[2 marks

04 Penicillin is an antibiotic drug.

04.1 Draw **one** line to link the person who discovered penicillin to the organism that it comes from.

person	organism
Mendel	fungus
Fleming	willow
Wallace	foxglove

[2 marks]

04.2 Some strains of bacteria have developed resistance to penicillin.

As more bacteria develop resistance, there is more pressure on scientists to produce new drugs.

Here are four steps carried out in the testing of new drugs.

Double-blind trials on patients	
Varying doses given to healthy volunteers	
Testing on live animals	
Low doses given to healthy volunteers	

Write the numbers **1, 2, 3** or **4** in the boxes to show the order of these steps in drug testing.

[3 marks]

Turn over for the next question

05 Sunflower plants, such as the ones shown in **Figure 4**, can grow well in the UK.

They often grow up to three metres tall.

Figure 4

05.1 Suggest why it is an advantage for sunflowers to be taller than the other plants growing around them.

....

....

....

.... **[3 marks]**

05.2 Growing tall means that sunflower plants have to transport water several metres up to the leaves from the roots.

Describe how they do this.

....

....

....

.... **[3 marks]**

Figure 5 shows an aloe plant.

Aloe is a plant that grows in desert areas.

In these areas there is little food for animals and very little water.

Figure 5

Table 1 gives some differences between the leaves of aloe plants and sunflowers.

Table 1

Plant	Thickness of Waxy Cuticle (micrometres)	Number of Stomata (per mm^2)
aloe	14.8	25
sunflower	6.1	150

05.3 Explain how aloe is adapted to living in desert conditions.

Use **Figure 5** and the data in **Table 1**.

....

....

....

....

....

....

....

....

[6 marks]

Turn over for the next question

06 Racehorses and athletes are both trained to run in races.

Figure 6

Horses can be tested to see how fit they are.

A horse's heart rate is measured when it is running at different speeds.

Some results for a horse are shown in **Table 2**.

Table 2

Speed of the Horse (Kilometres per Hour)	Heart Rate (Beats per Minute)
30	162
35	170
39	180
45	192

06.1 Plot the data from **Table 2** on the graph in **Figure 7**.

Finish the graph by drawing the line of best fit.

Figure 7

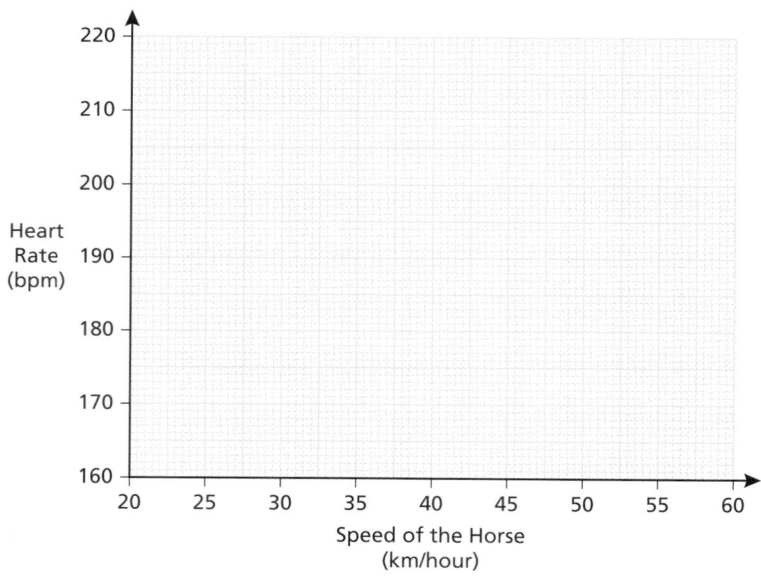

[4 marks]

06.2 Describe the pattern shown in the graph.

_____

_____

_____ **[2 marks]**

06.3 Above 200 heart beats per minute, a horse starts to use **anaerobic** respiration.

Use the graph to estimate the maximum speed at which this horse can run **without** using anaerobic respiration.

Show on the graph how you work out your answer.

Maximum speed = _____ km/hour **[2 marks]**

06.4 Write the word equation to show the reaction for **anaerobic** respiration in horses and athletes.

_____ **[2 marks]**

06.5 Explain why athletes and horses run better when they use **aerobic** rather than **anaerobic** respiration.

_____

_____

_____

_____ **[3 marks]**

Turn over for the next question

07 A student investigates the digestion of fats (lipids) by the enzyme lipase.

They find that when lipase digests fats, the pH of the solution changes from pH 8 to pH 6.

07.1 Why did the digestion of the fats change the pH of the solution?

Tick **one** box.

Amino acids are formed ☐

Fatty acids are formed ☐

Fats are alkaline ☐

Glucose is formed ☐

[1 mark

The student sets up three test tubes containing various liquids.

Figure 8 shows the contents of the tubes.

Figure 8

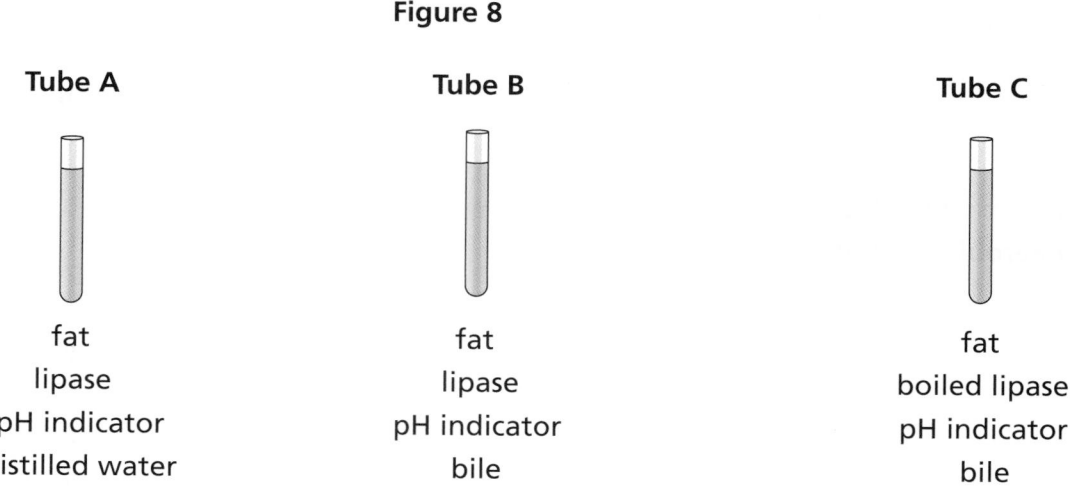

Tube A	Tube B	Tube C
fat	fat	fat
lipase	lipase	boiled lipase
pH indicator	pH indicator	pH indicator
distilled water	bile	bile

The student times how long it takes for the indicator to change colour after the lipase is added.

The results are shown in **Table 3**.

Table 3

Tube	Time Taken (minutes)
A	8
B	1
C	No change after 20 minutes

07.2 The indicator in tube **B** changes colour much faster than the indicator in tube **A**.

Explain why.

_____

_____

_____

_____ [3 marks]

07.3 Explain why the colour did not change in tube **C**.

_____

_____

_____ [2 marks]

07.4 Explain why lipase can digest fats (lipids) but amylase or protease cannot.

_____

_____

_____

_____ [3 marks]

Turn over for the next question

08 A student wants to find out the concentration of the cell contents of potato.
They design an experiment involving osmosis.

08.1 What is meant by the term 'osmosis'?

_____

_____

_____ **[3 marks**

The student cuts cylinders from a potato and weighs each cylinder.
They then place each cylinder in a test tube.
Each test tube contains a different concentration of sugar solution.
After several hours the student removes the cylinders from the solutions and reweighs them.
They then calculate the percentage change in mass for each cylinder.

Figure 9 shows the student's results.

Figure 9

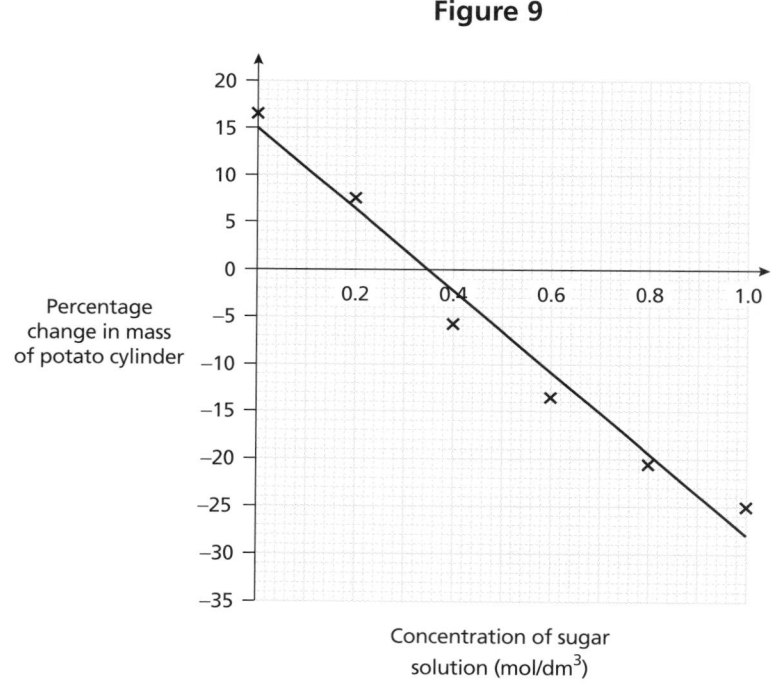

Percentage change in mass of potato cylinder

Concentration of sugar solution (mol/dm^3)

08.2 The potato cylinders in 0.4 to 1.0 mol/dm³ sugar solution all lost mass.

Explain why.

_____

_____

_____ **[3 marks]**

08.3 Use **Figure 9** to estimate the concentration of the cell contents of the student's potato.

Concentration = _____ mol/dm³ **[1 mark]**

08.4 **Table 4** shows possible variables in the student's experiment.

Put the letter **I**, **D** or **C** in the table to show the type of variable.

I = independent variable
D = dependent variable
C = any variables that should have been controlled

Table 4

The concentration of the sugar solution	
The volume of the sugar solution	
The change in mass of the potato cylinder	
The time that each cylinder was left to soak	

[4 marks]

END OF QUESTIONS

Biology Practice Exam Paper 1

There are no questions printed on this page

Collins

GCSE
COMBINED SCIENCE   H
Biology: Paper 2 Higher Tier

Materials

Time allowed: 1 hour 15 minutes

For this paper you must have:

- a ruler
- a calculator.

Instructions

Answer **all** questions in the spaces provided.

Do all rough work on the page. Cross through any work you do not want to be marked.

Information

- There are **70** marks available on this paper.
- The marks for each question are shown in brackets [ ].
- You are expected to use a calculator where appropriate.
- You are reminded of the need for good English and clear presentation in your answers.
- When answering questions 02.4 and 08.3 you need to make sure that your answer:
  - is clear, logical, sensibly structured
  - fully meets the requirements of the question
  - shows that each separate point or step supports the overall answer.

Advice

In all calculations, show clearly how you work out your answer.

Biology Practice Exam Paper 2

01   This question is about skin sensitivity.

Tim presses Kate's skin with two pin heads, as shown in **Figure 1**.

Sometimes Tim presses both pin heads on to Kate's skin and sometimes just one.
Every time he presses her skin, Kate says the number of points she feels.
Tim writes down how many times she calls the correct number of points.
He does this 20 times.

Tim does this experiment with the pins different distances apart and on three parts of the body.
The graph in **Figure 2** shows his results.

Figure 1

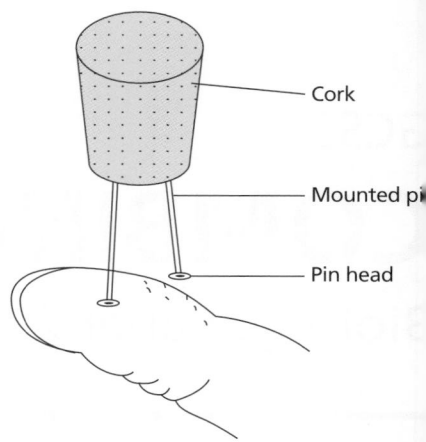

Cork

Mounted pin

Pin head

Figure 2

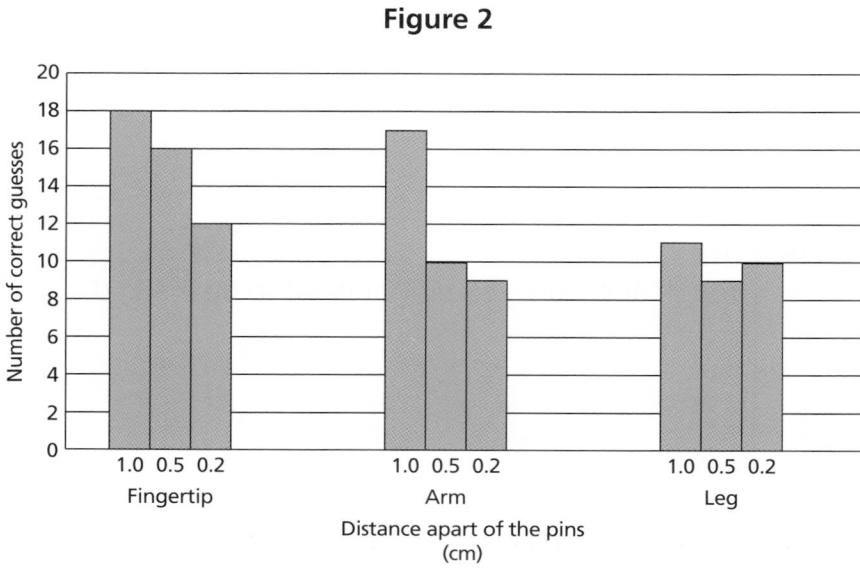

01.1 Name the part of Kate's body where her skin was most sensitive to the stimulus.

_____ **[1 mark**

01.2 Explain why Tim touched Kate's skin 20 times in each trial.

_____

_____ **[2 marks**

01.3 What do the results from Kate's arm show?

_____

_____ **[2 marks]**

01.4 Kate's response to the pins was not a reflex action.

Explain how you can tell this.

_____

_____ **[2 marks]**

01.5 **Figure 3** shows a reflex arc.

Figure 3

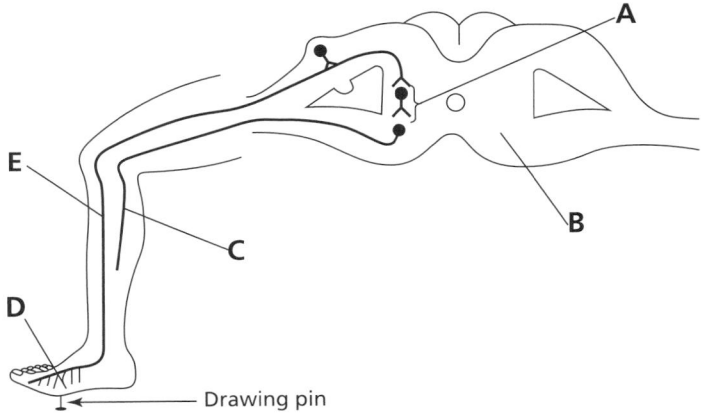

Complete **Table 1** by matching a letters on the diagram to each part of the reflex arc. The first one has been done for you.

Table 1

Part of Reflex Arc	Letter
relay neurone	A
receptor	
sensory neurone	
spinal cord	
motor neurone	

[3 marks]

Turn over for the next question

02   Polar bears (scientific name *Ursus maritimus*) live in the Arctic.
      They need to move about on the ice and they only eat seals.

Figure 4

02.1 Describe how polar bears are adapted for hunting and eating seals on ice.

....

.... **[2 marks]**

Alaskan bears (*Ursus arctos)* live in Alaska, south of the Arctic.
They catch fish for food.

Figure 5

02.2 What do their scientific names tell you about how closely related the Alaskan bear
         and the polar bear are?

....

.... **[2 marks]**

02.3 The temperature in Alaska is getting warmer.

Alaskan bears are now living in some of the same areas as polar bears.

Which of these statements are true?

Tick **two** boxes.

Polar bears and Alaskan bears will compete for food. ☐

Polar bears will mate with Alaskan bears and the offspring will be a new species. ☐

Polar bears and Alaskan bears may mate and produce sterile hybrids. ☐

Polar bears and Alaskan bears cannot mate because they are in different kingdoms. ☐

The habitats of the polar bears and the Alaskan bears may overlap. ☐ **[2 marks]**

02.4 The level of carbon dioxide in the air is increasing.

Explain why scientists think this is happening and how it might be causing the temperature in the Arctic to increase.

_____

_____

_____

_____

_____

_____

_____

[6 marks]

Turn over for the next question

03   Copper is an important element in many organisms. However, in high concentrations it is poisonous

Figure 6 shows waste soil from a copper mine piled up next to a river.

Figure 6

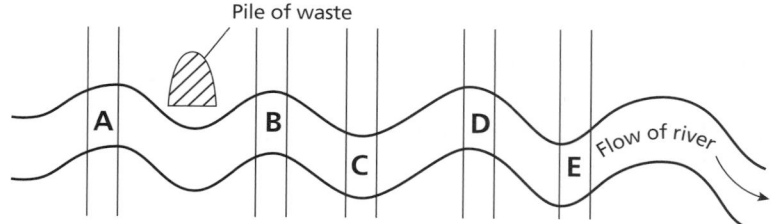

A scientist investigated copper concentrations in the river.
They took samples of water from locations **A**, **B**, **C**, **D** and **E**.
They then measured the concentration of copper in each sample.

The results are shown in **Table 2**.

Table 2

Sample Location	Concentration of Copper (Micrograms per Litre)
A	1
B	16
C	10
D	7
E	3

03.1 Suggest an explanation for the results.

....

.... **[2 marks]**

03.2 The scientist noticed that there were fewer plants growing on the river bank at location **B** than at location **A**.

Suggest an explanation for this difference.

....

.... **[2 marks]**

The scientist takes sample plants from each location and continues to grow them in the laboratory.

The scientist waters the plants regularly with water containing copper.

The survival rate of the plants is shown in **Table 3**.

Table 3

Location That Plants Were Taken From	Survival (%)
A	1
B	100
C	90
D	55
E	20

03.3 Why is there such a large difference in the survival rate of plants from location **A** compared with those from location **B**?

....

....

[2 marks]

03.4 The scientist explained the results by using natural selection.

Explain how natural selection could be used to explain why 100% of the plants from location **B** could survive the watering.

....

....

....

....

....

[4 marks]

Turn over for the next question

Biology Practice Exam Paper 2

04 In 1866, Gregor Mendel published a paper on genetics.

Figure 7

Mendel carried out breeding experiments with pea plants.
He crossed tall pea plants with dwarf pea plants.
He collected the seeds and planted them.
He then self-pollinated the plants that grew and again planted the seeds that were made.

His results are shown in **Figure 8**.

Figure 8

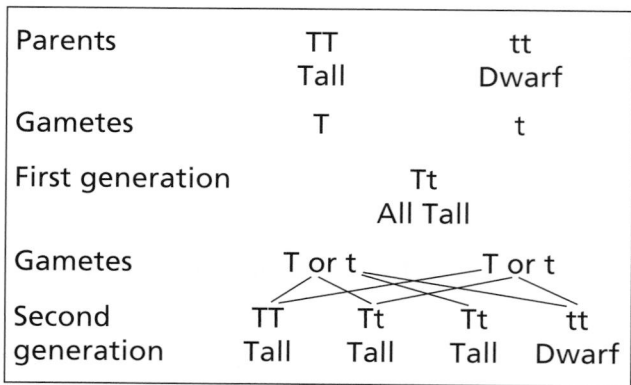

Parents	TT Tall	tt Dwarf
Gametes	T	t
First generation	Tt All Tall	
Gametes	T or t	T or t
Second generation	TT Tall Tt Tall	Tt Tall tt Dwarf

04.1 Complete the sentences about Mendel's results.

Use words from the box.

> dominant   dwarf   genotypes   heterozygous   homozygous   phenotypes   recessive   tall

The genotype of all the plants in the first generation is _____ .

The phenotype of these plants is _____ .

This is because the tall allele is _____ over the dwarf allele.

The second generation contains plants that have three

different _____ .

[4 marks]

04.2 To test his ideas, Mendel crossed a plant from the first generation with a dwarf plant.

Draw a genetic diagram to show this cross.
Write down the ratio of phenotypes that this cross would produce.

Ratio: .... **[4 marks]**

Mendel then self-pollinated the plants that were Tt **or** TT.
He looked at the first 10 offspring.
If they were **all** tall, he said that the parent plant was TT.
If **any** were dwarf, he said that the parent plant was Tt.

04.3 Explain why Mendel's assumptions may have been wrong in a small number of cases.

....

.... **[2 marks]**

Turn over for the next question

05   Here is an advert of a DNA testing kit.

BABY TESTING KIT

Do you want to know the sex of your baby before it is born?

This testing kit can answer that question.

When a woman is pregnant, a very small amount of the baby's DNA is in the mother's blood.

This means a small drop of the mother's blood can be tested to see if it has any DNA from a Y chromosome.

This will tell you if you are going to have a boy or a girl.

05.1 The test involves finding out if any DNA in the mother's blood is from a Y chromosome.

Explain how this can be used to tell the sex of the baby.

....

....   [2 marks]

05.2 At present, scientists do this test on cells taken directly from the baby inside the mother's uterus.

They hope that in the future they will be able to carry out the same tests using samples of the mother's blood.

Suggest why scientists think that this would be a better method.

....

....   [2 marks]

06 *Acetabularia* is a single-celled organism that lives in the sea.

It has an unusual shape, as shown in **Figure 9**.

Figure 9

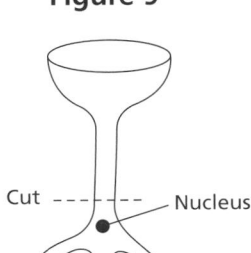

06.1 When *Acetabularia* reproduces, the nucleus and then the cytoplasm divides into two.

What is this type of reproduction called?

.... **[1 mark]**

06.2 *Acetabularia* produced by this type of reproduction always have the same shaped cap as the parent.

Why is this?

.... **[1 mark]**

06.3 A scientist called Hammerling performed some experiments on *Acetabularia*.
He cut the organism into two as shown on **Figure 9**.
He found that the bottom part of the organism survived and grew a new top (cap).

Explain these results.

....

....

....

.... **[3 marks]**

Turn over for the next question

07 This question is about the causes of infertility in humans.

Table 4 shows some of the problems that cause infertility.

Table 4

Problem	Percentage of Infertile Couples with this Problem	Percentage Success Rate of Treatment
blocked fallopian tubes	13	20
irregular ovulation	16	75
no ovulation	7	95
low sperm production	15	10
no sperm production	21	10
unknown cause	28	-

Use the information in **Table 4** and your biological knowledge to answer the following questions.

07.1 What is the total percentage of infertile couples where the problem is known to be in the **male**?

Percentage = _____ [1 mark]

07.2 Treatment of which problem produces the largest **number** of pregnant women among the affected infertile couples?

Explain how you worked out your answer.

_____

_____ [2 marks]

Figure 10 shows some of the hormones released from the pituitary gland and some of the effects that they have on the ovary.

Figure 10

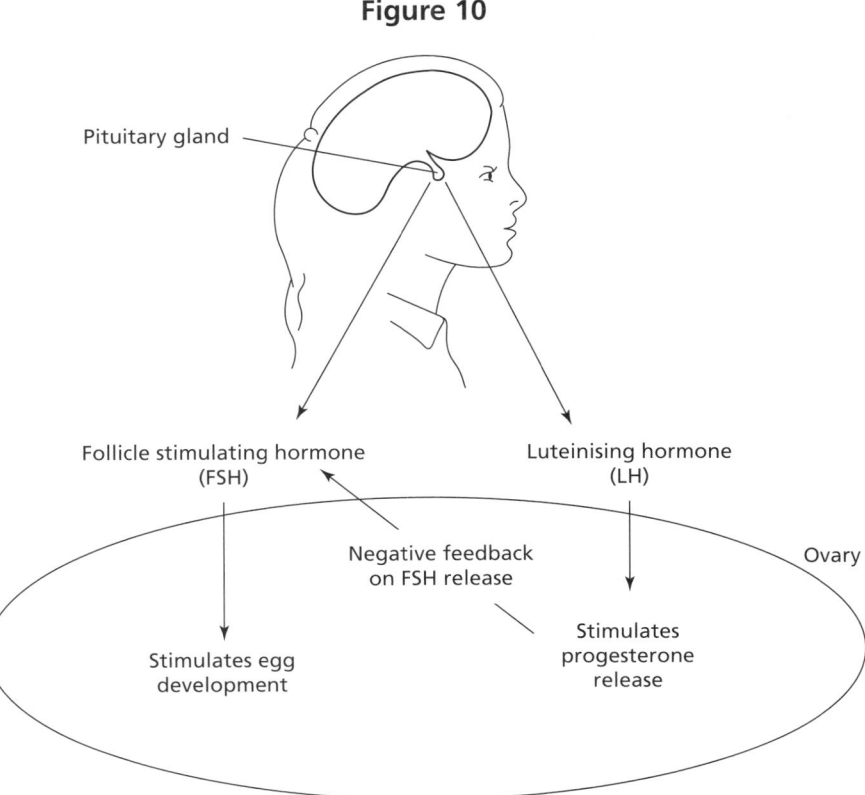

07.3 Progesterone has a negative feedback effect on FSH.

What does this mean?

[1 mark]

07.4 Some women with fertility problems are injected with a drug that prevents the pituitary gland from making LH.

Explain why this might help them to get pregnant.

[3 marks]

08 This question is about hormones and homeostasis.

08.1 What is a hormone?

....

.... **[2 marks]**

08.2 **Table 5** contains information about three different hormones.

Table 5

Hormone	Gland that Releases Hormone	Function of Hormone
		controls how much water is reabsorbed in the kidney
thyroxine		
	adrenal gland	

Complete the table by writing in the six blank boxes. **[6 marks]**

The article in **Figure 11** appeared in a recent newspaper.

Figure 11

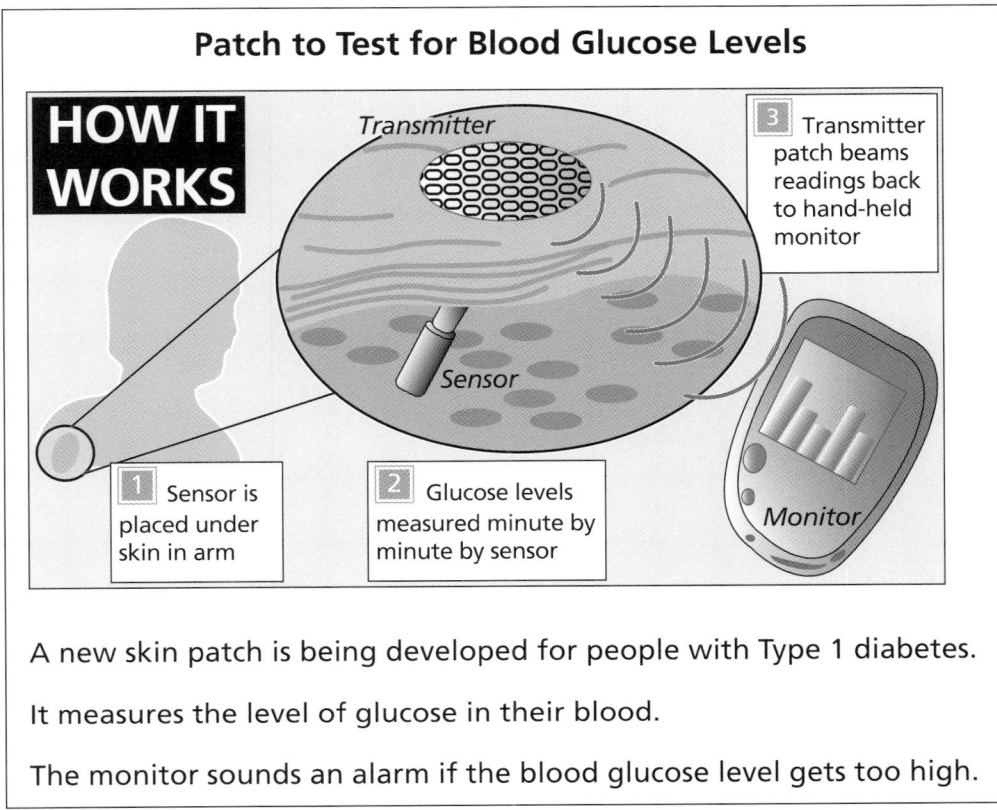

Patch to Test for Blood Glucose Levels

HOW IT WORKS

Transmitter

3 Transmitter patch beams readings back to hand-held monitor

Sensor

1 Sensor is placed under skin in arm

2 Glucose levels measured minute by minute by sensor

Monitor

A new skin patch is being developed for people with Type 1 diabetes.

It measures the level of glucose in their blood.

The monitor sounds an alarm if the blood glucose level gets too high.

08.3 It is very important for people with Type 1 diabetes to know if their blood glucose level gets too high.

Explain why.

_____

_____

_____

_____

_____ **[4 marks]**

END OF QUESTIONS

There are no questions printed on this page

![Collins]

GCSE
COMBINED SCIENCE
Chemistry: Paper 1 Higher Tier

Materials

Time allowed: 1 hour 15 minutes

For this paper you must have:

- a ruler
- a calculator
- the periodic table (see page 450).

Instructions

- Answer **all** questions in the spaces provided.
- Do all rough work on the page. Cross through any work you do not want to be marked.

Information

- There are **70** marks available on this paper.
- The marks for each question are shown in brackets.
- You are expected to use a calculator where appropriate.
- You are reminded of the need for good English and clear presentation in your answers.
- When answering question 05.2 you need to make sure that your answer:
  - is clear, logical, sensibly structured
  - fully meets the requirements of the question
  - shows that each separate point or step supports the overall answer.

Advice

- In all calculations, show clearly how you work out your answer.

Chemistry Practice Exam Paper 1

01 This question is about the atomic model.

Figure 1 shows a model of a helium atom.

Figure 1

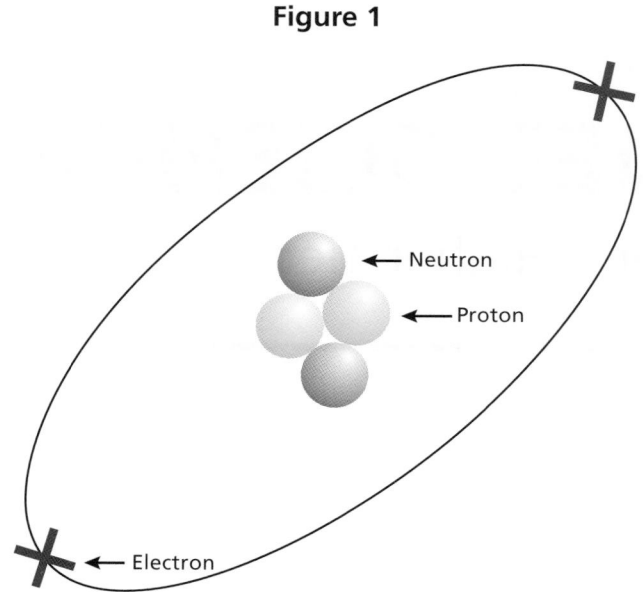

01.1 What are the relative electric charges on the particles in an atom?

Tick **one** box.

Proton	Neutron	Electron	
0	−1	+1	☐
+1	0	−1	☐
−1	0	0	☐
+1	+1	−1	☐ [1 mark]

01.2 A helium atom has an overall neutral charge.

State why.

....

.... **[1 mark]**

01.3 What is the mass number and atomic number of helium?

Tick **one** box.

Mass Number	Atomic Number	
2	2	☐
6	2	☐
6	4	☐
4	2	☐ [1 mark]

01.4 Why is helium an unreactive element?

Tick **one** box.

It has an equal number of protons and neutrons. ☐

Elements with two electrons in their outer shell are unreactive. ☐

It is a gas at room temperature and pressure. ☐

It has a full outer shell of electrons. ☐ [1 mark]

01.5 An isotope of helium has only one neutron in its atoms.

Which statements about how an atom of this isotope compares to the atom in **Figure 1** are true?

Tick **two** boxes.

It has the same atomic number. ☐

It has a higher mass. ☐

It has a different atomic number. ☐

It has a different mass number. ☐

It has the same mass number. ☐ [2 marks]

Turn over for the next question

02 Iron is found in the Earth as the compound iron oxide (Fe_2O_3).

02.1 How many atoms are in one molecule of Fe_2O_3?

Answer: _____ **[1 mark**

Iron is extracted from iron(III) oxide using carbon in the following reaction:

$$2Fe_2O_3 + 3C \rightarrow 4Fe + 3CO_2$$

02.2 Calculate the relative formula mass (M_r) of carbon dioxide (CO_2).

Relative atomic masses (A_r): carbon = 12; oxygen = 16

Answer: _____ **[1 mark**

02.3 Both oxidation and reduction take place when iron is extracted from iron(III) oxide.

Draw **one** line from each type of change to the name of the substance that is changed in that way.

Change		Substance
		iron
oxidation		iron(III) oxide
		carbon
reduction		carbon dioxide

[2 marks]

02.4 Gold does not have to undergo an extraction process.

Explain why.

....

....

....

....

[2 marks]

Turn over for the next question

Chemistry Practice Exam Paper 1

03   **Table 1** shows some of the properties of elements in Group 7 of the periodic table.

Table 1

Element	Density (g/cm³)	Melting point (°C)	Boiling point (°C)
Fluorine	0.0017	−219.6	−188.1
Chlorine	0.0032	−101.5	−34.0
Bromine	3.1028	−7.3	58.8
Iodine	4.9330	113.7	184.3

03.1 State **one** trend in the properties of the Group 7 elements as shown in **Table 1**.

_____

_____ **[1 mark]**

03.2 Astatine is found in Group 7, below iodine.

Use the information in **Table 1** to estimate the melting point of astatine.

Answer: _____ °C   **[1 mark]**

Group 7 elements all exist as molecules containing two atoms.

03.3 Complete the dot and cross diagram in **Figure 2** to show the covalent bonding in a molecule of chlorine.

Show the outer shell electrons only.

Figure 2

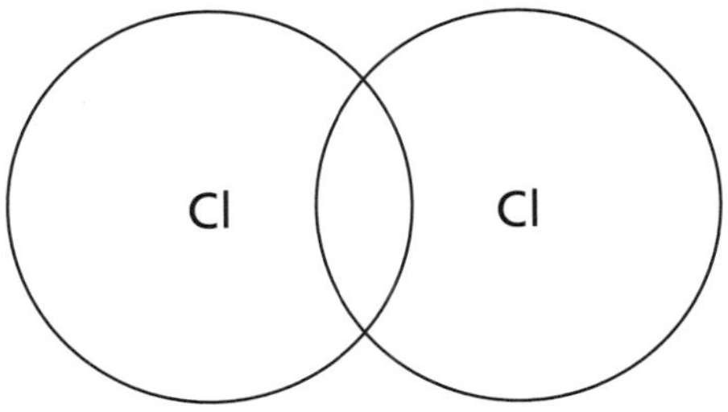

[2 marks]

03.4 What is the formula for a molecule of bromine?

Answer: _____ [1 mark]

03.5 Which statement correctly describes what happens when a Group 7 element boils?

Tick **one** box.

Covalent bonds form ☐

Intermolecular forces form ☐

Covalent bonds break ☐

Intermolecular forces break ☐ [1 mark]

Question 3 continues on the next page

03.6 The Group 7 elements all react with Group 1 elements to form ionic compounds.

Explain why all the Group 7 elements share this chemical property.

....

.... [1 mark]

03.7 The Group 1 element sodium reacts with chlorine to form sodium chloride.
Figure 3 shows the outer electrons in an atom of sodium and in an atom of chlorine.

Figure 3

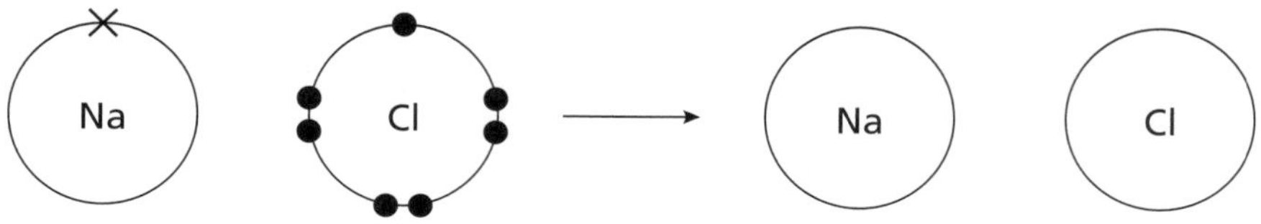

Complete the diagram to show the arrangement of electrons in sodium chloride.
You should give the formula of each ion formed. [5 marks]

03.8 The Group 7 elements become less reactive as you go down the group.

Explain why.

....

....

....

....

....

.... [4 marks]

4 A student was asked to produce copper(II) sulfate crystals. They used the method shown in **Figure 4.**

Figure 4

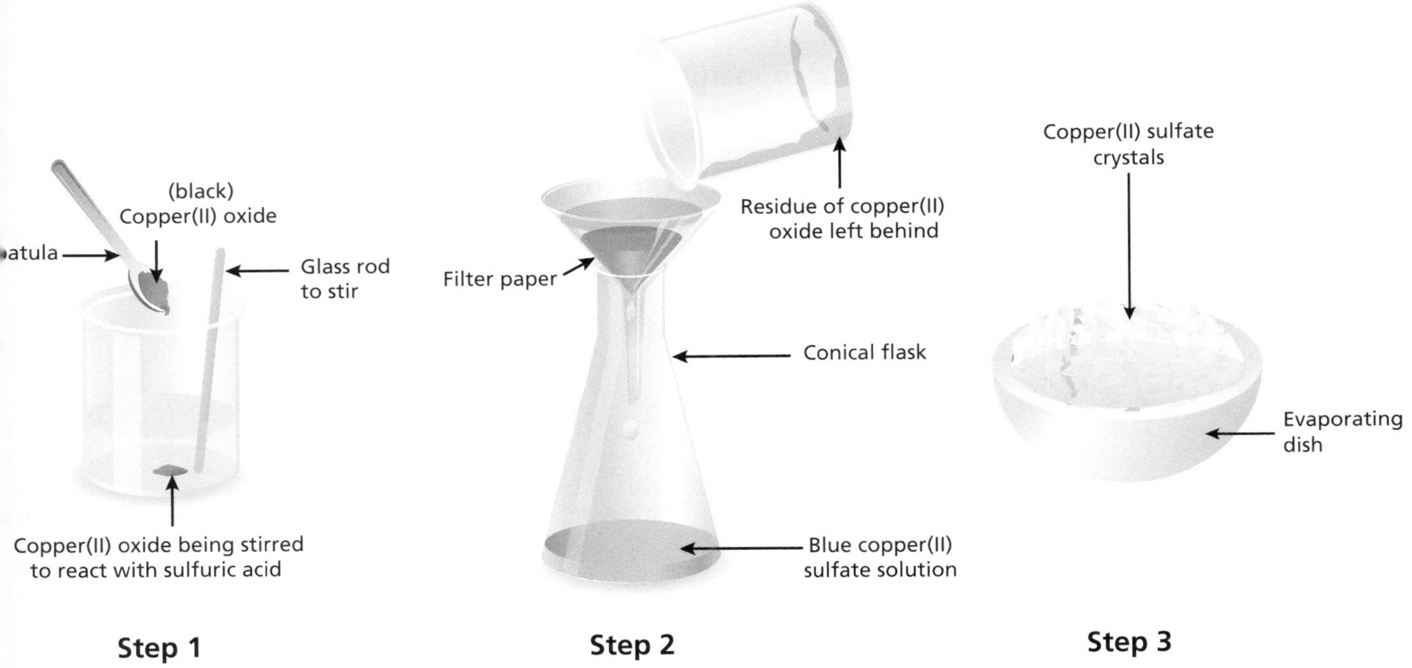

Step 1 — (black) Copper(II) oxide, spatula, Glass rod to stir, Copper(II) oxide being stirred to react with sulfuric acid

Step 2 — Filter paper, Residue of copper(II) oxide left behind, Conical flask, Blue copper(II) sulfate solution

Step 3 — Copper(II) sulfate crystals, Evaporating dish

04.1 Give the name of the separation process used in **Step 2** and explain why it was used.

[2 marks]

04.2 Identify one hazard in **Step 1** or **Step 2** and suggest a method of reducing the risk.

Hazard:

Way of reducing the risk:

[2 marks]

Question 4 continues on the next page

The equation for the reaction is:

$$CuO(s) + H_2SO_4(aq) \rightarrow CuSO_4.5H_2O\ (aq) + H_2O(l)$$

04.3 The student used 5.2 g of copper(II) oxide.

Calculate the maximum mass of copper(II) sulfate crystals ($CuSO_4.5H_2O$) that can be produced.

Relative atomic masses (A_r): H = 1; C = 12; O = 16; S = 32; Cu = 63.5

Answer: .... **g** **[4 marks**

05 **Figure 5** shows the structure of graphite.

Figure 5

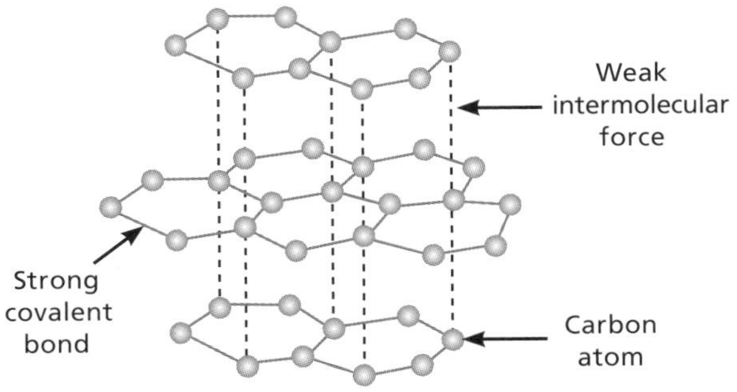

Weak intermolecular force

Strong covalent bond

Carbon atom

05.1 Graphite and diamond are both giant covalent compounds.

State **one difference** and **one similarity** between the structures of diamond and graphite.

Similarity: ....

Difference: .... **[2 marks]**

05.2 Graphene is a single layer of graphite.
Its properties include being an electrical conductor, strong and transparent.
In the future, it could be used to make touch-screens for electronic devices like mobile phones.

Explain why graphene has these properties in terms of its structure and why it is a good choice of material for a touch-screen.

....

....

....

....

....

....

.... **[6 marks]**

Turn over for the next question

06 A student was asked to carry out the electrolysis of copper(II) sulfate solution using inert graphite electrodes.

06.1 Complete **Figure 6** to show the apparatus they should use.
Label the following on your completed diagram:

- Anode
- Cathode.

Figure 6

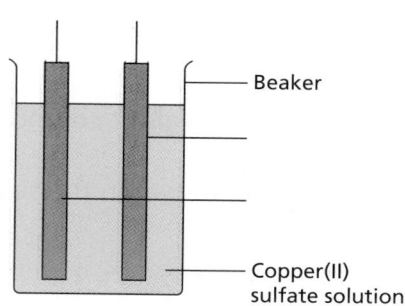

Beaker

Copper(II)
sulfate solution

[3 marks]

06.2 Hydrogen (H^+) ions and copper (Cu^{2+}) ions are both attracted to the cathode but only one product is formed.

Predict the name of the product formed at the cathode.
Explain why only this product is formed.

_____

_____ **[2 marks]**

06.3 Complete the balanced half equation for the reaction at the anode.

$$4OH^-(aq) \rightarrow \underline{\quad} H_2O(l) + O_2(g) + \underline{\quad} e^-$$ **[2 marks]**

06.4 Is the reaction at the anode an example of reduction or oxidation?
Give a reason for your answer.

_____

_____

_____ **[2 marks]**

7 A student investigated how the mass of magnesium changes when it reacts with dilute hydrochloric acid.

Figure 7 shows the apparatus they used.

Figure 7

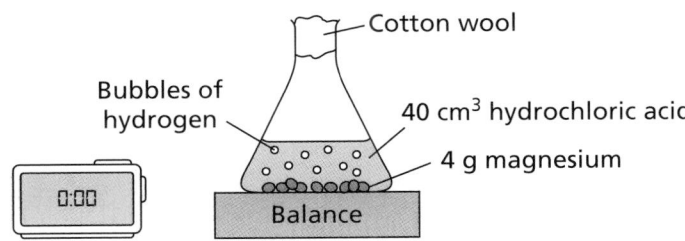

07.1 Complete the balanced symbol equation for the reaction by adding state symbols.

Mg_____ + 2HCl_____ → MgCl$_2$(aq) + H$_2$_____   **[1 mark]**

07.2 State **one** function of the cotton wool.

_____

_____   **[1 mark]**

Table 2 shows the student's results.

Table 2

Time in s	Total loss of mass in g
0	0.00
10	1.25
20	2.30
30	3.00
40	3.45
50	3.70
60	3.87
70	4.00
80	4.00
90	4.00

Question 7 continues on the next page

07.3 On **Figure 8**:

- Plot the results from **Table 2**.
- Draw a line of best fit.

Figure 8

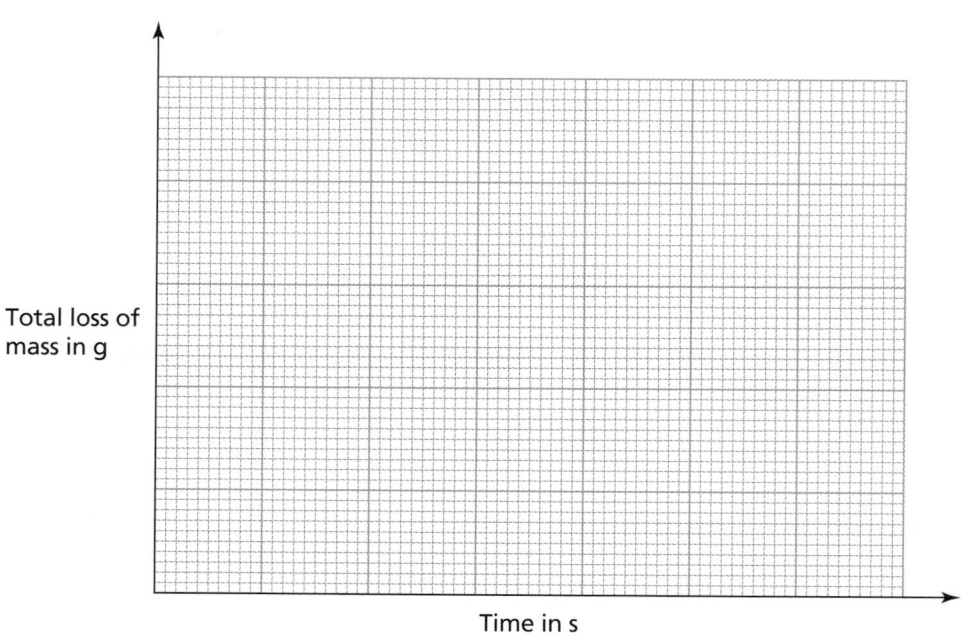

Total loss of mass in g

Time in s

[4 marks]

07.4 At what time did the mass loss stop changing?

Answer: .... s [1 mark]

07.5 The student observed that there was still magnesium in the flask at this time.

Explain why the mass loss stopped changing.

....

.... [1 mark]

07.6 Explain, in terms of particles, why the mass changes during the reaction.

....

....

....

.... [2 marks]

07.7 Zinc is less reactive than magnesium.

Sketch a line on the graph in **Figure 8** to show the results you would expect if the experiment was repeated using zinc of the same surface area.

Label this line **A**. [2 marks]

Turn over for the next question

08 This question is about the reaction of hydrogen with oxygen.
The equation for the reaction is:

$$2H_2(g) + O_2(g) \rightarrow 2H_2O(g)$$

Figure 9 shows the displayed formulae for the reaction.

Figure 9

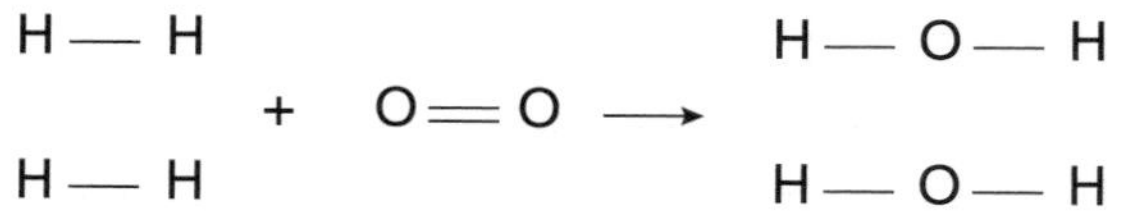

```
H — H                    H — O — H
        +   O = O  ⟶
H — H                    H — O — H
```

The bond enthalpies for the reaction are shown in **Table 3**.

Table 3

	H–H	O=O	O–H
Energy (kJ/mol)	432	495	467

08.1 Calculate the energy transferred in the reaction.

Answer: .... kJ/mol   **[3 marks]**

08.2 Explain, in terms of bond energies, why the reaction is exothermic.

....

....

....

[2 marks]

END OF QUESTIONS

Collins

GCSE
COMBINED SCIENCE

H

Chemistry: Paper 2 Higher tier

Materials

Time allowed: 1 hour 15 minutes

For this paper you must have:

- a ruler
- a calculator
- the periodic table (see page 450).

Instructions

- Answer **all** questions in the spaces provided.
- Do all rough work on the page. Cross through any work you do not want to be marked.

Information

- There are **70** marks available on this paper.
- The marks for each question are shown in brackets.
- You are expected to use a calculator where appropriate.
- You are reminded of the need for good English and clear presentation in your answers.
- When answering questions 03 and 05.2 you need to make sure that your answer:
  - is clear, logical, sensibly structured
  - fully meets the requirements of the question
  - shows that each separate point or step supports the overall answer.

Advice

- In all calculations, show clearly how you work out your answer.

Chemistry Practice Exam Paper 2

01 This question is about atmospheric pollutants from fuels.

Methane (CH_4) is a hydrocarbon which is used as a fuel.

The equation shows the reaction for the complete combustion of methane:

$$CH_4 + 2O_2 \rightarrow CO_2 + XH_2O$$

01.1 **X** represents what number?

Tick **one** box.

1 ☐

2 ☐

3 ☐

4 ☐

[1 mark]

If there is a limited amount of oxygen then methane will undergo incomplete combustion. One product is carbon monoxide.

01.2 Which other product may be produced in this reaction?

Tick **one** box.

carbon particles ☐

sulfur dioxide ☐

nitrogen oxide ☐

hydrogen ☐

[1 mark]

01.3 Give **two** reasons why it is difficult to detect the toxic gas carbon monoxide.

....

....

....

[2 marks]

01.4 When methane is burned at high temperatures, oxides of nitrogen are produced by the reaction between nitrogen and oxygen from the air.

Which equation correctly shows the production of nitrogen dioxide?

Tick **one** box.

$N + O_2 \rightarrow NO_2$ ☐

$2N_2 + O_2 \rightarrow 2N_2O$ ☐

$N + O \rightarrow NO$ ☐

$N_2 + 2O_2 \rightarrow 2NO_2$ ☐   **[1 mark]**

01.5 Increased amounts of pollutants in the air cause problems.

Draw **one** line from each pollutant to the problem it causes.

Pollutant	**Problem**
	global warming
sulfur dioxide	global dimming
	acid rain
carbon particles	destruction of ozone layer

[2 marks]

Turn over for the next question

Chemistry Practice Exam Paper 2

02 Cracking can be carried out in the laboratory using the apparatus shown in **Figure 1**.

Figure 1

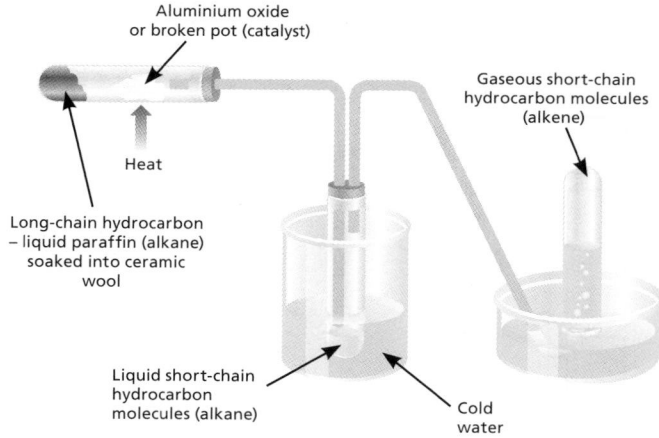

One example of a cracking reaction is:

$$C_{10}H_{22} \rightarrow C_7H_{16} + M$$

02.1 State the formula and name of **M**.

Formula: ....

Name: .... **[2 marks]**

02.2 Describe a test to show that the gas produced is an alkene.

Give the expected result of the test.

....

....

....

[2 marks]

An alkene called ethene is used to make a polymer called poly(ethene).

There are two types of poly(ethene): low density (LD) and high density (HD).

Table 1 shows the properties of the two types of poly(ethene).

Table 1

Property	LD Poly(ethene)	HD Poly(ethene)
Density	0.91–0.94 g/cm³	0.95–0.97 g/cm³
Flexibility	High	Low
Strength	Low	High

02.3 A manufacturer wants to make plastic buckets.

Suggest what type of poly(ethene) they should use.
Give a reason for your choice.

[2 marks]

Turn over for the next question

03 An increase in average global temperature is a cause of climate change.

Explain the effects of global climate change on the environment, humans and wildlife.

_____

_____

_____

_____

_____

_____

_____

_____

_____

_____

_____

_____

_____

[6 marks]

4 Humans need clean drinking water.

Water that is safe to drink is called potable water.

04.1 What is the name of the process that produces potable water from seawater?

_____ **[1 mark]**

04.2 Potable water cannot be described as pure.

Explain why.

_____

_____

_____ **[2 marks]**

04.3 A student is given the following equipment:

A beaker of water, a Bunsen burner, a tripod and gauze, and a thermometer.

Describe how the student can use the equipment to test if the water sample is pure.

You should state what they will observe if the water is pure in your answer.

_____

_____

_____ **[2 marks]**

04.4 Name **one** sterilising agent used to kill microorganisms in water.

_____ **[1 mark]**

Question 4 continues on the next page

04.5 Fluoride is sometimes added to water.

A sample of drinking water contains 1.35 mg of fluoride per dm³ of water.

Calculate the mass of fluoride in 250 cm³ of water.

Give your answer to 2 decimal places.

1000 cm³ = 1 dm³

Mass = ... mg   **[3 marks**

5   Steel and aluminium are used to make drinks cans.
    They are both limited resources.

05.1 Explain what is meant by 'limited resources'.

....

.... **[1 mark]**

05.2 The metal from drinks cans can be recycled.

The following steps are used:
1. The cans are collected and sorted.
2. They are melted down and then cooled to form blocks of metal.
3. The blocks are rolled into thin sheets, which can be used to make new products.

Evaluate the use of recycling metal cans as a method of reducing the use of
limited resources.

....

....

....

....

....

....

.... **[5 marks]**

06 Magnesium reacts with dilute hydrochloric acid:

$$Mg(s) + 2HCl(aq) \rightarrow MgCl_2(aq) + H_2(g)$$

A student investigated how the volume of hydrogen produced over time changes when magnesium is reacted with two different concentrations of dilute hydrochloric acid.

This is the method used:

1. Measure 20 cm^3 of 0.5 mol/dm^3 hydrochloric acid using a measuring cylinder.
2. Pour the acid into the conical flask.
3. Add 2 g of magnesium strip.
4. Place the bung into the flask.
5. Measure the volume of gas every 30 seconds until the reaction is complete.
6. Repeat using 1 mol/dm^3 hydrochloric acid.

The student reached the end of **Step 2** and set up the apparatus as shown in **Figure 2**.

Figure 2

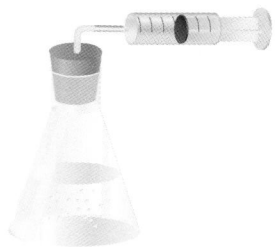

06.1 Identify what the student should do before continuing with the method.
Describe what could happen if the student continued without making any changes.
Explain how this would affect the results.

_____

_____

_____

_____

_____

[3 marks

Question 6 continues on the next page

The student corrected the error.

Their results are shown in **Table 2**.

Table 2

Time in s	Total volume of hydrogen in cm³	
	0.5 mol/dm³ acid	1 mol/dm³ acid
0	0.0	0.0
30	8.2	14.1
60	14.4	25.5
90	20.0	33.6
120	25.1	36.8
150	29.3	37.6
180	33.7	38.0
210	36.2	38.0
240	37.8	38.0
270	38.0	38.0

06.2 On **Figure 3**:
- Plot both sets of results on the grid.
- Draw two lines of best fit.

Figure 3

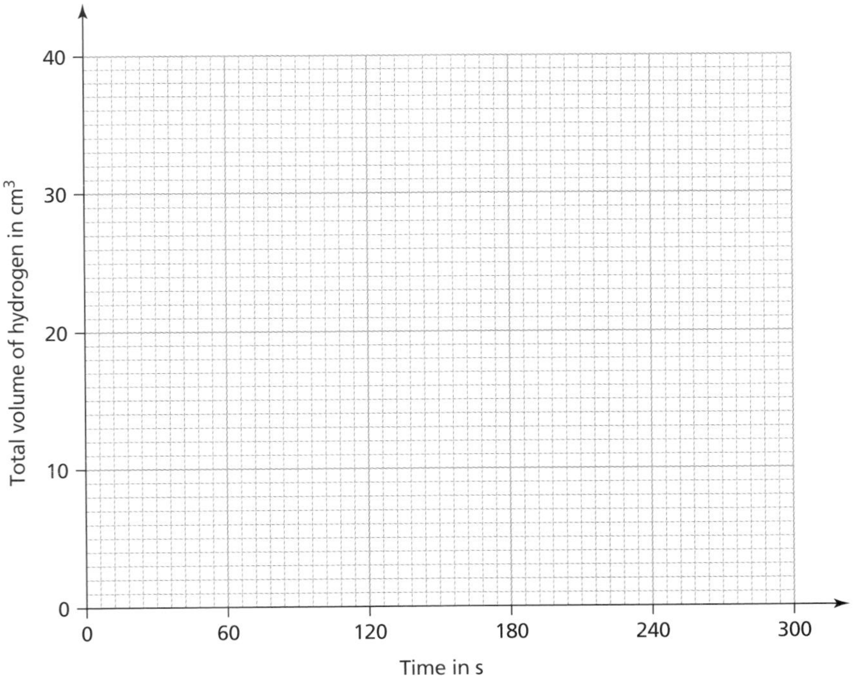

[4 marks]

06.3 How does the concentration of acid affect the rate of the reaction?

....

.... [1 mark]

06.4 Explain why, in terms of particles, the concentration of acid affects the rate of the reaction.

....

....

....

....

....

.... [3 marks]

The student decided to research how the temperature of the acid affected the rate of reaction.

Figure 4 shows a graph the student found in a text book.

Figure 4

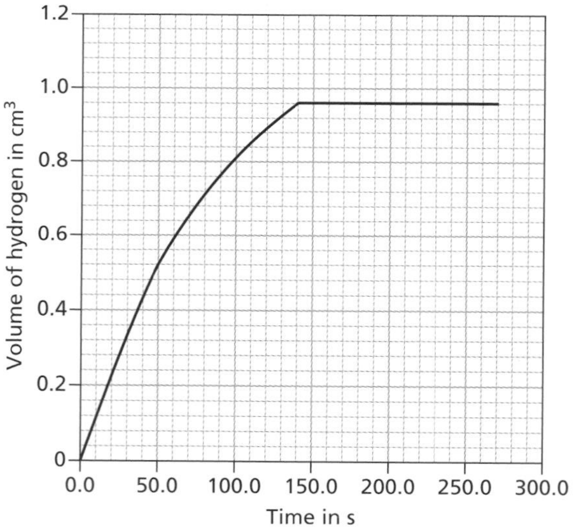

06.5 State the time at which the reaction finishes.

Answer: .... s [1 mark]

Question 6 continues on the next page

06.6 Use **Figure 4** to calculate the rate of the reaction at 50 seconds.
Give your answer to one significant figure.
Give the unit.

Rate of reaction = _____ **[6 marks]**

07 This question is about how the amounts of the different gases in the atmosphere have changed.

Figure 5 shows how the levels of oxygen in the atmosphere have changed since the Earth was formed.

Figure 5

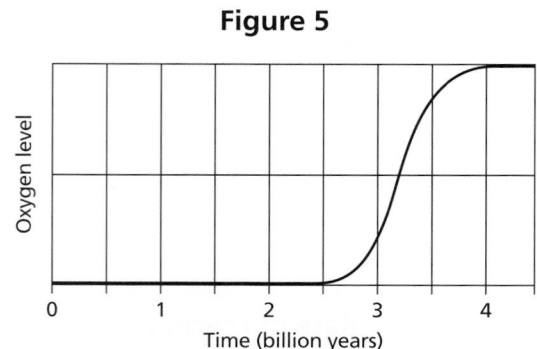

07.1 Use the graph to state when oxygen first started to be produced.

Answer: _____ billion years   **[1 mark**

07.2 The air today is approximately one-fifth oxygen.

Calculate the approximate volume of oxygen in 200 cm³ of air.

Answer: _____ cm³   **[1 mark**

07.3 Explain how the amount of oxygen in the air increased to the amount found today.

_____

_____

_____

_____

_____   **[3 marks]**

Question 7 continues on the next page

Another gas found in the air today is carbon dioxide.

07.4 Describe how carbon dioxide helps to maintain temperatures on Earth.

[3 marks]

07.5 In what way has the amount of carbon dioxide in the atmosphere changed over the last 100 years?

[1 mark]

07.6 Describe **one** way in which human activity has brought about this change to the amount of carbon dioxide in the atmosphere today.

[2 marks]

08 Paper chromatography can also be used to identify substances.

Figure 6 shows the results from chromatography carried out on a mixture.

Figure 6

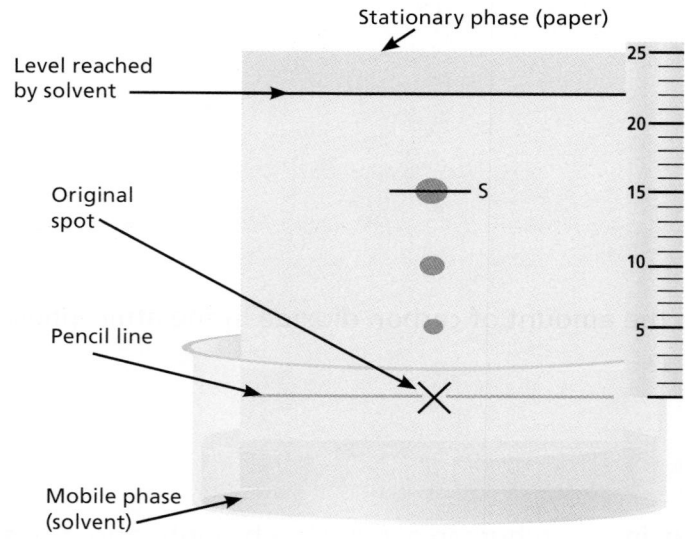

08.1 How many substances are present in the mixture?

Answer: .... **[1 mark**

08.2 To identify a substance, its R_f value must be calculated.

Calculate the R_f value of **S**.
Give your answer to 2 decimal places.

Answer: .... **[3 marks**

08.3 Explain how paper chromatography separates the substances in a mixture.

....

....

....

[3 marks]

END OF QUESTIONS

Collins

GCSE
COMBINED SCIENCE
Physics: Paper 1 Higher Tier

H

Materials

Time allowed: 1 hour 15 minutes

For this paper you must have:

- a ruler
- a calculator
- the Physics Equation Sheet (page 449).

Instructions

Answer **all** questions in the spaces provided.

Do all rough work on the page. Cross through any work you do not want to be marked.

Information

- There are **70** marks available on this paper.
- The marks for each question are shown in brackets [ ].
- You are expected to use a calculator where appropriate.
- You are reminded of the need for good English and clear presentation in your answers.
- When answering questions 01.2 and 04.5 you need to make sure that your answer:
- – is clear, logical, sensibly structured
- – fully meets the requirements of the question
- – shows that each separate point or step supports the overall answer.

Advice

In all calculations, show clearly how you work out your answer.

01 **Figure 1** shows what happens to each 100 joules of energy from coal that is burned in a power station.

Figure 1

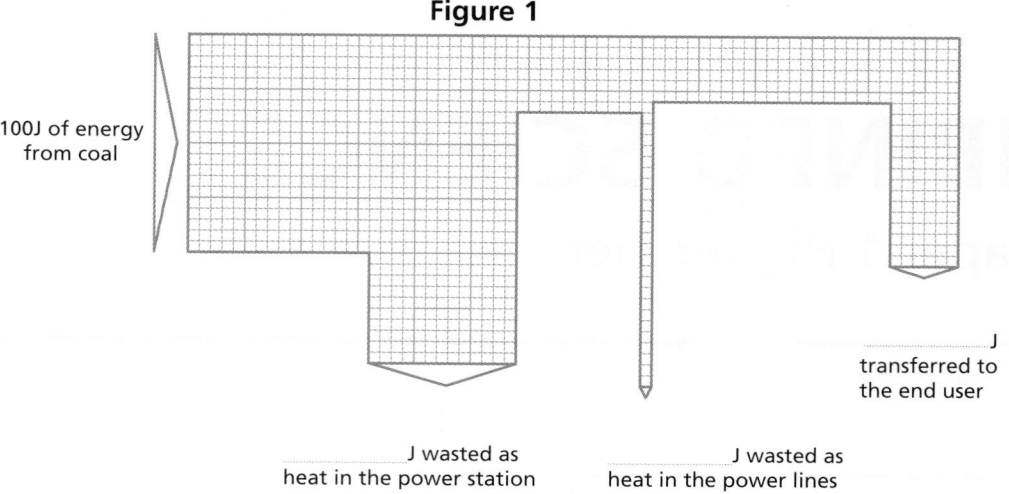

100J of energy from coal

_____ J wasted as heat in the power station

_____ J wasted as heat in the power lines

_____ J transferred to the end user

01.1 Add the missing figures to the diagram.

[3 marks]

01.2 For the same cost, the electricity company could:
- install new power lines that only waste half as much energy as the old ones
  OR
- use a quarter of the heat wasted at the power station to heat schools in a nearby town.

Which of these two things do you think they should do?
Give a reason for your answer.

_____

_____

_____

_____

_____

[4 marks]

01.3 Calculate the efficiency of the coal powered station in **Figure 1**.

Efficiency = _____ %   **[1 mark]**

02   A gas burner is used to heat some water in a pan.

By the time the water starts to boil:

- 60% of the energy released has been transferred to the water
- 20% of the energy released has been transferred to the surrounding air
- 13% of the energy released has been transferred to the pan
- 7% of the energy released has been transferred to the gas burner itself.

02.1   Some of the energy released by the burning gas is wasted.

What happens to this wasted energy?

....

....   **[2 marks]**

02.2   What percentage of the energy from the gas is wasted?

Percentage = ... %   **[1 mark]**

02.3   How efficient is the gas burner at heating water?

Efficiency = ... % .   **[1 mark]**

Turn over for the next question

03 A book weighs 6 newtons.

A librarian picks up the book from the ground and puts it on a shelf that is 2 metres high.

03.1 Calculate the work done on the book.

Work done = _____ J   **[2 marks]**

03.2 The next person to take the book from the shelf accidentally drops it.
The book falls 2 m to the ground.

Calculate how much gravitational energy it loses as it falls.
The gravitational field strength is = 10 N/kg.

Answer = _____ J   **[2 marks]**

03.3 All of the book's gravitational energy is converted to kinetic energy when it falls.

Calculate the velocity with which the book hits the floor.

Velocity = _____ m/s   **[3 marks]**

4   Electricity can be produced from a number of different energy resources.

04.1   Complete **Table 1** to show the energy transfers that occur for different resources.

Table 1

Device	Energy resource	Useful energy transfer from resource	
Coal-fired power station	Coal	chemical →	electrical
Hydroelectric power station	Stored water	→	electrical
Solar panel	Sun	→	electrical
Wind turbine	Wind	→	electrical
Gas-fired power station	Gas	→	electrical

[4 marks]

04.2   State which of the five energy resources in **Table 1** are not renewable.

_____
[1 mark]

04.3   Give another non-renewable energy resource.

_____
[1 mark]

04.4   Give a renewable energy source **not** listed in **Table 1**.

_____
[1 mark]

Question 4 continues on the next page

04.5 State and explain the advantages and disadvantages of using nuclear power stations to produce electricity.

....

....

....

....

....

....

[4 marks

5 A student carries out an experiment to investigate the current through component **X**.

A circuit is set up as shown in **Figure 2**.

The current is measured when different voltages are applied across component **X**.

Figure 2

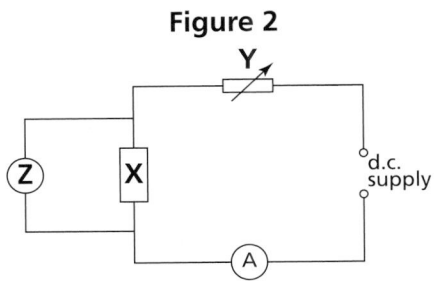

05.1 Name the components labeled **Y** and **Z** in the circuit.

Y = ... [1 mark]

Z = ... [1 mark]

05.2 What is the role of component **Y** in the circuit?

.... [1 mark]

Table 2 shows the measurements obtained in this experiment.

Table 2

Voltage (V)	−0.6	−0.4	−0.2	0	0.2	0.4	0.6	0.8
Current (mA)	0	0	0	0	0	50	100	150

05.3 Name the independent variable in this experiment.

Independent variable = ... [1 mark]

05.4 Name the dependent variable in this experiment.

Dependent variable = ... [1 mark]

Question 5 continues on the next page

05.5 Plot a graph on the axes in **Figure 3** using the data from **Table 2**.

Figure 3

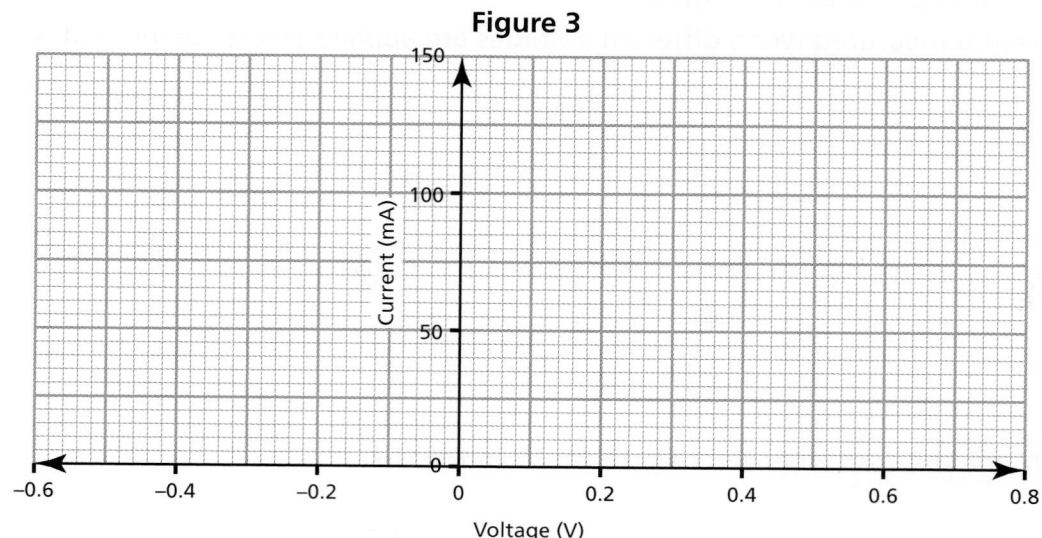

Voltage (V)

[3 marks]

05.6 The student looks at their measurements and decides that there are no anomalous results

Are they correct?
You must explain your answer.

....

....

[1 mark]

05.7 Use the shape of the graph to name component **X**.

Component **X** = ... [1 mark]

6 There are many isotopes of the element strontium (Sr).

06.1 What do the nuclei of different strontium isotopes have in common?

_____ [1 mark]

06.2 When the nucleus of a strontium-90 atom decays, it emits radiation and changes into a nucleus of yttrium-90

$$^{90}_{38}\text{Sr} \rightarrow\, ^{90}_{39}\text{Y} + \text{Radiation}$$

What type of decay is this?

Answer _____ [1 mark]

06.3 Give a reason for your answer to **06.2**.

_____ [1 mark]

Strontium-90 has a half-life of 30 years.

06.4 What is meant by the term 'half-life'?

_____ [1 mark]

06.5 After formation in the nuclear reactor, strontium-90 is stored as radioactive waste.

For how many years does strontium-90 have to be stored before its radioactivity has fallen to $\frac{1}{8}$ of its original level?

Answer _____ [2 marks]

Turn over for the next question

07 A car that is moving has kinetic energy.
The faster a car goes, the more kinetic energy it has.

    07.1 The kinetic energy of a car was 472 500 J when travelling at 30 m/s.

        Calculate the total mass of the car.
Give the unit.

        Mass = _____ **[4 marks**

There is a government road safety campaign to reduce the speed at which people drive in residential areas.
It uses the slogan 'Kill your speed, not a child'.
The scientific reason for this is that kinetic energy is transferred from the vehicle to the person it knocks down.

    07.2 A bus and car are travelling at the same speed.
The bus is likely to cause more harm to a person who is knocked down than the van would.

        Explain why.

        _____

        _____

    [2 marks

07.3 A car and its passengers have a total mass of 1200 kg.
The car is travelling at 8 m/s.

Calculate the increase in kinetic energy when the car increases its speed to 14 m/s.

Increase _____ **J**   **[3 marks]**

07.4 Explain why the increase in kinetic energy is much greater than the increase in speed.

_____

_____   **[1 mark]**

Turn over for the next question

08 In order to jump over the bar, a high jumper must raise his mass above the ground by 1.25 m. The high jumper has a mass of 65 kg.

The gravitational field strength is 10 N/kg.

08.1 The high jumper just clears the bar.

Calculate the gain in his gravitational potential energy.

Gain = _____ J **[2 marks**

08.2 Calculate the minimum vertical speed the high jumper must reach in order to jump over the bar.

Use your answer to **08.1** and the formula for kinetic energy.

Minimum vertical speed = _____ **m/s** **[3 marks**

09 The circuit diagram in **Figure 4** shows a circuit used to supply electricity for car headlights.

Figure 4

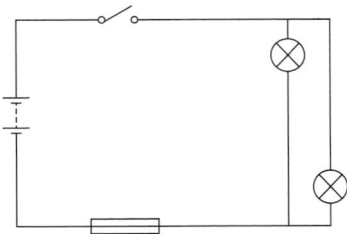

The current through the filament of one car headlight is 2 A.
The potential difference supplied by the battery is 12 V.

09.1 What is the potential difference across each headlight?

Potential difference = _____ V   [1 mark]

09.2 Work out the total current through the battery.

Current = _____ A   [1 mark]

09.3 Calculate the resistance of each headlight filament when in use.

Resistance = _____ Ω [2 marks]

Question 9 continues on the next page

09.4 How does the total resistance of the circuit compare to the resistance of each individual bulb?

_____ **[1 mark**

09.5 Calculate the power supplied to each of the two headlights of the car.

Power = _____ **W** **[2 marks**

09.6 The fully charged car battery can deliver 96 kJ of energy at 12 V.

How long can the battery keep both the headlights fully on?

Length of time = _____ **s** **[2 marks**

END OF QUESTIONS

Collins

GCSE
COMBINED SCIENCE   H

Physics: Paper 2 Higher Tier

Materials   Time allowed: 1 hour 15 minutes

> **For this paper you must have:**
> - a ruler
> - a calculator
> - a protractor
> - the Physics Equation Sheet (page 449).

Instructions

- Answer **all** questions in the spaces provided.
- Do all rough work on the page. Cross through any work you do not want to be marked.

Information

- There are **70** marks available on this paper.
- The marks for each question are shown in brackets [ ].
- You are expected to use a calculator where appropriate.
- You are reminded of the need for good English and clear presentation in your answers.
- When answering questions 04.1 and 06.5 you need to make sure that your answer:
  - is clear, logical, sensibly structured
  - fully meets the requirements of the question
  - shows that each separate point or step supports the overall answer.

Advice

- In all calculations, show clearly how you work out your answer.

Physics Practice Exam Paper 2

01 A student used a lever system to investigate how the force of attraction between a coil and an iron rocker varied with the current in the coil.

She supported a coil vertically and connected it to an electrical circuit as shown in **Figure 1**.

The weight of the iron rocker is negligible.

Figure 1

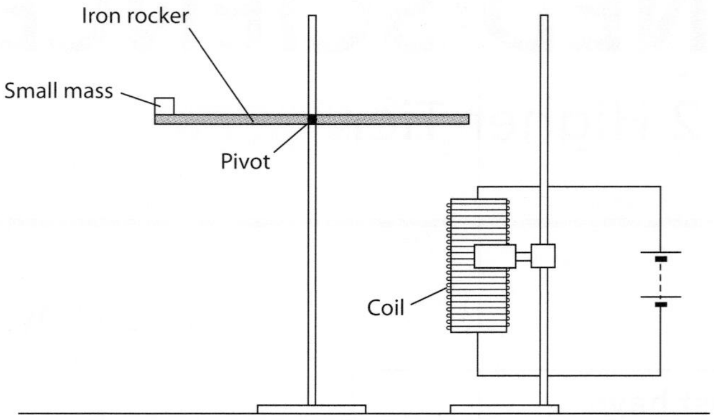

01.1 Why is it important that the rocker in this experiment is made of iron?

.... **[1 mark]**

01.2 The student put a small mass on the end of the rocker and adjusted the current in the coil until the rocker balanced.

To keep the rocker balanced, how will the current through the coil need to change as the size of the mass is increased?

.... **[1 mark]**

01.3 Explain your answer to **01.2**.

....

....

.... **[2 marks]**

01.4 A second student set up the same experiment and put an iron core inside the coil.

How will this affect the size of the mass that can be balanced?
You must explain you answer.

....

....

.... **[2 marks]**

Turn over for the next question

Physics Practice Exam Paper 2

02 A group of students investigate circular motion. **Figure 2**
They swing a bung attached to a string around in circle.
The string is attached to a force meter, which measures
the centripetal force (the force that keeps an object on
a circular path).

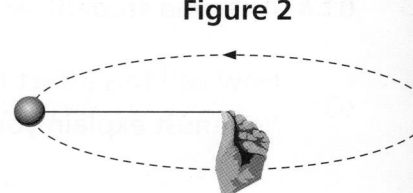

The students record how the reading on the force meter
changes to determine how the force affects the speed of the orbit.

02.1 In which direction does the centripetal force act on the rubber bung?

_____ **[1 mark**

02.2 In this experiment, what provides the centripetal force?

_____ **[1 mark**

02.3 One student swung the rubber bung around in a circle at constant speed.
A second student timed how long it took the rubber bung to complete 10 rotations.

Give **two** variables that are important to control in this experiment.

_____

_____ **[2 marks**

02.4 The Moon orbits the Earth in a circular path.

direction	resistance	speed	velocity

Use words from the box to complete the sentences.
You may use each word once, more than once or not at all.

The Moon's _____ is constant but its _____ changes.

This is because its _____ changes. **[3 marks**

02.5 What force provides the centripetal force needed to keep the Moon in its orbit
around the Earth?

_____ **[1 mark**

3 Radio waves, ultraviolet waves, visible light and X-rays are all types of electromagnetic radiation.

03.1 Choose wavelengths from the list below to complete **Table 1**.

3×10^{-8} m 1×10^{-11} m 5×10^{-7} m   1500 m

Table 1

Type of Radiation	Wavelength (m)
Radio waves	
Ultraviolet waves	
Visible light	
X-rays	

[3 marks]

03.2 Give **two** properties that are common to all the waves in **Table 1**.

....

.... [2 marks]

03.3 State the name and a use of **one** type of electromagnetic wave **not** listed in **Table 1**.

....

.... [2 marks]

03.4 Radio waves can be used to control remote control cars.

Calculate the frequency of radio waves of wavelength 300m.
(The velocity of electromagnetic waves is 3×10^8 m/s.)

Frequency = ... Hz   [3 marks]

Question 3 continues on the next page

The graph in **Figure 3** shows the speed of a remote-controlled vehicle during a race.

Figure 3

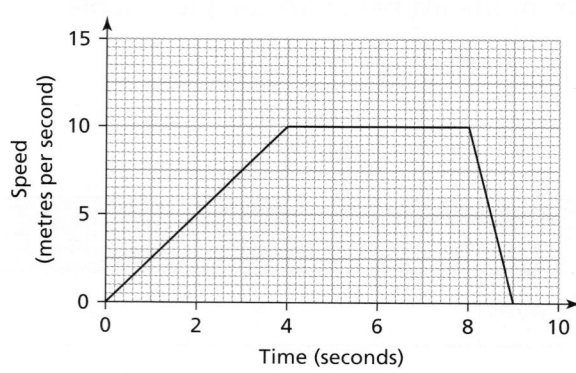

03.5 Calculate the acceleration during the first four seconds.

Acceleration = _____ m/s² **[3 marks]**

03.6 What is the maximum speed reached by the vehicle?

Maximum speed = _____ m/s **[1 mark]**

03.7 How far does the vehicle travel between 4 and 8 seconds?

Distance = _____ m **[2 marks]**

4   The distance–time graph in **Figure 4** represents the motion of a car during a race.

Figure 4

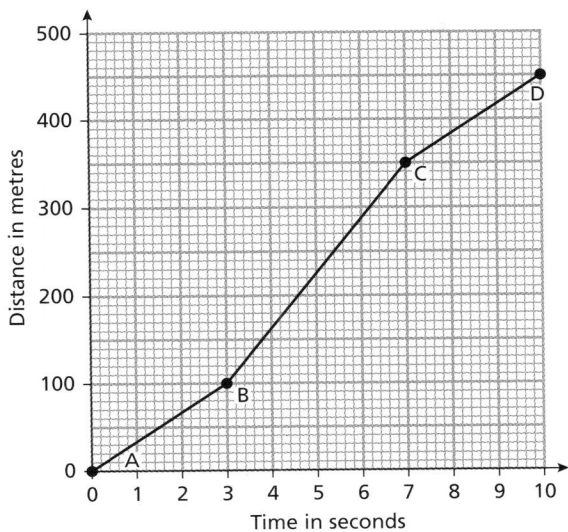

04.1   Describe the motion of the car between points **A** and **D**.
You should calculate the maximum speed reached by the car and include this in your answer.

_____

_____

_____

_____

_____

_____

_____

                                                      [6 marks]

Question 4 continues on the next page

04.2 At the start of the race, the car accelerates from rest to a speed of 30 m/s in 6 seconds.

Calculate the acceleration of the car.

Acceleration = _____ m/s² **[3 marks**

5 A car of mass 1200 kg has an engine thrust of 3500 N and experiences a resistive force of 2000 N.

05.1 Calculate the acceleration of the car.

Acceleration = _____ m/s² **[4 marks]**

05.2 Explain why the car reaches a top speed even though the thrust force remains constant at 3500 N.

_____

_____

_____ **[2 marks]**

The driver of the car notices a hazard.
They apply the brakes to come to a complete stop in a certain distance.
This stopping distance is made up of the thinking distance and the braking distance.

05.3 What is meant by the term 'thinking distance'?

_____

_____ **[1 mark]**

05.4 State **two** factors that affect thinking distance.

_____

_____

_____ **[2 marks]**

Question 5 continues on the next page

05.5 The braking distance of a car depends on the speed of the car and the braking force applied.

State **one** other factor that affects the braking distance.

....

.... **[1 mark**

05.6 During braking, the temperatures of the brakes increase.

Explain why.

....

....

....

.... **[2 marks**

6 **Figure 5** shows two waves.

Figure 5

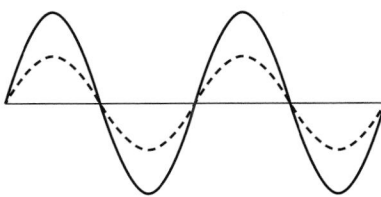

06.1 Name **one** wave quantity that is the same for both waves.

_____ [1 mark]

06.2 Name **one** wave quantity that is different for the two waves.

_____ [1 mark]

06.3 The waves shown in **Figure 5** are transverse waves.

Which of the following types of wave is **not** a transverse wave?
Draw a ring around the correct answer.

gamma rays   **sound**   **visible light**   [1 mark]

06.4 A student studies waves in a ripple tank.
Every second, eight waves pass a fixed point in the tank.
The waves have a wavelength of 0.015 m.

Calculate the speed of the water waves.

Wave speed = _____ m/s   [3 marks]

Question 6 continues on the next page

06.5 Measuring the wavelength of moving water waves is difficult.
To improve accuracy, the student decides to use a stroboscope and a ruler.

Describe the procedure that the student should use and outline **one** hazard they need to be aware of.

_____

_____

_____

_____

_____

[4 marks

7 **Figure 6** shows two lines of the magnetic field pattern around a current-carrying wire.

Figure 6

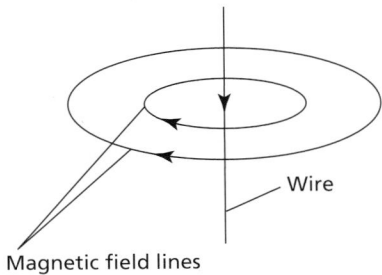

07.1 The direction of the current is reversed.

How does this affect the lines in the magnetic field pattern?

.... **[1 mark]**

07.2 The size of the current through the wire is increased.

How does this affect the lines in the magnetic field pattern?

.... **[1 mark]**

Question 7 continues on the next page

07.3 Fleming's left hand rule can be used to identify the direction of a force acting on a current-carrying wire in a magnetic field.

Figure 7

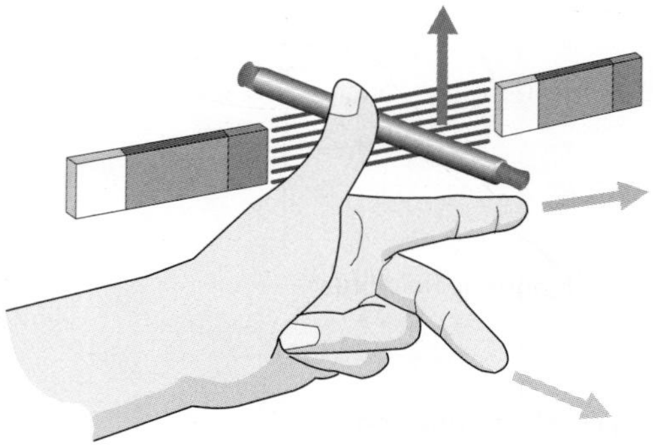

Complete the table below to show what is indicated by each part of the hand.

Part of hand	Indicates the direction of:
First finger	
Second finger	
Thumb	

[3 marks]

Figure 8 illustrates a pair of magnets.

The arrow shows the direction of the current in a wire passing between the magnets.

Figure 8

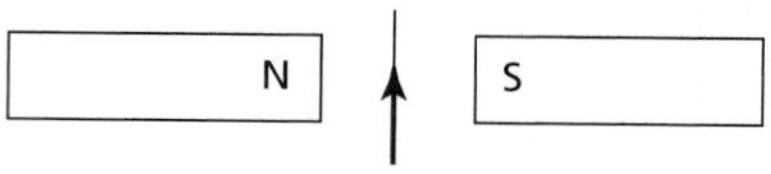

Current

07.4 In which direction does the force on the wire act?

[1 mark]

07.5 Give **three** changes that would **decrease** the force acting on the wire.

[3 marks]

END OF QUESTIONS

Physics Equation Sheet

1	(final velocity)^2 − (initial velocity)^2 = 2 × acceleration × distance	$v^2 - u^2 = 2as$
2	elastic potential energy = 0.5 × spring constant × (extension)^2	$E_e = \frac{1}{2}ke^2$
3	change in thermal energy = mass × specific heat capacity × temperature change	$\Delta E = mc\Delta\theta$
4	period = $\dfrac{1}{\text{frequency}}$	
5	force on a conductor (at right-angles to a magnetic field) carrying a current = magnetic flux density × current × length	$F = BIl$
6	thermal energy for a change of state = mass × specific latent heat	$E = mL$
7	potential difference across primary coil × current in primary coil = potential difference across secondary coil × current in secondary coil	$V_s I_s = V_p I_p$

The Periodic Table

Key

relative atomic mass
atomic symbol
name
atomic (proton) number

1	**H** hydrogen 1

1	2											3	4	5	6	7	0
																	4 **He** helium 2
7 **Li** lithium 3	9 **Be** beryllium 4											11 **B** boron 5	12 **C** carbon 6	14 **N** nitrogen 7	16 **O** oxygen 8	19 **F** fluorine 9	20 **Ne** neon 10
23 **Na** sodium 11	24 **Mg** magnesium 12											27 **Al** aluminum 13	28 **Si** silicon 14	31 **P** phosphorus 15	32 **S** sulfur 16	35.5 **Cl** chlorine 17	40 **Ar** argon 18
39 **K** potassium 19	40 **Ca** calcium 20	45 **Sc** scandium 21	48 **Ti** titanium 22	51 **V** vanadium 23	52 **Cr** chromium 24	55 **Mn** manganese 25	56 **Fe** iron 26	59 **Co** cobalt 27	59 **Ni** nickel 28	63.5 **Cu** copper 29	65 **Zn** zinc 30	70 **Ga** gallium 31	73 **Ge** germanium 32	75 **As** arsenic 33	79 **Se** selenium 34	80 **Br** bromine 35	84 **Kr** krypton 36
85 **Rb** rubidium 37	88 **Sr** strontium 38	89 **Y** yttrium 39	91 **Zr** zirconium 40	93 **Nb** niobium 41	96 **Mo** molybdenum 42	[98] **Tc** technetium 43	101 **Ru** ruthenium 44	103 **Rh** rhodium 45	106 **Pd** palladium 46	108 **Ag** silver 47	112 **Cd** cadmium 48	115 **In** indium 49	119 **Sn** tin 50	122 **Sb** antimony 51	128 **Te** tellurium 52	127 **I** iodine 53	131 **Xe** xenon 54
133 **Cs** caesium 55	137 **Ba** barium 56	139 **La*** lanthanum 57	178 **Hf** hafnium 72	181 **Ta** tantalum 73	184 **W** tungsten 74	186 **Re** rhenium 75	190 **Os** osmium 76	192 **Ir** iridium 77	195 **Pt** platinum 78	197 **Au** gold 79	201 **Hg** mercury 80	204 **Tl** thallium 81	207 **Pb** lead 82	209 **Bi** bismuth 83	[209] **Po** polonium 84	[210] **At** astatine 85	[222] **Rn** radon 86
[223] **Fr** francium 87	[226] **Ra** radium 88	[227] **Ac*** actinium 89	[261] **Rf** rutherfordium 104	[262] **Db** dubnium 105	[266] **Sg** seaborgium 106	[264] **Bh** bohrium 107	[277] **Hs** hassium 108	[268] **Mt** meitnerium 109	[271] **Ds** darmstadtium 110	[272] **Rg** roentgenium 111	[285] **Cn** copernicium 112	[286] **Uut** ununtrium 113	[289] **Fl** flerovium 114	[289] **Uup** ununpentium 115	[293] **Lv** livermorium 116	[294] **Uus** ununseptium 117	[294] **Uuo** ununoctium 118

* The Lanthanides (atomic numbers 58–71) and the Actinides (atomic numbers 90–103) have been omitted.
Relative atomic masses for **Cu** and **Cl** have not been rounded to the nearest whole number.

Answers

Topic-Based Questions

Page 286 Cell Structure

1. a) B [1]
   b) A [1]
   c) C [1]
2. a) B [1]; it has a nucleus / chloroplasts [1]
   b) Eukaryotic cells are much larger than prokaryotic [1]
   c) Using flagella [1]; which whip around [1]

Page 287 Investigating Cells

1. a) To act as a stain [1]; because some of the structures are transparent [1]
   b) i) 30 micrometres [1]
      ii) $\frac{20}{0.03}$ [1]; = 667, magnified 667 times (×667) [1]
2. a) 0.11mm [1]
   b) 10 micrometres [1]

Page 288 Cell Division

1. a) Mitosis [1]
   b) P, Q, (S), R [2] (1 mark for one in the correct place)
2. a) Stem cells can divide to produce different types of cells [1]; these cells can replace lost or defective cells or be grown into tissues [1]
   b) Any one of: the stem cells may come from cloned embryos [1]; embryos may be destroyed in the process [1]; concerned that the long-term effects are not known [1]

Page 289 Transport In and Out of Cells

1. a) 6 × (10 × 10) = 600 [1]
   b) Diffusion [1]
   c) Block C [1]; it has the largest surface area [1]; so there is more surface for the dye to diffuse across [1]
2. a) Osmosis always involves movement of water particles, whereas diffusion is movement of particles [1]; osmosis involves a cell membrane and diffusion does not [1]
   b) Active transport is movement of particles against a concentration gradient / from low to high concentration, rather than from high to low concentration as with diffusion [1]; active transport requires energy from respiration (diffusion does not require an input of energy from the cell) [1]

Page 290 Levels of Organisation

1. a) cells, tissues, organs, systems [1]
   b) Top to bottom: organ [1]; cell [1]; organ [1]; tissue [1]
   c) Muscle cells are specialised for contraction. [1]
2. Phloem cells join end to end [1]; the end walls have holes to let sugar through [1]; xylem cells join end to end with end walls broken down [1]; to form hollow tubes / strengthened with lignin [1]

Page 291 Digestion

1. a) The rate of reaction increases to a peak / optimum [1]; and then decreases [1]
   b) 40°C [1]
2. a) Pancreas / small intestine [1]
   b) i) There are more than three fatty acids / it contains sugar / no glycerol is present [1]
      ii) Olestra is a different shape to fats [1]; so it cannot fit into the enzyme's active site [1]

Page 292 Blood and the Circulation

1. 3, 1, 2, 4 [3] (2 marks for two correct; 1 mark for one correct)
2. a) Haemoglobin combines with oxygen at the lungs, when in high concentration [1]; and forms oxyhaemoglobin [1]; it releases oxygen at the tissues, as the concentration is low there, to go back to haemoglobin [1]

   haemoglobin + oxygen ⇌ oxyhaemoglobin

   b) Lower surface area to take up or release oxygen [1]; shape means they get stuck in capillaries causing pain [1]

Page 293 Non-Communicable Diseases

1. a) i) The heart stops beating (Accept an accurate explanation of why) [1]
      ii) It restricts blood flow to the heart [1]; so not enough oxygen and glucose are supplied to cells in the heart muscle for respiration [1]
   b) Having an atheroma does not seem to depend on level of physical activity [1]; increased exercise reduced incidence of heart disease [1]; suggesting atheromas are not the only factor in causing heart disease [1]

Page 294 Transport in Plants

1. a) water [1]; transpiration [1]; windy [1]
   b) i) A smaller change in mass due to less transpiration [1]; because less water is taken up by the roots [1]
      ii) A smaller change in mass due to less transpiration [1]; as some of the stomata have been blocked [1]
2. Any two of: water is transported through the xylem, sugars are transported in phloem [1]; water is moved upwards only, sugars move all over [1]; water movement does not need energy from respiration, sugars move by active transport [1]

Page 295 Pathogens and Disease

1. a) Four correctly draw lines [3] (2 marks for two lines; 1 mark for one line)
   rose black spot – fungus
   salmonella – bacterium
   measles – virus
   malaria – protozoan
   b) vector [1]; *Plasmodium* [1]; host [1]
2. a) Any two of: infected needles [1]; sexual activity [1]; across the placenta [1]
   b) The virus attacks white blood cells [1]; making the immune system less effective [1]
3. To prevent the spread of the disease [1]; to destroy the spores [1]

Page 296 Human Defences Against Disease

1. Nose: hairs trap particles [1]; skin: sebum kills bacteria [1]; stomach: acid kills microorganisms [1]
2. A pathogen has antigens on its surface [1]; lymphocytes make antibodies [1]; that attach to antigens on pathogens (with a specific fit) [1]; phagocytes ingest pathogens [1]

Page 297 Treating Diseases

1. a) HIV is a virus [1]; TB is a bacterial infection [1]
   b) Many TB bacteria are resistant [1]; and are not killed by antibiotics [1]
   c) Any two of: finish the dose [1]; only prescribe if necessary [1]; rotate antibiotics used [1]
   d) Use of a placebo [1]; patient does not know if they are having the placebo / real drug [1]; neither does the doctor [1]

Page 298 Photosynthesis

1. a) carbon dioxide + water [1]; ⟶ glucose + oxygen [1]
   b) i) Any two of: root tip [1]; shoot tip [1]; fruits [1]; seeds [1]; storage organs [1];
      ii) Any two of: starch [1]; for storage [1] OR lipids [1]; for storage [1] OR proteins [1]; for growth [1] OR cellulose [1]; for cell walls [1]
2. a) Some of the mass did come from the water [1]; but some also comes from carbon dioxide / photosynthesis [1]
   b) He showed that carbon dioxide was needed as well as water [1]

Answers

Page 299 Respiration and Exercise

1. a) oxygen [1]; → carbon dioxide [1]; + water [1] (The two products can be given in any order)
   b) i) In the race / after training, lactic acid does not start to increase so early [1]; and does not increase as much [1]; lactic acid causes muscle cramps, and because there is less she can run more efficiently [1]
      ii) The lactic acid is sent to the liver [1]; where it is broken down [1]; using oxygen [1]

Page 300 Homeostasis and the Nervous System

1. a) Insulation [1]; increases the speed of the impulse [1]
   b) Transmitter molecules are released when an impulse reaches a synapse [1]; they diffuse across the synapse [1]; and stimulate an impulse in the next neurone [1]
   c) i) The drug mimics the neurotransmitter [1]; and stimulates an impulse in the next neurone [1]
      ii) Neurotransmitter molecules will not be broken down [1]; so they continue to stimulate the next neurone [1]

Page 301 Hormones and Homeostasis

1. a) Type 1   [1]
   b) Insulin   [1]
   c) Any two of: blood glucose level too high [1]; glucose starts to pass out in urine [1]; coma and / or death [1]
   d) i) Type 2   [1]
      ii) Treat by modifying diet [1]; rather than by insulin injections [1]
2. Less water is reabsorbed back into the body [1]; urine is less concentrated / higher volume [1]; which could lead to dehydration [1]

Page 302 Hormones and Reproduction

1. a) Hormone A: Oestrogen [1]; inhibits FSH release / stimulates LH release / repairs lining of uterus [1]; Hormone B: progesterone [1]; maintains the lining of the uterus / inhibits FSH and LH [1]
   b) i) X marked just after peak of Hormone A   [1]
      ii) Y marked at start or end of the cycle   [1]
   c) i) Stimulates follicle development [1]; ready for ovulation [1]
      ii) Causes release of an egg   [1]
2. The pill contains both oestrogen and progesterone [1]; it inhibits FSH [1]; so that eggs do not develop / no ovulation [1]

Page 303 Sexual and Asexual Reproduction

1. a) It produces runners [1]; that touch the ground and root, growing into new plants [1]
   b) Any two of: it produces flowers [1]; pollen is transferred from plant to plant [1]; it makes seeds [1]
2. Egg cell: meiosis [1]; eye cell: 38 and mitosis [1]; sperm cell: 19 [1]; leg cell: 38 [1]

Page 304 Patterns of Inheritance

1. a) i) Sachin and Rose   [1]
      ii) Sara, Tia and Rohit   [1]
   b) Sachin: gg (genotype) [1]; male with Gaucher disease (phenotype) [1]; Tia: female with normal enzyme (phenotype) [1]
   c) Probability: 50% / $\frac{1}{2}$ / 1 : 1 [1];

	G	g	
	G	g	[1];
g	Gg	gg	[1];
g	Gg	gg	[1]

   d) Any two of: She might prefer not know until it is born [1]; to avoid having to decide whether to have a termination [1]; the test may increase the risk of miscarriage [1] (Accept any other sensible reason)

Page 305 Variation and Evolution

1. Genetic: Bill and Ben have brown eyes [1]; Environment: Ben has a scar [1]; Genetics and Environment: Bill is 160cm tall [1]; Ben's body mass is 60kg [1]
2. a) Darken / turn black [1]
   b) The dark moths are better camouflaged [1]; so fewer are eaten by birds [1]

Page 306 Manipulating Genes

1. Selective breeding [1]; genetic engineering [1]
2. a) (D), A, B, C [2] (1 mark for one in the correct place)
   b) The process uses hormones to cut DNA. [1]; The insulin molecules made are a mixture of human and bacterial insulin. [1]

Page 307 Classification

1. a) Organisms that can mate with each other [1]; to produce fertile offspring [1]
   b) i) Bobcat [1]; ocelot [1]
      ii) They are both in the same genus [1]; Felis [1]
2. a) i) 160 million years ago   [1]
      ii) Archaeopteryx   [1]
   b) Any two of: new disease [1]; new predators [1]; more successful competitors [1]; catastrophic event, e.g. meteor [1]

Page 308 Ecosystems

1. a) i) To spread their weight / stop them sinking in the sand   [1]
      ii) To shade them from the sun / store of food or fat / produce water from respiration (by breaking down the fat store) [1]
   b) population [1]; habitat [1]; community [1]; competed [1]; biotic [1]
   c) i) Use a quadrat [1]; placed at random [1]; count cacti present in quadrat [1]; repeat many times [1]; calculate the mean number per m^2 [1]; multiply results up by the area [1]
      ii) Camels move around [1]; so may count them more than once / not at all [1]

Page 309 Cycles and Feeding Relationships

1. A = Photosynthesis [1];
   B = Respiration [1];
   C = Decomposition [1];
   D = Combustion / burning [1];
   E = Eating / feeding [1]
2. a) Predator–prey graph   [1]
   b) Rabbit numbers increase, so there is more food available for the foxes [1]; more foxes survive and numbers increase [1]; this means more rabbits are eaten, so the fox numbers drop again (as less food is available) [1]

Page 310 Disrupting Ecosystems

1. a) There is no pollution / no industry / no cars   [1]
   b) Any two of: in summer, there is more sunlight [1]; so more photosynthesis [1]; and more carbon dioxide taken in by plants [1]
   c) Greenhouse effect / greenhouse gas [1]; allows heat from the Sun through the atmosphere [1]; but absorbs / traps energy reflected back from the Earth's surface, leading to a rise in temperature [1]
2. a) Frog   [1]
   b) When some fossil fuels are burned [1]; acidic gases are released, e.g. sulfur dioxide gas [1]; which dissolve in water in the atmosphere, which then falls as acid rain [1]

Page 311 Atoms, Elements, Compounds and Mixtures

1. a) $2Cu + O_2 \rightarrow 2CuO$ [1]
   b) 18.9 – 15.6 = 3.3g   [1]
   c) 0.1g   [1]

The resolution of the balance is the degree of accuracy. In this case, the measurements are given to the nearest 0.1g.

2. a) Place the salt solution in the round bottomed flask [1]; heat the solution [1]; and collect the liquid that distils at 100°C [1]
   b) **Any one of:** Burns from the Bunsen burner / hot water / steam [1]; equipment may crack so risk of cuts [1]

Page 312 Atoms and the Periodic Table

1. a) Thomson showed that the atom could be divided up into simpler substances [1]; the old model was no longer correct [1]
   b) Electron [1]
2. a) Based on Thomson's model, the alpha particles would have all gone through [1] OR There would have been no deflection [1] (Accept: It went against Thomson's model)
   b) To reduce the effects of errors [1]; to check repeatability [1] (Accept: To make sure they were correct / accurate [1])
   c) **Any three of:** small nucleus [1]; nucleus in the centre of the atom [1]; nucleus has a positive charge [1]; electrons in orbit around the nucleus [1] (Accept a labelled diagram)

Page 313 The Periodic Table

1. C [1]

   The number of electrons in the outer shell is the same as the group that the element is in. The number of electron shells is the same as the row (period) that the element is in.

2. a) X [1]; because it does not conduct electricity [1]
   b) W [1]; because it is a liquid at room temperature [1]; and it conducts electricity [1]
   c) (lithium) hydroxide [1]; hydrogen [1]

Page 314 States of Matter

1. a) g [1]; s [1]

   Gas = g, solid = s, liquid = l, aqueous / dissolved in water = aq

   b) 650°C [1]
   c) −183°C [1]
   d) **Any six of:** Oxygen has a low boiling point [1]; magnesium oxide has a high boiling point [1]; oxygen has weak forces of attraction between its particles / molecules [1]; magnesium has strong forces between its ions / particles [1]; oxygen is a simple molecule [1]; magnesium oxide is an ionic compound [1]; only a small amount of energy is needed to boil oxygen / turn it from a liquid into a gas [1]; a lot of energy is needed to boil magnesium oxide / turn it from a liquid into a gas [1]

Page 315 Ionic Compounds

1. a) calcium chloride [1]; sodium carbonate [1]

   Ionic compounds contain both metal and non-metal elements. Look at the periodic table if you are not sure whether an element is a metal or a non-metal. The metals are on the left side, the non-metals on the right.

   b) They cannot conduct electricity when solid because the ions cannot move [1]; they can conduct electricity when in solution because the ions are free to move about [1]; and carry the charge [1]
2. Electrons are transferred from magnesium to bromine [1]; the magnesium atom loses two electrons [1]; forming Mg^{2+} / 2+ ions [1]; the two bromine atoms each gain one electron [1]; forming Br^- / 1– ions [1]

Page 316 Covalent Compounds

1. The forces between bromine molecules are stronger [1]
2. a) A shared pair of electrons drawn between H and Br [1]; no additional hydrogen electrons and three non-bonding pairs shown on bromine [1] (second mark dependent on first)
   b) HBr [1]
3. High melting point – Strong covalent bonds between many carbon atoms [1]; Does not conduct electricity when molten – There are no charged particles that are free to move [1]

   The covalent bonds between atoms are very strong. The force of attraction between molecules is called the intermolecular force and is much weaker.

Page 317 Metals and Special Materials

1. High melting point: strong force of attraction between positive ions and negative electrons / metallic bond is strong [1]; so, a lot of energy is needed to break metallic bonds (to melt metal) [1]; thermal conductivity: electrons are delocalised [1]; and are free to move through the metal and transfer energy [1]; malleability: the ions are arranged in layers / have a regular arrangement [1]; the layers are able to slide over each other easily [1]

Page 318 Conservation of Mass

1. a) $Zn(s) + 2H^+(aq) \rightarrow Zn^{2+}(aq) + H_2(g)$ [1]

   In an ionic equation, only the ions that change (gain or lose electrons) are shown. They must be balanced.

   b) Gas is produced in the flask (increasing the pressure inside) [1]; the flask may crack / the bung may be forced out [1] (Accept: 'it is not safe' for 1 mark)
   c) The mass reading will decrease [1]; because the hydrogen particles (atoms) in the hydrochloric acid are rearranged [1]; to form hydrogen gas, which leaves the flask into the air [1]

Page 319 Amount of Substance

1. a) $2 \times 2 \times 2 = 8cm^3$ [1]
   b) mass = density × volume [1]; $10.49 \times 8 = 83.92g$ [1]
   c) moles = $\frac{mass}{A_r}$ [1]; $\frac{83.92}{108} = 0.78$ moles [1]
   d) 0.5mol of oxygen (O_2) [1]; 12g of carbon (C) [1]

Page 320 Reactivity of Metals

1. a) magnesium [1]; sodium [1]
   b) i) Carbon [1]
      ii) Zinc(II) oxide [1]
2. Oxidation: $Zn \rightarrow Zn^{2+} + 2e^-$ [1]; Reduction: $Cu^{2+} + 2e^- \rightarrow Cu$ [1]

   Remember, OIL RIG: Oxidation Is Loss (of electrons); Reduction Is Gain (of electrons)

Page 321 The pH Scale and Salts

1. 4 [1]
2. a) Add excess copper(II) oxide to acid (accept alternatives, e.g. 'until no more will react') [1]; filter (to remove excess copper(II) oxide) [1]; heat filtrate to evaporate some water or heat to point of crystallisation [1]; leave to cool (so crystals form) [1]
   b) **Any one of:** wear apron [1]; use eye protection [1]; tie hair back [1]
   c) Copper(II) chloride [1]

Page 322 Electrolysis

1. a) Aluminium is more reactive than carbon [1]
   b) 2 [1]; 4 [1]
   c) It requires a lot of electricity to melt the aluminium oxide / keep it molten [1]; so aluminium oxide is dissolved in molten cryolite [1]; to reduce the melting point, so less electricity is used [1]

Page 323 Exothermic and Endothermic Reactions

1. a) It reduces the movement of heat to and from the surroundings [1]; which could affect the accuracy of the results [1] (Accept 'it is an insulator' for 1 mark)

Answers

b) **Any two of:** type of acid [1]; concentration of acid [1]; surface area of metal [1]; temperature of acid [1]; volume of acid [1]; mass of metal [1]

c) The more reactive the metal [1]; the more exothermic the reaction [1]

Page 324 Measuring Energy Changes

1. a) 366 × 2 [1]; = 732 [1]
   b) 436 + 193 [1]; = 629 [1]
   c) Endothermic [1]; because the amount of energy needed to break the bonds in the reactants is greater than the energy released from forming new bonds [1]

Page 325 Rate of Reaction

1. a) Repeat using different concentrations of acid [1]; for example, $0.5mol/dm^3$, $1mol/dm^3$, $1.5mol/dm^3$, $2mol/dm^3$ [1] (Accept any other concentrations within a sensible range)
   b) **Any one of:** wear eye protection [1]; wear apron [1]; do not heat mixture over 50°C [1]
   c) The time taken for the cross to disappear will decrease as concentration of acid increases [1]; as the (acid) concentration increases so does the number of (acid) particles in a given volume [1]; so they collide more often with the sodium thiosulfate particles [1]; resulting in more successful collisions and a reaction taking place (sulfur forming) [1]

Page 326 Reversible Reactions

1. a) Reversible (reaction) [1]
   b) Forward reaction [1]; the backward reaction is endothermic [1]; because heat is needed to decompose the ammonium chloride [1]
   c) The forward and reverse reactions are taking place at the same rate. [1]; The amounts of reactants and products are constant. [1]
   d) More ammonium chloride will be produced [1]; until equilibrium is reached again [1]

Page 327 Alkanes

1. a) A molecule / compound [1]; that only contains carbon and hydrogen [1]
   b) The different molecules / hydrocarbons condense [1]; at a place in the column just below their boiling point [1]
2. O_2 [1]; 2 [1]

Page 328 Cracking Hydrocarbons

1. a) 2 [1]
   b) C_8H_{18} only [1]
2. Add bromine water [1]; if it goes colourless it is an alkene [1]; if it stays orange / brown / does not change colour then it is an alkane [1]

Page 329 Chemical Analysis

1. a) Paper [1]
   b) Pencil [1]; the ink from the pen would 'run' on the paper / dissolve in the solvent and affect the results [1]
   c) **B** and **C** [1]; there is only one spot [1]
   d) 18cm (distance moved by **X**) and 28cm (distance moved by solvent) [1]; $\frac{18}{28}$ [1] = 0.64286 = 0.64 to 2 d.p. [1] (award 2 marks if calculation is correct but the answer is not given to 2 d.p.)

Page 330 The Earth's Atmosphere

1. C [1]
2. $\frac{0.05}{4.2} \times 100$ [1]; = 1.2% [1]
3. a) Evidence [1]; that supported their theory [1]
   b) **Any three of:** it happened a long time ago [1]; nobody was alive to record how it happened [1]; there is little evidence available [1]; there is evidence to support both theories [1]

Page 331 Greenhouse Gases

1. a) Carbon dioxide OR methane [1]; **plus:** increased burning of fossil fuels / increased deforestation (for carbon dioxide) OR more animal farming (digestion, waste decomposition) / decomposition of rubbish in landfill sites (for methane) [1]
   b) From: 0.6°C [1]; To: 3.6°C [1] (If no units are included only award 1 mark)
   c) **Any two of:** complex systems [1]; many different variables [1]; may only be based on parts of evidence [1]

Page 332 Earth's Resources

1. a) Water that is safe to drink [1]
   b) It contains dissolved substances [1]; it is a mixture as it contains more than just water molecules [1]
   c) **Any one of:** adding chlorine [1]; adding ozone [1]; using UV light [1]
   d) To remove the water from the salt [1]
   e) The sea water needs to be heated [1]; which requires a large amount of energy [1]

Page 333 Using Resources

1. **Any six of:** wood pulp is from trees, a renewable resource [1]; trees should be replanted before wood pulp can be considered a sustainable resource [1]; clay, chalk and titanium oxide are quarried which can have negative environmental impacts [1]; paper production uses a lot of water [1]; transportation of plastic bags uses less fuel [1]; plastic is longer lasting / can be reused many times [1]; (plastic longer lasting) so plastic may use less finite resources [1]; (plastic may use less finite resources) and so plastic may have a lower energy requirement [1]; paper is biodegradable so spends less time in landfill [1]; paper is more likely to be recycled which lowers raw material usage [1]

> For this type of question, you need to give your answer in a clear and logical way, using good English and correct grammar and punctuation. Typically, you might expect to receive 5–6 marks for a clear description of the advantages and disadvantages of both types of bag, with logical links; 3–4 marks if a number of relevant points are made, but the logic is unclear; 1–2 marks for fragmented points, with no logical structure.

Page 334 Forces – An Introduction

1. a) Weight / gravity [1]; non-contact [1]; air resistance (or drag) [1]; contact [1]
   b) weight = mass × gravitational field strength / $W = mg$ [1]; weight = 120 000 × 10 = 1 200 000N [1]
   c) A correctly drawn horizontal arrow [1]; a vertical arrow that is half the length of the horizontal arrow [1]; and a diagonal arrow showing the total resultant force [1]

Resultant Force

2. a) Friction [1]
   b) Friction is a contact force [1]; lifting the glider off the track means there is no contact, so no friction [1]

Page 335 Forces in Action

1. a) Remove the weights [1]; if it returns to its original shape it was behaving elastically [1]
   b) i) 8.0 (cm) [1]
      ii) The extension appears to be linear [1]; and is increasing by 4cm each time [1]

c) force [1]; 0N [1]; 6N [1]
d) It allows you to check for errors / anomalies [1]; you can calculate a mean (average) result [1]

Page 336 Forces and Motion

1. a) Zero [1]

> Displacement is the distance from the start point. The person returns home – it is a circular journey – so the total displacement at the end of their journey is zero.

b) B [1]
c) Stationary [1]
d) A is travelling away from home [1]; D is travelling in the opposite direction (back towards home) [1]; D is travelling slightly faster than A [1]

2. a) speed = $\dfrac{\text{distance travelled}}{\text{time}}$ / $v = \dfrac{s}{t}$ [1]; = $\dfrac{180}{6}$ = 30m/s [1]

b) i) 50 – 40 = 10m/s [1]
   ii) 50 + 40 = 90m/s [1]

> The velocity changes from 50m/s to the east to 40m/s to the west.

Page 337 Forces and Acceleration

1. a) The mass of the system [1]
   b) The force accelerating the trolley (provided by the hanging masses) [1]
   c) The acceleration of the trolley [1]

> The independent variable is the one deliberately changed and the dependent variable is the one being measured.

2. a) acceleration = $\dfrac{\text{change in velocity}}{\text{time taken}}$ / $a = \dfrac{\Delta v}{t}$ [1]; acceleration = $\dfrac{16 - 4}{8}$ = 1.5 [1]; m/s² [1]
   b) resultant force = mass × acceleration / $F = ma$ [1]; force = 68 × 1.5 = 102N [1]
3. a) acceleration = $\dfrac{33}{11}$ [1]; = 3m/s² [1]
   b) force = 950 × 3 [1]; = 2850N [1]

Page 338 Terminal Velocity and Momentum

1. They could change position e.g. dive head first [1]; to become more aerodynamic / reduce air resistance [1]
2. a) momentum = mass × velocity / $p = mv$ [1]; = 5000 × 1340 [1] = 6 700 000kg m/s [1]
   b) 6 700 000 in one second from 1 engine [1]; so 6 700 000 × 5 × 10 = 335 000 000kg m/s momentum

to the exhaust gases [1]; and the same amount of momentum is gained by the the rocket [1]

> Momentum is conserved, so the momentum gained by the rocket is equal to the momentum given to the fuel.

c) momentum = mass × velocity / $p = mv$ [1]; 335 000 000 = 3 000 000 × v [1]; $v = \dfrac{335\,000\,000}{3\,000\,000}$ = 111.67m/s [1]
d) As the rocket burns fuel, it becomes lighter [1]; resultant force = mass × acceleration / $F = ma$, so acceleration (a) increases [1]; at a higher altitude, there is less air resistance [1]; so the resultant force increases [1]

Page 339 Stopping and Braking

1. a) 1.4 seconds [1]
   b) 1.4 × 15 [1]; = 21m [1]
   c) 4 – 1.4 = 2.6 seconds [1]
   d) momentum = mass × velocity / $p = mv$ [1]; 1300 × 15 = 19 500kg m/s [1]
   e) It would be longer [1]
   f) A correctly drawn graph line that starts horizontally at 15m/s [1]; then starts sloping downwards between 0.2s and 0.8s [1]; and has the same gradient on the downslope as the original line [1]

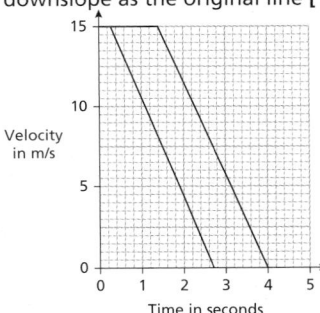

Velocity in m/s (vertical axis), Time in seconds (horizontal axis)

g) **Any two of:** The down slope would start at the same point [1]; but have a shallower gradient [1]; and take a longer total time to stop [1]

Page 340 Energy Stores and Transfers

1. a) 100 – 20 = 80 degree change [1]; energy = 2 × 4200 × 80 [1]; = 672 000J [1]
   b) 672 000 = 2.5 × 4200 × temp change [1]; temp change = $\dfrac{672\,000}{(2.5 \times 4200)}$ = 64°C [1]; final temp = 20 + 64 = 84°C [1];
2. a) gravitational potential energy = mass × gravitational field strength × height / $E_p = mgh$ [1]; $E_p = 0.1 \times 10 \times 0.05$ [1]; = 0.05J [1]

b) kinetic energy = 0.5 × mass × (speed)² / $E_k = \frac{1}{2}mv^2$ [1]; 0.05 = 0.5 × 0.1 × v² [1]; $v^2 = \dfrac{0.05}{(0.5 \times 0.1)}$ = 1, v = 1m/s [1]

Page 341 Energy Transfers and Resources

1. light [1]; electrical / thermal [1]; heat [1]
2. a) Start temperature of water [1]; thickness of fleece [1]
   b) Fleece M [1]; because it cools the slowest, so has the lowest thermal conductivity [1]; and insulates the best [1]

Page 342 Waves and Wave Properties

1. a) Half a wave per second [1] (Accept: 1 wave every 2 seconds)
   b) wave speed = frequency × wavelength / $v = f\lambda$ [1]
   c) speed = $\dfrac{\text{distance}}{\text{time}}$ / $v = \dfrac{s}{t}$, $v = \dfrac{50}{10}$ [1]; = 5m/s [1]
   d) $v = f\lambda$, $\lambda = \dfrac{5}{0.5}$ [1]; = 10m [1]
   e) They will go slower [1]

2. In longitudinal waves, the particles oscillate [1]; parallel to the direction of energy transfer / wave motion [1]; in transverse waves, the oscillation is at right-angles to the direction of energy transfer / wave motion [1]

Page 343 Electromagnetic Waves

1. Can be shown on the diagram to help explain but must include 'refraction' in the answer for full marks, e.g. Light rays from the pin [1]; are refracted when they leave the water [1]; away from the normal and into the eye [1]

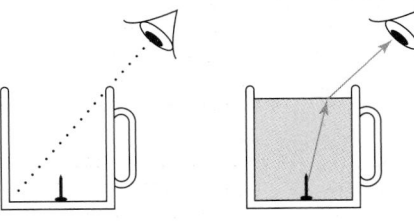

2. frequency [1]; wavelength [1]

Page 344 The Electromagnetic Spectrum

1. a) Microwaves [1]
   b) Accept any sensible answer, e.g. X-rays for photographing bones OR gamma rays for sterilisation OR UV for sunbeds [2] (1 mark for wave, 1 mark for use)
2. a) X-rays [1]
   b) It can penetrate soft tissue [1]; but is blocked by bone [1]
   c) i) Skin cancer [1]
      ii) It has more / higher frequency energy [1]; and is ionising [1]

Answers

iii) Any two of: they think it looks good / healthy **[1]**; they don't think it will happen to them **[1]**; they don't think it is that risky **[1]** (Accept any other sensible answer)

Page 345 An Introduction to Electricity

1. current **[1]**; charge **[1]**; greater **[1]**; current **[1]**
2. a) potential difference = current × resistance / $V = IR$ **[1]**
   b) $230 = 5 \times R$ **[1]**; $R = \frac{230}{5} = 46\Omega$ **[1]**
   c) charge = current × time / $Q = It$ **[1]**
   d) $Q = 5 \times 120$ **[1]**; $= 600C$ **[1]**
3. energy **[1]**; greater **[1]**; current **[1]**
4. a) Closed switch **[1]**
   b) Battery **[1]**
   c) Fuse **[1]**
   d) Light dependent resistor / LDR **[1]**

Page 346 Circuits and Resistance

1. a) To adjust the resistance of the circuit **[1]**; and change the voltage across the component **[1]**
   b) Series **[1]**
   c) Parallel **[1]**
2. Four correctly drawn lines **[3]** (2 marks for two correct lines; 1 mark for one correct line)
   Light dependent resistor (LDR) – Resistance decreases as light intensity increases.
   Thermistor – Resistance decreases as temperature increases.
   Diode – Has a very high resistance in one direction.
   Filament light – Resistance increases as temperature increases.

Page 347 Circuits and Power

1. a) energy transferred = power × time / $E = Pt$ **[1]**; $E = 1600 \times 120$ **[1]**; $= 192\,000J$ **[1]**
   b) $\frac{192\,000}{100} \times 10$ **[1]**; $= 19\,200J$ **[1]**
   (accept any equivalent method)
2. a) $V = IR$, $V = 2 \times 3$ **[1]**; $= 6V$ **[1]**
   b) $18 = 6 + V$ **[1]**; $V = 18 - 6 = 12V$ **[1]**

> The total potential difference in a series circuit is shared across the components.

   c) $18 = 2 \times R$ **[1]**; $R = \frac{18}{2} = 9\Omega$ **[1]**

Page 348 Domestic Uses of Electricity

1. a) 1.5V **[1]**; d.c. **[1]**
   b) 230V **[1]**; a.c. **[1]**
2. When a device is switched off, the live wire before the switch can still be at a non-zero potential **[1]**; touching this would create a potential difference between the wire and the ground **[1]**; this would make current flow through the person **[1]**; which would cause an electric shock **[1]**

Page 349 Electrical Energy in Devices

1. a) Kinetic **[1]**
   b) It is dissipated / lost **[1]**; to the surroundings **[1]**
2. a) Energy input = electrical **[1]**; useful energy output = kinetic **[1]**
   b) It disappears. **[1]**
   c) 500J **[1]**
3. The output from the generators goes through a step-up transformer **[1]**; this increases the voltage and also reduces the current **[1]**; the low current stops the cables from becoming hot **[1]**; which means less energy is lost during transmission **[1]**

Page 350 Magnetism and Electromagnetism

1. If free to move, the magnet will rotate so that the north pole of the magnet **[1]**; points to the Earth's north pole **[1]**
2. a) They will repel **[1]**
   b) It will be attracted to the magnet **[1]**
3. When the switch is pressed current flows in the electromagnet **[1]**; this magnetises the magnet **[1]**; which attracts the armature, causing the hammer to hit the gong **[1]**; the movement of the armature breaks the circuit, switching off the magnet **[1]**; the armature springs back and remakes the circuit, which starts the cycle again **[1]**

Page 351 The Motor Effect

1. a) force on a conductor (at right-angles to a magnetic field carrying a current = magnetic flux density × current × length / $F = BIl$, $F = (0.3 \times 10^{-3}) \times 2 \times 0.05$ **[1]**; $= 0.00003N$ **[1]**
   b) There will be no force on the wire **[1]**
2. a) 1st finger = direction of field **[1]**; 2nd finger = direction of current **[1]**; thumb = direction of force **[1]**
   b) Increase the current **[1]**; use stronger magnets **[1]**
   c) Reverse the current **[1]**; reverse the magnets / magnetic field **[1]**

Page 352 Particle Model of Matter

1. a) The energy required **[1]**; to change 1kg of a substance from a solid to a liquid **[1]**
   b) thermal energy for a change of state = mass × specific latent heat / $E = mL$, $E = 0.012 \times (2.3 \times 10^6)$ **[1]**; $= 27\,600J$ **[1]**
2. a) The substance is condensing **[1]**; from a gas to a liquid **[1]**
   b) The particles are slowing down **[1]**; and the substance is cooling **[1]**

Page 353 Atoms and Isotopes

1. a) Electron, –1 **[1]**; neutron, 0 **[1]**; proton, +1 **[1]**
   b) It has the same number of protons **[1]**; as electrons **[1]**
   c) ion **[1]**; positive **[1]**
2. Path **A** is a long way from the nucleus and the alpha particle goes straight through **[1]**; Path **B** is close to the positive nucleus so the alpha particle is deflected **[1]**; Path **C** comes very close to the nucleus and the alpha particle is repelled back the way it came **[1]**

> Two positively charged particles will repel each other.

Page 354 Nuclear Radiation

1. a) An unstable atom that gives out radiation **[1]**
   b) Beta decay **[1]**
   c) It is stable / non-radioactive **[1]**
   d) Sodium **[1]**
2. Becquerel **[1]**
3. Gamma, beta, alpha **[1]**

Page 355 Half-Life

1. a) No longer a risk **[1]**; because it has a half-life of just 8 days **[1]**; so would have completely decayed to the same level as background radiation in the last 30 years **[1]**
   b) 24 days is 3 half-lives **[1]**; $256 \to 128 \to 64 \to 32$, so a count rate of 32 remains **[1]**
   c) To fall to $\frac{1}{8}$ takes 3 half-lives, so $(3 \times 30 =)$ 90 years **[1]**; (1986 + 90 =) 2076 **[1]**
   d) $^{40}_{19}$potassium **[1]**; $\to {}^{40}_{20}$caesium $+ {}^{0}_{-1}$e **[1]**

Pages 357-372 Biology Practice Exam Paper 1

01.1 Bacterium **[1]**
01.2 $\frac{70}{14\,000}$ **[1]**; $= 0.005$ **[1]**
01.3 2, 4, 8, 16, 32, 64 (idea of doubling) **[1]**; 64 **[1]**
02.1 **Any two of:** fever **[1]**; vomiting **[1]**; abdominal cramps **[1]**; diarrhoea **[1]**
02.2 Incidences higher in the summer **[1]**; because food not kept at a cold enough temperature in summer / references to barbeques and undercooked food **[1]**
03.1 Right ventricle **[1]**
03.2 A build-up of fatty material inside the coronary arteries **[1]**; narrows them down and reduces the flow of blood **[1]**; resulting in a lack of oxygen for the heart muscle **[1]**; **Any three risk factors from:** saturated fat in diet **[1]**; smoking **[1]**; lack of exercise **[1]**; stress **[1]**; genetic factors **[1]**
03.3 Will have the same antigens / tissue type **[1]**; so no chance of rejection **[1]**

03.4 Stem cells removed from embryos [1]; the embryos are destroyed / mention of ethics of disposing of human embryos [1]

04.1 Fleming [1]; fungus [1]

04.2 4, 3, 1, 2 [3] (1 mark for 4 before 3; 1 mark for 3 before 1; 1 mark for 1 before 2)

05.1 To outcompete other plants [1]; and get more light [1]; so more photosynthesis / they can produce more food [1]

05.2 Water is transported in the xylem [1]; it is pulled up [1]; due to loss of water / transpiration from the leaves [1]

05.3 The presence of spines [1]; mean animals are less likely to eat it [1]; plus lower numbers of stomata [1]; so less water loss [1]; the thick waxy cuticle [1]; means less loss of water through the epidermis [1]

06.1 Correctly plotted points [3]; straight line of best fit [1]

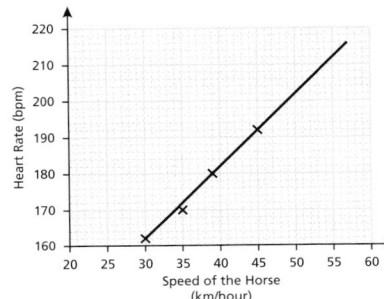

06.2 As speed increases heart rate increases [1]; Straight line (for these speeds) / use of figures [1]

06.3 A horizontal line should be drawn on the graph from 200bpm on the y-axis to the line of best fit and a vertical line should be drawn down from this point to the x-axis [1]; accept 48, 49 or 50 km/hour [1]

06.4 glucose [1]; → lactic acid [1]

06.5 No lactic acid is made [1]; so muscles are not fatigued [1]; and more energy released [1]

07.1 Fatty acids are formed [1]

07.2 Bile is present [1]; which emulsifies the fats [1]; giving a larger surface area for lipase to work on [1]

07.3 The lipase had been boiled [1]; so the enzyme had been denatured [1]

07.4 Amylase only works on starch or carbohydrates / protease only works on protein [1]; lipase is the wrong shape to work on starch / protein [1]; reference to the lock and key model and specific fit [1]

08.1 The movement of water [1]; from a dilute solution to a concentrated solution [1]; through a selectively permeable membrane [1]

08.2 The cylinders had lost water [1]; by osmosis [1]; because the contents of the potato cells were more dilute than the solution [1]

08.3 Answer from intercept on graph in the range of 0.35–0.38mol/dm³ [1]

08.4 **Top to bottom: I [1]; C [1]; D [1]; C [1]**

Pages 373–388 Biology Practice Exam Paper 2

01.1 Fingertip [1]

01.2 To make sure the results are reliable [1]; because she has a 50/50 chance of guessing right [1]

01.3 Less sensitive than fingertip / more sensitive than leg [1]; she can tell the difference between one pin and two at 1cm only [1]

01.4 She is thinking about it / it is not involuntary [1]; it does not involve protecting her body [1]

01.5 **Top to bottom: (A), D, E, B, C [3]** (2 marks for two correct; 1 mark for one correct)

02.1 **Any two of:** sharp teeth to catch seals [1]; white fur for camouflage [1]; sharp claws to catch seals [1]; forward facing eyes to judge distance [1]

02.2 They are in the same genus (*Ursus*) [1]; but are in different species [1]

02.3 Polar bears and Alaskan bears may mate and produce sterile hybrids. [1]; The habitats of the polar bears and the Alaskan bears may overlap. [1]

02.4 Increased burning of fossil fuels [1]; and deforestation are causing the increase [1]; carbon dioxide is a greenhouse gas [1]; it traps long wavelength energy from the sun [1]; that is being reflected back by the Earth's surface [1]; leading to an increase in temperature [1]

03.1 Copper from waste is passing into the river [1]; the concentration is low before the waste pile / high directly after the waste pile / decreases in concentration as you move away from the waste pile [1]

03.2 At **B** the concentration of copper is higher [1]; plants at **B** are being poisoned by the copper [1]

03.3 Plants at **B** have adapted to living in high copper concentrations [1]; plants at **A** have not experienced high concentrations before [1]

03.4 All plants have a different resistance to copper [1]; the ones that are most resistant survive [1]; they reproduce and pass on the genes for resistance [1]; and over many generations the plants become more resistant [1]

04.1 heterozygous [1]; tall [1]; dominant [1]; genotypes [1]

04.2 1:1 [1]; (Accept a final answer of 50% or $\frac{1}{2}$)

	T	t	
t	Tt	tt	[1];
t	Tt	tt	[1]

04.3 **Any two of:** two Tt plants could produce 10 tall plants [1]; it is unlikely, but could happen by chance [1]; 10 plants is not enough, more crosses need to be done [1]

05.1 If Y chromosome is present then it is a boy [1]; as boys have XY sex chromosomes [1]

05.2 Easier to obtain mother's blood than baby's cells [1]; less risk / less likely to cause damage or miscarriage [1]

06.1 Asexual [1]

06.2 They have exactly the same genes [1]

06.3 The bottom part contained the nucleus [1]; this means in has the genetic material / DNA / chromosomes [1]; and can make new proteins [1]

07.1 15 + 21 = 36% [1]

07.2 Irregular ovulation [1]; 75% of 16 is the highest figure [1]

07.3 If levels of progesterone increase less FSH is released / if levels of progesterone fall then more FSH is released [1]

07.4 Less progesterone is released [1]; so more FSH is released [1]; and eggs are stimulated to develop [1]

08.1 A chemical messenger [1]; carried in the blood [1]

08.2 **From top to bottom:** ADH [1]; pituitary gland [1]; thyroid [1]; controls metabolic rate [1]; adrenaline [1]; gets the body ready for action [1]

08.3 **Any four of:** they cannot produce insulin [1]; therefore they cannot reduce their glucose levels [1]; glucose may start to pass out in the urine [1]; high levels can lead to a coma [1]; they need to know to inject themselves with insulin [1]

Pages 389–404 Chemistry Practice Exam Paper 1

01.1 Proton +1, Neutron 0, Electron −1 [1]

01.2 It has equal numbers of protons and neutrons / equal numbers of positive and negative charges [1]

01.3 Mass number 4, Atomic number 2 [1]

Answers

The atomic number is the number of protons (and also electrons). The mass number is the total number of particles in the nucleus (protons plus neutrons).

01.4 It has a full outer shell of electrons. [1]

01.5 It has the same atomic number. [1]; It has a different mass number. [1]

02.1 5 [1]

02.2 (12 + 16 + 16) = 44 [1]

02.3 oxidation–carbon [1]; reduction–iron(III) oxide [1]

02.4 It is found pure in the ground [1]; because it is unreactive [1]

03.1 **Any one of:** the density increases as you go down the group [1]; the melting point increases as you go down the group [1]; the boiling point increases as you go down the group [1] (Accept alternative answers, e.g. the density decreases as you go up the group)

03.2 Any answer in the range 200–230°C [1]

03.3 One shared pair of electrons drawn between the two atoms [1]; three non-bonding pairs drawn on each atom [1] (the second mark will only be awarded if the first mark is achieved)

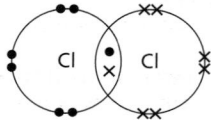

03.4 Br_2 [1]

03.5 Intermolecular forces break [1]

03.6 They all have the same number of electrons (seven) in their outer shell [1]

03.7 Sodium ion has no electrons drawn [1]; labelled Na^+ / + ion [1]; chlorine ion has eight electrons drawn [1]; seven represented by dots and one a cross [1]; labelled Cl^- / 1– ion [1]

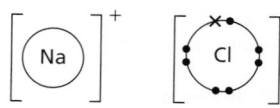

03.8 The atoms get larger [1]; the outer shell gets further from the nucleus [1]; the attraction between the nucleus and electrons gets weaker [1]; so an electron is less easily gained [1]

04.1 Filtration [1]; to remove the excess copper(II) oxide [1]

04.2 Hazard: **any one of:** chemical on skin [1]; chemical in eyes [1]; cuts from broken glass [1]; Way of reducing the risk: **any one of (as appropriate to hazard):** wash hands [1]; wear eye protection [1]; ask teacher to clear away broken glass [1]

04.3 M_r CuO = 63.5 + 16 = 79.5 [1]; M_r $CuSO_4.5H_2O$ = 63.5 + 32 + (16 × 4) + 5 × (2 × 1 + 16) = 249.5 [1]; moles CuO = $\frac{5.2}{79.5}$ = 0.0654088 [1]; mass $CuSO_4.5H_2O$ = 0.0654088 × 249.5 = 16.3g [1]

05.1 Similarity: **any one of:** they both are made up of carbon atoms [1]; both contain strong covalent bonds between carbon atoms [1] Difference: **any one of:** graphite is made up of layers, diamond has no layers [1]; graphite contains weak intermolecular forces, diamond only contains strong covalent bonds [1]; graphite has delocalised (free) electrons, diamond does not [1]

05.2 Electrical conductivity comes from delocalised electrons, which are able to move through the structure [1]; this is useful for touch-screens, as they need to be able to conduct electricity to work [1]; strength comes from strong covalent bonds between carbon atoms [1]; this is useful for touch-screens so they do not crack / shatter when dropped [1]; graphene is transparent because it is only one atom thick [1]; this is useful for touch-screens so you can see the light coming through from the display underneath [1]

06.1 Battery / cell added and joined to electrodes [1]; electrode connected to positive side of cell labelled anode [1]; electrode connected to negative side of cell labelled cathode [1]

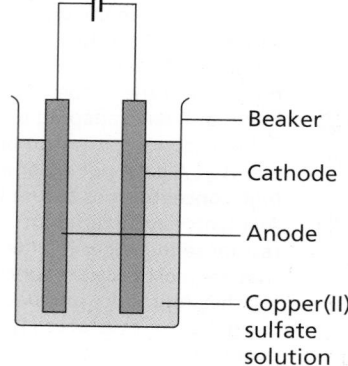

— Beaker

— Cathode

— Anode

— Copper(II) sulfate solution

06.2 Copper [1]; it is less reactive than hydrogen [1]

06.3 2 [1]; 4 [1]

06.4 Oxidation [1]; because electrons are lost from the (hydroxide) ions [1]

07.1 **In order:** (s), (aq), (g) [1]

07.2 To let the gas out / to stop the acid spraying [1]

07.3 Sensible scales, using at least half the grid for the points [1]; all points correct [2]; connected by smooth curve [1]

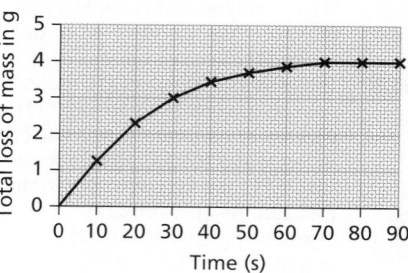

07.4 70s [1]

07.5 All of the acid had reacted [1]

07.6 Hydrogen gas is formed [1]; which escapes into the air [1]

07.7 Less steep line to right of original line [1]; finishes at same overall mass loss [1]

08.1 Bonds broken = (432 × 2) + 495 = 1359kJ/mol [1]; bonds formed = 467 × 4 = 1868kJ/mol [1]; bonds broken – bonds formed = –509kJ/mol [1]

08.2 Less energy is required to break the bonds in the reactants [1]; than is produced when the bonds are formed in the products [1]

Pages 405–420 Chemistry Practice Exam Paper 2

01.1 2 [1]

01.2 carbon particles [1]

01.3 It is colourless [1]; and odourless [1]

01.4 $N_2 + 2O_2 \rightarrow 2NO_2$ [1]

Di- means two, so nitrogen dioxide contains one nitrogen atom and two oxygen atoms.

01.5 sulfur dioxide – acid rain [1]; carbon particles – global dimming [1]

02.1 Formula: C_3H_6 [1]; Name: Propene [1]

02.2 Add bromine water [1]; turns orange to colourless [1]

02.3 HD poly(ethene) [1]; because it is stronger [1]

03 **Effects on the environment: any two of:** melting of ice caps [1]; sea level rise, which may cause flooding and coastal erosion [1]; changes in

amount, timing and distribution of rainfall [1]; desertification in some regions [1]; more frequent and severe storms [1]

Effects on people: any two of: flooding of homes [1]; migration of people from affected areas [1]; temperature and water stress [1]; lack of food in some regions [1]

Effects on wildlife: any two of: changes to distribution of species [1]; extinction of some species [1]; temperature and water stress [1]; lack of food in some regions [1]

04.1 Desalination / distillation [1]

04.2 It contains other substances [1]; which are dissolved [1] (Accept 'it's a mixture' for 1 mark)

04.3 Heat the water (with the Bunsen burner) [1]; if the water is pure it will start to boil at 100°C [1]

04.4 Chlorine / ozone / ultraviolet light [1]

04.5 $\frac{250}{1000}$ × 1.35 [1]; = 0.3375 [1]; = 0.34mg (to 2 decimal places) [1]

05.1 They are finite / non-renewable / may run out [1]

05.2 **Any five valid points, e.g.** recycling reduces the amount of metal being mined (metals will last longer) [1]; mining, processing metals and recycling all require energy, which mostly comes from the use of finite resources [1]; collecting and transporting cans uses petrol / diesel [1]; sorting cans and rolling metal blocks requires electrical machinery [1]; melting the cans requires a lot of energy [1]; overall, recycling consumes less energy than producing new cans [1]

06.1 Push the plunger of the syringe down to remove any air in it [1]; the plunger may be pushed out before the end of the reaction [1]; so volume of gas produced not accurately measured [1]

06.2 All points plotted correctly for two sets of data [2]; two lines of best fit drawn [2]

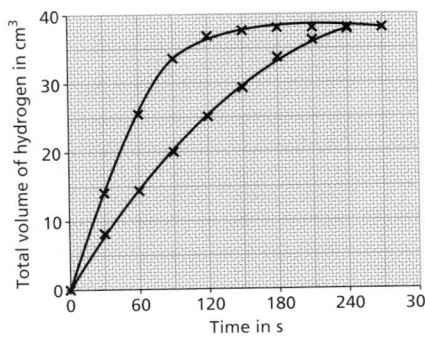

06.3 The higher the concentration, the faster the rate of reaction [1]

06.4 As the concentration increases so does the number of particles of acid in a given volume [1]; so there are more frequent collisions / more collisions per second with magnesium particles [1]; so rate increases / reaction speeds up [1]

06.5 140s [1]

06.6 Tangent drawn at 50s [1]; volume of hydrogen calculated, e.g. 0.64 − 0.36 = 0.28 [1]; time calculated, e.g. 60 − 30 = 30 [1]; gradient calculated to give rate of reaction, e.g. $\frac{0.28}{30}$ = 0.009333 [1]; = 0.009 [1]; cm³/s [1] (Accept 0.008)

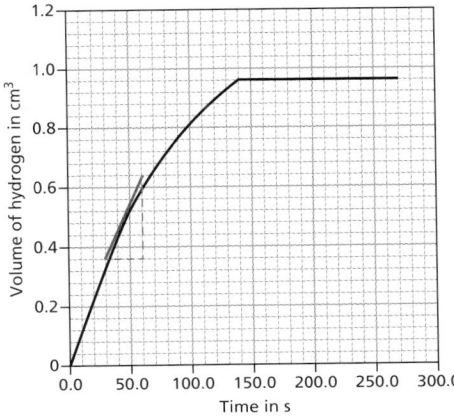

07.1 2.5 billion years [1]

07.2 40cm³ [1]

07.3 Algae evolved and starting producing oxygen [1]; via photosynthesis [1]; land plants evolved to increase amounts of oxygen in the atmosphere further [1]

07.4 Carbon dioxide allows short wavelength radiation to pass through [1]; the atmosphere to the Earth's surface [1]; carbon dioxide absorbs outgoing long wavelength radiation [1]

07.5 Increased [1]

07.6 Increased burning of fossil fuels [1]; in vehicle engines/power stations [1] OR increased deforestation [1]; so fewer trees to absorb carbon dioxide from the air [1]

08.1 3 [1]

08.2 15cm (distance moved by S) and 22cm (distance moved by solvent) [1]; $\frac{15}{22}$ = 0.68182 [1]; = 0.68 [1]

08.3 Mobile phase / solvent moves through the paper [1]; and carries different compounds different distances [1]; depending on their attraction for the paper and the solvent [1]

Pages 421–434 Physics Practice Exam Paper 1

01.1 From left to right: 65J [1]; 5J [1]; 30J [1]

01.2 Heat the schools [1]; because it saves the most energy [1]; half of 5J is 2.5J [1]; but a quarter of 65J is over 15J [1]

01.3 efficiency =

$\frac{\text{useful output energy transfer}}{\text{useful input energy transfer}}$

× 100%, = $\frac{30}{100}$ × 100% = 30% [1]

02.1 It spreads out / is dissipated [1]; into the surroundings [1]

02.2 20 + 13 + 7 = 40% [1]

02.3 60% [1]

03.1 work done = force × distance / $W = Fs$, $W = 6 × 2$ [1]; = 12J [1]

03.2 weight = mass × gravitational field strength / $W = mg$, $m = \frac{6}{10}$ = 0.6kg [1]; gravitational potential energy = mass × gravitational field strength × height / $E_p = mgh$, $E_p = 0.6 × 10 × 2 = 12J$ [1]

03.3 kinetic energy = 0.5 × mass × (speed)² / $E_k = \frac{1}{2}mv^2$ [1]; $v^2 = \frac{12}{0.5 × 0.6}$ = 4 [1]; $v = \sqrt{40}$ = 6.3m/s [1]

04.1 Top to bottom: gravitational [1]; light [1]; kinetic [1]; chemical [1]

04.2 Coal and gas [1]

04.3 Nuclear / oil [1]

04.4 Wave / tidal / geothermal / biomass [1]

04.5 Nuclear power stations have high output [1]; and are reliable [1]; **plus any two of:** however they produce nuclear waste [1]; are expensive to build and decommission [1]; have a risk of explosion if something went wrong [1]; and are not renewable [1]

05.1 Y = variable resistor [1]; Z = voltmeter [1]

05.2 To control the voltage applied to component X / adjust the resistance [1]

05.3 The voltage [1]

05.4 The current [1]

05.5 Accurately plotted voltage [1]; and current [1]; with all points connected [1]

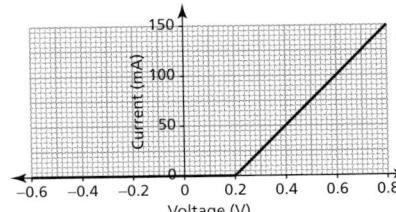

Answers

05.6 Yes, they are correct as all the points fit the line **[1]**

An anomalous result would be significantly higher or lower than the other results or would not fit the pattern.

05.7 A diode **[1]**

06.1 They have the same number of protons **[1]**

06.2 Beta decay **[1]**

06.3 The mass number has not changed during emission, but the proton number has increased by one **[1]**

06.4 The time it takes for half of the radioactive isotopes to decay / for the count rate to halve **[1]**

06.5 $\frac{1}{8}$ means that 3 half-lives have passed **[1]**; so is (30 × 3 =) 90 years **[1]**

07.1 kinetic energy = 0.5 × mass × (speed)² / $E_k = \frac{1}{2}mv^2$, $m = \frac{E_k}{\frac{1}{2}v^2}$ **[1]**; $m = \frac{472500}{450}$ **[1]**; = 1050 **[1]**; final answer = 1050kg (unit must be correctly stated for mark) **[1]**

07.2 The bus has more mass **[1]**; so it has more kinetic energy **[1]**

07.3 kinetic energy = 0.5 × mass × (speed)² / $E_k = \frac{1}{2}mv^2$ **[1]**; increase in E_k = 0.5 × 1200 × (14² − 8²) **[1]**; = 79 200J **[1]**

07.4 Because the kinetic energy depends on the square of the speed **[1]**

08.1 gravitational potential energy = mass × gravitational field strength × height / $E_p = mgh$, E_p = 65 × 10 × 1.25 **[1]**; = 812.5J **[1]**

08.2 kinetic energy = 0.5 × mass × (speed)² / $E_k = \frac{1}{2}mv^2$, 812.5 = 0.5 × 65 × v² **[1]**; v² = 25 **[1]**; $v = \sqrt{25}$ = 5m/s **[1]**

09.1 12V **[1]**

09.2 2 + 2 = 4A **[1]**

09.3 potential difference = current × resistance / $V = IR$, $R = \frac{12}{2}$ **[1]**; = 6Ω **[1]**

09.4 It is less / half **[1]**

09.5 power = potential difference × current / $P = VI$, P = 12 × 2 **[1]**; = 24W **[1]**

09.6 $\frac{96\,000}{(24 \times 2)}$ **[1]**; = 2000s **[1]**

Pages 435–448 Physics Practice Exam Paper 2

01.1 So it will be attracted by the magnetic coil **[1]**

01.2 It will need to increase **[1]**

01.3 A bigger mass makes a bigger force pulling down on the left, so a bigger force is needed on the right **[1]**; a bigger current will increase the strength of the magnetic field created by the coil **[1]**

01.4 It will increase the mass that can be balanced **[1]**; because the iron core will make the magnet stronger **[1]**

02.1 Towards the centre of the circle **[1]**

02.2 The tension of the string **[1]**

02.3 The radius of the circle **[1]**; and the mass of the bung **[1]**

02.4 speed **[1]**; velocity **[1]**; direction **[1]**

02.5 Gravity / the gravitational attraction of the Earth on the Moon **[1]**

03.1 **From top to bottom:** 1500m, 3 × 10⁻⁸m, 5 × 10⁻⁷m, 1 × 10⁻¹¹m **[3]** (2 marks for two in the correct position; 1 mark for one correct)

03.2 **Any two of:** wave speed **[1]**; can travel through a vacuum **[1]**; transverse waves **[1]**

03.3 **Any pair from:** infrared **[1]**; used for cooking / remote control **[1] OR** Microwaves **[1]**; used for communication / cooking **[1] OR** gamma rays **[1]**; used for sterilising equipment **[1]**

03.4 wave speed = frequency × wavelength / $v = f\lambda$ **[1]**; $f = \frac{3 \times 10^8}{300}$ **[1]**; = 1 × 10⁶Hz **[1]**

03.5 acceleration = $\frac{\text{change in velocity}}{\text{time}}$ / $a = \frac{\Delta v}{t}$ **[1]**; $a = \frac{10}{4}$ **[1]**; = 2.5m/s² **[1]**

03.6 10m/s **[1]**

03.7 distance travelled = speed × time / $s = vt$ **[1]**; $s = 10 \times 4 = 40$m **[1]**

04.1 speed = $\frac{\text{distance travelled}}{\text{time}}$ / $v = \frac{d}{t}$ **[1]**; $= \frac{(350 - 100)}{(7 - 3)}$ **[1]**; = 6.5m/s **[1]**; the car travels at constant speed for the first 3 seconds **[1]**; it then speeds up to 6.5m/s for 4 seconds **[1]**; before slowing again for the next 3 seconds **[1]**

To gain full marks in this type of question, your response must be written using correct grammar and punctuation. Your ideas must in a sensible order and use the appropriate scientific words.

04.2 acceleration = $\frac{\text{change in velocity}}{\text{time taken}}$ / $a = \frac{\Delta v}{t}$ **[1]**; $= \frac{30}{6}$ **[1]**; = 5m/s² **[1]**

05.1 resultant force = mass × acceleration / $F = ma$ **[1]**; F = 3500 − 2000 = 1500 **[1]**; $a = \frac{1500}{1200}$ **[1]**; = 1.25 m/s² **[1]**

05.2 As the speed increases, the resistive forces increase **[1]**; until they balance the engine thrust / it reaches terminal velocity **[1]**

05.3 The distance a car travels between the driver seeing the hazard and applying the brakes **[1]**

05.4 **Any two of:** fatigue **[1]**; age **[1]**; alcohol / drugs **[1]**; distractions **[1]** (Accept named distractions, e.g. using a mobile phone)

05.5 Road conditions (Accept a specific adverse condition, e.g. ice, snow, rain, etc.) **[1]**

05.6 As the brakes are applied friction occurs between the brakes and the wheels **[1]**; this takes kinetic energy from the car, which is converted to heat **[1]**

06.1 Wavelength **[1]**

06.2 Amplitude **[1]**

06.3 Sound **[1]**

06.4 wave speed = frequency × wavelength / $v = f\lambda$ **[1]**; v = 8 × 0.015 **[1]**; = 0.12m/s **[1]**

8 waves per second means 8Hz.

06.5 Set up the tank and switch on the stroboscope and adjust the scopes flashing speed so that the waves (or shadows of the waves) appear stationary **[1]**; measure the wavelength with a ruler **[1]**; if using shadows, scale the value based on the total size of shadow compared to the total size of tank **[1]**; flashing lights are a hazard as they can trigger photosensitive epilepsy **[1]**

07.1 They are reversed **[1]**

07.2 They would be closer together / more of them in the same space **[1]**

07.3 First finger: field **[1]**; second finger: current **[1]**; thumb: movement / force **[1]**

07.4 Into the paper / downwards at 90° to the surface of the page **[1]**

07.5 **Any three of:** a weaker current **[1]**; using weaker magnets **[1]**; moving the magnets further apart **[1]**; changing the angle of the wire so not at right-angles to the field **[1]**

Notes

Notes